BY THE EDITORS OF CONSUMER GUIDE®

HOW THINGS WORK

PUBLICATIONS INTERNATIONAL, LTD.

CONTRIBUTING WRITERS:

Brent Butterworth
Dennis Eskow
Sandy Fritz
Lee Green
Richard Koff
Evan Powell
James Rogers
Mort Schultz
Bruce Strasser

ILLUSTRATIONS:

Bill Whitney

CONSULTANTS:

American Dental Association
Peter Bussolini
Dr. Peter Cavanagh
Dr. Alvin Davis
Dr. Michael Gozola, C.P.
Lee Green
Mark Mitros, M.D.
Thomas Orszulak, M.D.
Roger Stube
Steve Toth

Louis Weber, C.E.O.
Publications International, Ltd.
7373 North Cicero Avenue
Lincolnwood, Illinois 60646

Manufactured in U.S.A.

h g f e d c b a

Library of Congress Catalog Card Number: 89-69865

ISBN: 0-517-69335-6

This edition published by Beekman House
Distributed by Crown Publishers, Inc.
225 Park Avenue South
New York, New York 10003

FOREWORD

How Things Work is a peek into the world of machinery and inventions. It explains, with words and pictures, the operation of hundreds of devices both everyday and exotic. It's a book you can browse through, discovering the inside workings of the traffic light, the television set, the EKG machine, or the door chime. It's also a book you can refer to any time you want to know how something works.

Some devices, such as transistors and motors, are used in many inventions. And certain principles, such as electricity, are so widespread that all our technology depends on them. You'll find many articles on this type of subject in the first chapter, "What Makes Machines Tick," which describes four fundamental principles behind how things work. **Energy** is what makes things happen; anything that moves, makes a noise, produces heat, gives off light, or generates electricity requires energy. **Electricity** is a natural phenomenon. The ability to use it not only gives us machines and motors that we can turn on with a switch, but it also makes possible the transistors and microchips of our electronic age. **Mechanics**—screws, levers, wheels, and gears—are the building blocks of machinery. **Waves**—sound waves, light waves, radio waves—surround us always, whether we can see or hear them or not, and inventors have put them to use in the radio, the microwave oven, the X-ray machine, and many other devices.

Although most of the topics in this book rely on these fundamental principles, it's not necessary to read "What Makes Machines Tick" before you look something up. As you read about a topic, you're likely to find cross references to pages in "What Makes Machines Tick." Those pages will contain explanations that relate to the topic you're reading about.

The topics in the book are arranged by areas of interest. "At Work," for example, is a chapter devoted to office equipment, with articles on what happens inside the photocopier, how the automatic scanner at the supermarket works, and other subjects. "Health and Personal Care" contains articles on the equipment and techniques you see in doctors' offices and hospitals; "Sight and Sound" will tell you about stereo equipment, cameras, musical instruments, and more. Look through the table of contents to see the types of devices included in each chapter. You can investigate a whole subject area or you can leaf through the book, lighting on topics that interest you.

As a further help, you'll find a glossary—"Terms to Know"—where you can get a concise definition of an unfamiliar term. And "A Time Line of Discoveries" is a thumbnail sketch of the history of inventions from the wheel to the compact disc.

When you're curious about the operation of some gadget—or if you're stumped when the kids ask, "How does that work?"—you'll reach for *How Things Work* to get a clear answer.

C O N T

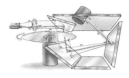

E N T S

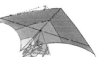

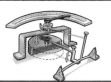

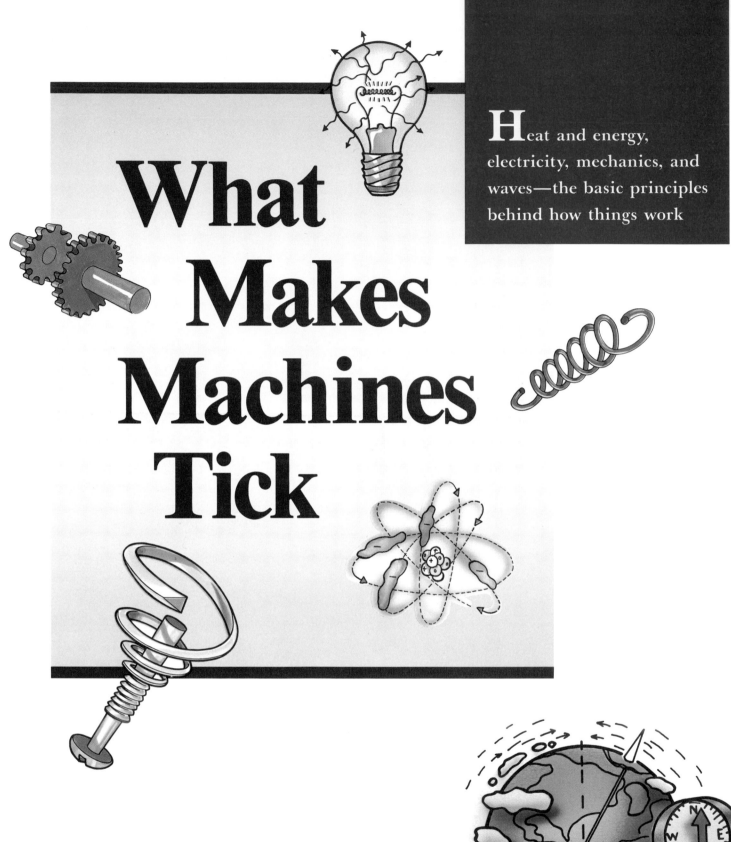

What Makes Machines Tick

Heat and energy, electricity, mechanics, and waves—the basic principles behind how things work

FRICTION

Whenever you try to move any object that's touching another object, you'll encounter some resistance to that movement. That resistance is called **friction.** Energy is required to overcome friction, but friction is not necessarily a disadvantage. The friction between the wheels of a car and the roadway not only keep the car from skidding, but enable the car to move in the first place.

Any movement involves some friction. Friction is present not only when one surface is sliding against another, but also when one of the surfaces is rolling against the other and when an object is moving through a liquid or gas. That is why it is impossible for an airplane to glide from New York to Los Angeles.

Some of the energy you put into moving an object against frictional resistance is converted to heat. The faster an object moves, and the greater the friction, the more heat will be produced. This is clearly shown by the old Boy Scout feat of starting a fire by rubbing two sticks together.

SCOOTER

A scooter has no gas tank or engine, but it needs energy to overcome friction. The "engine" of the scooter consists of the rider's muscles and digestive system. The "fuel" is the chemical energy contained in food.

Friction at the axles slows down the scooter while it's coasting and makes the rider work harder. Although friction can't be eliminated completely, it can often be reduced—by oiling the contacting surfaces or by using low-friction ball or roller bearings.

Friction at the tires makes the wheels roll in the direction that they're pointed, rather than sliding off-course sideways.

Friction between the rider's foot and the ground provides a scooter's driving force. Most of the energy supplied goes to push the scooter, but some is converted to heat. (The sole of the rider's foot gets warm.)

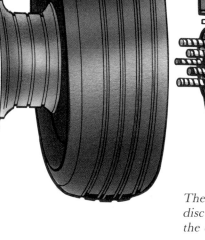

The pads, which provide one of the rubbing surfaces, do not rotate. They are attached to the calipers, which are attached to the car's suspension.

When you step on the brake pedal, the pistons inside the caliper push the pads against the rotating disc, slowing the disc and producing heat. After a time, this friction wears down the pads, which then need to be replaced.

The disc, which provides the other one of the disc brake's functioning surfaces, is attached to the car's wheel and rotates with the wheel.

DISC BRAKE

Brakes use friction to convert the energy of a moving device to heat, stopping or slowing the motion. But excess heat produced when braking can lower the frictional resistance, resulting in "brake fade." And in wet weather, water can get between the rubbing surfaces and act like a lubricant.

The automobile disc brake eliminates these problems. Its small, high-pressure friction area almost instantly boils off any water between the disc and the pressure pad, making braking failures due to wet brakes a thing of the past. And the disc brake's open construction provides a lot of cooling air, preventing overheating and brake fade.

COMBUSTION

*From the mixing chamber, the gas/air mixture goes to the **burner**, where it is ignited by the flame of a pilot light or, in more modern ranges, an energy-saving spark igniter.*

*The **burner control valve** adjusts the amount of gas that is supplied to the burner.*

*The lower pressure sucks air into the **mixing chamber**.*

*The gas passes through a **spud**—a smaller opening or restricting orifice—which increases its velocity, creating a lower pressure.*

Combustion consists of burning something to produce heat. This process converts chemical energy to thermal energy. Although other reactions also achieve this energy conversion, the most common one is a chemical reaction between a substance that acts as a fuel—usually one that contains carbon and hydrogen—with the oxygen in the air.

Most carbon-based compounds have an extremely large amount of stored chemical energy. The reason for this is that it took a great deal of energy to combine the atoms of carbon, hydrogen, and (sometimes) oxygen into a single molecule of the fuel oil. The more energy it took to produce the item to begin with, the more energy you can get out of it. Because a tree uses a lot of solar energy to grow, it can give off a lot of heat energy as fireplace wood. Long-dead plants and animals, which depended directly or indirectly on the sun's energy in order to grow, become fossil fuels—natural gas, coal, and oil.

When such fuels combine with oxygen, the large molecules are broken down into the products of combustion—carbon dioxide and

GAS RANGE BURNER

The burner on a gas range is a simple device that allows you to adjust the flame size while automatically mixing in the proper amount of air. The amount of air that is mixed with the gas is determined when the serviceman adjusts the air inlet opening into the mixing chamber. Once adjusted, the mixing of air and gas is automatic; if you open the burner control valve to increase the gas flow, the gas velocity through the spud increases, thus lowering the pressure further and pulling in more air.

water, compounds having a much lower energy content. The difference in energy levels between the original fuel molecule and the combustion products is released as heat.

To get the fuels to combine with oxygen, a source of ignition is needed. This is an energy source that raises the temperature of the fuel to the point where combustion can start. Matches, pilot lights, and engine spark plugs all serve as ignition sources.

A specific quantity of any fuel requires a specific amount of oxygen for complete combustion. If you supply less air than is needed, combustion will not be complete. Less heat will be produced, along with more smoke, soot, and other products of incomplete combustion (such as poisonous carbon monoxide). If too much air is supplied, combustion will be complete, but you will still get less heat, since the unneeded air cools off the reaction.

WOOD STOVE

A wood-burning stove also depends upon combustion. The fuel intake can't be regulated as finely as in a gas burner, so the heat output is regulated by adjusting the amount of air that is supplied for combustion.

*A wood stove's source of ignition must perform two tasks—it must raise the temperature of the wood high enough for combustion to start, and it must heat the air inside the stove sufficiently to create a draft in the flue (see **Convection**, page 10). This draft pulls fresh combustion air through the stove's inlet dampers, which can be adjusted to increase or reduce the rate of combustion.*

Flue

Flue damper

Wood

Inlet dampers

Loading door

Combustion chamber (firebox)

CONDUCTION AND CONVECTION

All substances are composed of molecules, and, no matter how solid a substance may appear, its molecules are in motion to some degree. In solid materials they vibrate back and forth as if they were connected by springs. In fluids they also drift about randomly, with little or no force keeping them in place. When you heat something, the energy it receives is stored in the form of increased molecular motion.

By heating a skillet on the stove, you increase the speed and distance of the molecules' vibration on the bottom of the skillet. These molecules transfer some of their movement to other, less active molecules by bouncing into them, much in the same way that a bowling ball causes the pins to scatter. The result is that heat, in the form of molecular motion, is transmitted along the edges of the skillet. Eventually the handle becomes too hot to touch. This process of moving heat by direct contact is called **conduction.**

The closer together the molecules are, the easier it is to conduct heat through the material. Substances like metals, which have closely packed molecular structures, are good conductors. Gases, with much greater spacing between the molecules, are poor conductors, or good **insulators.**

Even though gases are a poor conductor of heat, they can transmit thermal energy through a different mechanism—**convection.** When you heat a gas, the increase in molecular motion causes the molecules to occupy a larger space. The gas increases in volume, becoming less dense and more buoyant (see **Buoyancy,** page 236). As a result, it tends to rise, with cooler gas around it taking its place.

Convection is the reason that the upper floors of buildings are usually warmer than the lower floors and that scuba divers find the water colder as they dive deeper. Convection has given rise to the popular, but incorrect, saying that heat rises. It doesn't, but hot air does.

HOME INSULATION

Through convection, hot air inside the house reaches the outside walls and the ceiling, heating those surfaces. The heat then travels through the walls by conduction, until it reaches the outside surface of the house. Then it escapes to the outdoors by convection. The net result is that a lot of your fuel bill goes to heating the outdoors.

A layer of insulation with a high resistance to the flow of heat (high R-value) placed in the conductive path in the walls or ceiling slows down the flow of heat to the outside. Materials such as loosely packed fiberglass batting, which have air making up a large portion of their bulk, are used for insulation. An uninsulated wall typically has an R-value of 5; an insulated wall with 3½ inches of fiberglass has an R-value of 16, reducing heat loss by more than two-thirds.

HOT-AIR BALLOON

The increased buoyancy of heated air is colorfully illustrated by the hot-air balloon. A bottled-gas burner heats air that rises to fill a large balloon made of lightweight synthetic cloth. Since the hot air in the balloon is less dense than the cooler air outside, the balloon rises, with enough excess buoyancy to support a passenger-carrying basket beneath it. Hot-air balloons can be made to rise or sink by controlling the size of the burner flame—they can be steered, more or less, by causing the balloon to rise until you find an altitude with a wind blowing in the direction you want to go.

11

COMPRESSION

The molecules of a gas move more rapidly, and for greater distances, when you heat them. Thus the gas expands and increases in volume. If you heat a gas in a closed container without allowing it to expand, the increased speed of the gas molecules makes them hit the walls of the container harder. The result is a higher pressure inside the closed container.

Forcing more gas into the container achieves the same effect. You can start with a container full of gas and reduce the size of the container without allowing any of the gas to leak out. More molecules strike the container's walls, with greater speeds, resulting in a pressure increase.

Because it takes effort to increase pressure, the stored energy is increased. That energy can be recovered in several ways. A toy balloon illustrates one of these: If you pressurize the air inside it (blow it up), and then suddenly release your grip, the compressed air inside the balloon escapes in a rush, propelling the balloon across the room. (This illustrates the principle of thrust, which is used in rockets.) The stored energy is released as motion.

When you pump up a bicycle tire, both the pump and the tire get warm. Some of the energy that goes into compressing the air in the tire is converted to thermal energy.

The more gas you compress, and the greater the pressure you achieve, the more energy you store in it. Inflating an automobile tire to 30 pounds of pressure takes more work than inflating a bicycle tire to 30 pounds of pressure, because the automobile tire holds more air. If you used the same amount of energy inflating the bicycle tire that you used to inflate the automobile tire to 30 pounds of pressure, the bicycle tire would contain more pressure—and thus more stored energy.

STEAM ENGINE

The reciprocating steam engine was the invention that kicked off the Industrial Revolution. For many years, this type of engine was the major source of power for industry and transportation, but it has largely given way to electric motors and gasoline and diesel engines (see **Gasoline engine,** page 186; **Diesel engine,** page 188; **Motors,** page 26). The steam engine uses compression as a key element in converting the chemical energy in the fuel it burns to the mechanical energy that drives a locomotive or a mill.

1. Water is heated in a **boiler** to produce steam, which is a gas. The fuel is burned outside the engine itself; steam engines are **external combustion engines.**

2. The boiler is connected to a **steam chest** that acts as a temporary holding chamber for the steam.

3. A **valve** on the steam chest directs steam to one end of the **cylinder,** and the steam, having no place to expand, pressurizes the cylinder.

4. When the pressure becomes high enough, it forces the **piston** to move forward.

6. Another connecting rod moves a **slide valve rod** so that a valve on the opposite end of the cylinder is open.

8. The connecting rod and the flywheel convert the back-and-forth motion of the piston to the rotary motion of a turning wheel. This back-and-forth motion is the reason for the name **reciprocating** steam engine.

5. The piston moves a pivoting **connecting rod,** which turns the flywheel. The piston continues to move until it reaches the end of its travel.

7. At the same time, a **port** is opened that allows the spent steam that had already pushed the piston to escape through the exhaust pipe. Steam from the steam chest now enters the opposite end of the cylinder, thus moving the piston back to its original position.

RADIATION

Heat can travel from one place to another without the movement of molecules associated with conduction or convection. This is called radiation—or **electromagnetic radiation,** as it's often called to distinguish it from other forms of radiation (see **Electromagnetic Waves,** page 50).

When radiation waves strike a surface, they can be transmitted through it, reflected from it, or absorbed into it. When sunlight hits a window, the heat rays go right through the window to warm your face on the other side. If sunlight hits a mirror, the heat waves bounce off. If you wear a black shirt on a sunny day, the heat rays are absorbed by the shirt. Through absorbed radiation, the energy carried in the waves is transferred to the surface that does the absorbing. That's the way that a heat lamp warms your skin without the necessity of warming all the air between you and the lamp. Absorbed radiation is also the mechanism by which the sun's energy reaches the earth.

All surfaces, even if they are not hot, radiate heat. The hotter the surface is, the greater the amount of radiation. As a result, if your cool face is facing a heat lamp, both your face and the lamp are radiating heat and absorbing heat at the same time. But your face, which is cooler, absorbs much more heat than it radiates. Thus the heat is moved from the warmer heat lamp to your face.

STEAM RADIATOR

The old-fashioned home heating radiator delivers heat to the room both by convection and radiation. Hot water or steam, generated by a boiler fired by oil, gas, or coal, is circulated through the radiator's tubes, heating the radiator. The air in contact with the radiator is heated, which then moves through the room by convection. The radiator's surface also radiates heat, which strikes any surface in its path, such as the cat. As a result, the radiator can warm the cat even if the room is still cool. The radiation that misses the cat is eventually absorbed by the room's walls and furnishings, which, in turn, heats the air in the room.

The heater's portability lets you place the heat wherever you need it.

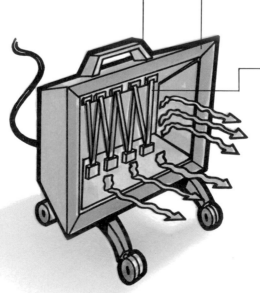

The hood reflects radiant energy from the hot wire of the resistance element and directs the heat.

An electric current passes through a wire (resistance element), heating it and causing it to glow red-hot. The hot wire radiates heat.

ELECTRIC SPACE HEATER

Electric space heaters are popular for use in poorly heated areas like garages, where they make it possible to warm a person without the necessity of heating the entire area.

ENERGY FROM ATOMS

The only way that you can obtain energy without a corresponding input of another form of energy is by converting matter to energy—nuclear power, of one form or another. All matter is made up of atoms, each of which is composed of electrons moving in orbit around a heavy nucleus, which is made up of protons and neutrons. Nuclear fission is a process in which the nuclei of heavy atoms are broken apart with a resulting release of energy. This is the reaction that powers nuclear power plants and atom bombs.

Heat can also be produced by fusion, a process in which the nuclei of light atoms are forced to combine. Normally it takes a great deal of energy to start the fusion process. In order to combine two nuclei, you have to force them close enough together to merge them. This isn't an easy task, because all atomic nuclei carry a positive charge and tend to repel each other. The closer they get, the more strongly they repel. Once a fusion chain reaction is started, though, it liberates much more energy than it took to get it started.

Fusion is the reaction that powers the sun and stars, as well as the thermonuclear devices we know as hydrogen bombs. It is tempting to try to construct a fusion power reactor to make electricity, since fusion reaction products are much less hazardous than those of a fission reactor. Deuterium (heavy hydrogen), the major fuel, is present in great quantities in the ocean and not difficult to extract. However, designing machines that can handle the high temperatures (millions of degrees) that make the fusion reaction possible have not yet proved practical.

Atoms resemble miniature solar systems, with tiny particles called electrons revolving around a much heavier nucleus. The nucleus is composed of protons, which have a positive electrical charge, and, for most elements, companion particles called neutrons. Since the neutrons have no electric charge, the nucleus as a whole has a positive charge from the protons. The electrons carry a negative electrical charge, which is what binds them in orbit around the nucleus.

A substance whose atoms are all the same structure, and which can't be broken down into simpler substances, is called an element. Each element has a different atomic structure. Hydrogen, the lightest element, has atoms that consist of one electron orbiting a nucleus containing only a proton. Uranium, the heaviest element occurring in nature, has a nucleus of 92 protons and more than 140 neutrons, which is orbited by a cloud of 92 electrons.

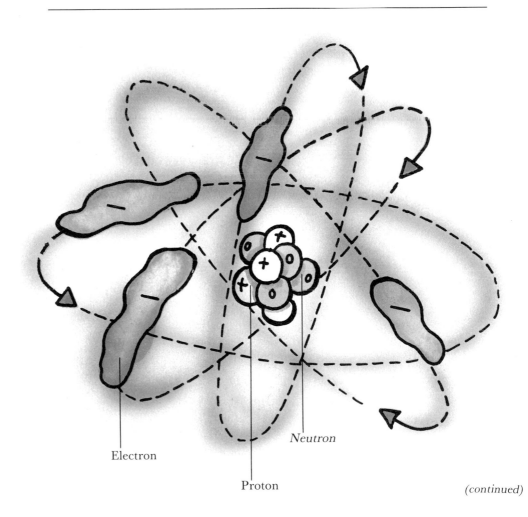

Electron

Proton

Neutron

(continued)

ENERGY FROM ATOMS

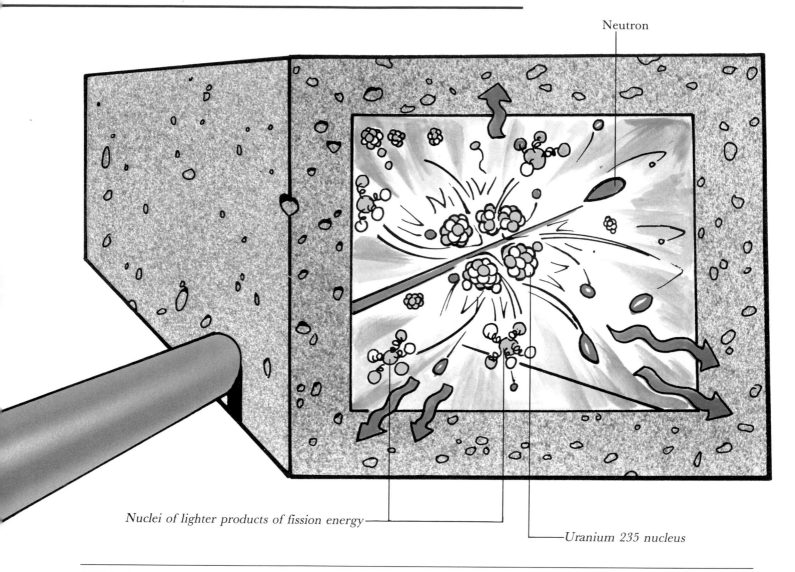

Neutron

Nuclei of lighter products of fission energy

Uranium 235 nucleus

FISSION

Because it has mass but no positive or negative electric charge, a freely traveling neutron can penetrate the outer precincts of an atom and approach the nucleus without being repelled by the electrical charge of the protons. That property makes it possible to use neutrons as "bullets" to shoot at an atom's nucleus. In a fission reaction, neutrons are used to bombard the nuclei of an atomic fuel. Uranium atoms, with their heavy nuclei, can be broken up into smaller atoms. An **isotope** (variant form) of uranium is frequently used to start fission reactions. This isotope, called uranium 235, has atoms that are inherently unstable; neutrons are spontaneously emitted from them. These neutrons provide the "bullets."

The emitted neutrons crash into neighboring uranium atoms. The impact of the neutrons splits some atoms into smaller, lighter atoms, creating smaller, lighter elements. In the process, more neutrons are released. Those neutrons collide with other uranium atoms nearby, causing those atoms to split. Each time a nucleus is split, some of its mass is converted into thermal energy, or heat. The chain reaction, if left unchecked, continues to throw off heat until all the nearby uranium is converted into lighter elements, such as boron, krypton, strontium, and xenon. Those smaller fractions are known as fission products.

Atomic power reactors use the heat of the reaction to generate steam for turbines, which spin generators to make electricity. The reaction process is controlled with movable carbon rods. Carbon is an element that can absorb neutrons, so that the speed and number of neutrons bouncing around the reaction chamber can be regulated by raising or lowering the rods.

MAGNETISM

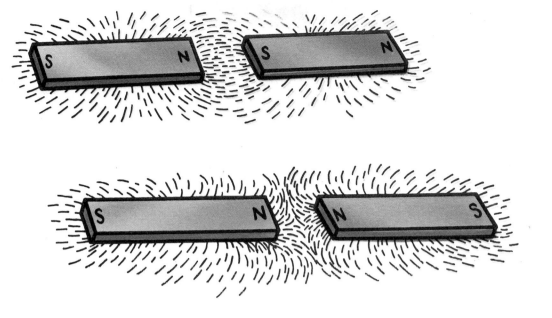

Magnetism is the power of repulsion or attraction that can be induced with an electric current in certain metals. Some metals, usually steel or steel alloys, retain those magnetic properties and are called permanent magnets. Other materials such as iron tend to lose the magnetic property almost immediately once the power is removed, making them useful for **electromagnets** (see page 24).

Magnetism can be visualized as lines of force flowing around and through materials that have this magnetic property. Every magnet has two opposite poles, called a north pole and a south pole. The magnetic lines of force flow out of the north pole toward the south pole, then through the magnet back to the north pole, so that they always form a continuous loop.

The property of magnetism is vital to the generation of electric currents, and magnetism is also involved in many of the devices such as motors, relays, and solenoids that put electricity to work. Fast, smooth, frictionless trains and automobiles are a dream with great promise for the transportation systems of tomorrow, and it will be superconductive electromagnets levitating them above the ground and zipping them to their destination that makes this dream a real possibility (see **Superconductors,** page 32).

The magnetic field around a magnet can be demonstrated by sprinkling iron filings on a sheet of white paper and placing a magnet beneath it. The filings align themselves along the lines of force. The magnetic field is strongest at the poles, where the lines of force converge. When opposite poles of two magnets are placed near each other, the lines of force flow from one magnet toward the other, tending to pull them together. When like poles of two magnets are placed near each other, the lines of force flowing from the north to the south poles of each magnet push against each other, and the magnets repel each other.

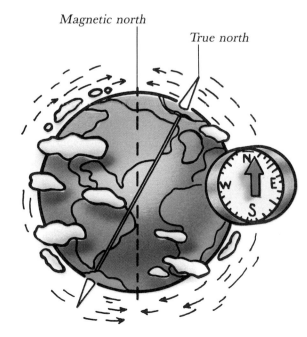

Magnetic north

True north

COMPASS

The best-known magnet of all is the Earth itself, and as in the bar magnet, the lines of force align themselves from north to south. This fact allows you to use a magnetized needle, pivoted in the center, to find north. The needle aligns itself along the Earth's lines of force, indicating the north and south magnetic poles. The Earth's magnetic poles are slightly apart from the geographic poles, so that in most locations a magnetic compass needle doesn't point exactly to the geographic North Pole.

THE ELECTRICAL CIRCUIT

An electric circuit can be compared in certain ways to a plumbing system. The water pipe is the equivalent of the conductor, or the wires in an electric circuit. The water is the equivalent of the electric current flowing through the wires.

Electricity occurs in nature—in thunderstorms and in friction when certain materials are rubbed together. Actually, everything contains electricity. All matter consists of atoms, which in turn have positively charged protons in a nucleus with negatively charged electrons orbiting outside. (See **Energy from atoms,** page 15.) The electrons in the outermost orbit are the least attracted to the nucleus, and by bumping these free electrons from atom to atom, electron flow, or electricity, is caused. The chain reaction is like pushing a marble into one end of a tube already filled with marbles, expelling another from the opposite end.

The atomic structure of any substance determines whether it will permit easy electron flow or not. Materials such as aluminum and copper, which have large numbers of free electrons, are **conductors** and encourage electron flow. For that reason, these are the most common materials used for electric wires. Those materials with few free electrons, such as plastic and glass, are **insulators,** which block the flow of electrons. Some materials, such as silicon, conduct electricity under one set of conditions but act as insulators under another set of conditions. These materials are called **semiconductors.**

*4. The water falling into the ladles makes the drill turn. How much power the water can divert to the drill depends on the pump's force and the rate at which the water travels through the system. In an electric circuit, the power consumption in **watts** of an appliance or other electrical device is equal to the current in amperes multiplied by the force in volts.*

5. Once the water has done its work and falls back into the container, it has lost its energy. An electric circuit must provide a path so that the electrons return to the power source after they have done their work. The returning electrons enter the positive pole of the electric circuit.

*1. The pump forces water through the pipe with a certain amount of pressure. This is analogous to **voltage,** which is the pressure or force that bumps electrons through the conductor. A voltmeter measures the force that makes the electrons flow in units of volts. The source of electrons in an electric circuit is the negative pole, as opposed to the positive pole. Most household receptacle circuits in the United States are rated at 115 volts; most automotive and boat circuits are 12 volts.*

3. If the opening through the pipe is made smaller, the flow of water is restricted. The restriction to the flow of electrons in an electric circuit is called **resistance,** which is measured in **ohms.** Insulators have a high resistance; conductors have a low resistance. Lowering the resistance in a circuit permits greater current flow.

2. Water flows through the pipes at so many gallons per second. The volume or flow rate of electrons through a conductor is the **current,** which is measured in **amperes,** or amps. Amps are measured with an ammeter.

(continued)

THE ELECTRICAL CIRCUIT

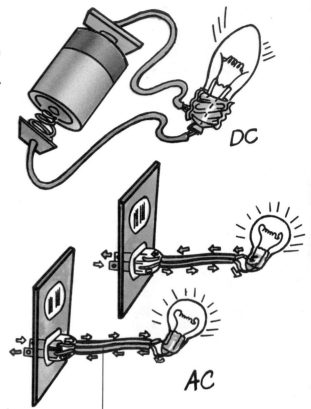

DC

AC

The current reverses itself 60 times per second.

DIRECT CURRENT AND ALTERNATING CURRENT

In a **direct current (DC)** power supply, the current flows in one direction only, from the negative terminal to the positive, and the voltage is of a constant value. The current from a battery is DC.

The current and voltage in **alternating current (AC)** circuits shifts between positive and negative, because the current repeatedly changes direction. This is the type of output from typical generators and is the type found in household circuits in the United States. In a 110/120-volt 60-cycle AC current,

the voltage rises to a peak of approximately +120 volts and then, as the current reverses direction, drops to −120 volts. This cycle occurs 60 times each second. Electric appliances, of course, continue to work no matter which direction the current flows in. This rise and collapse is important because an electric current moving through a wire produces a magnetic field. When the current reverses itself, so does the magnetic field. This principle is used in **induction motors** (see page 27).

SERIES CONNECTION

24V
5 amps capacity

PARALLEL CONNECTION

12V
10 amps capacity

12V battery, 5 amps capacity

SERIES AND PARALLEL CIRCUITS

Sources of electric power, such as batteries, can be combined in two basic ways to provide different results. In a **series** installation of two batteries, the positive pole of one battery is connected to the negative pole of another, leaving one positive pole and one negative pole to be connected into the circuit. This increases the voltage, but the current capacity remains the same. Batteries are connected in **parallel** by connecting both positive poles and both negative poles to the circuit. The voltage output remains the same, but the power capacity is increased. Parallel circuits are often used to provide the necessary power for starting diesel engines. Electrically wired components can also be connected in series and in parallel.

CONTINUITY TESTER

The primary purpose of a continuity tester is to tell you if a complete circuit exists. You attach one of the probes to one prong on the plug of an electric appliance and touch the other probe to the other prong. If the circuit is complete, a battery passes a small current through the tester and lights a lamp in the tester. If the light doesn't go on, the circuit is incomplete—there is a break in it somewhere.

BATTERIES

Batteries convert chemical energy into electrical energy. There are two basic types of batteries—storage batteries, or "wet-cell" batteries, and dry-cell batteries.

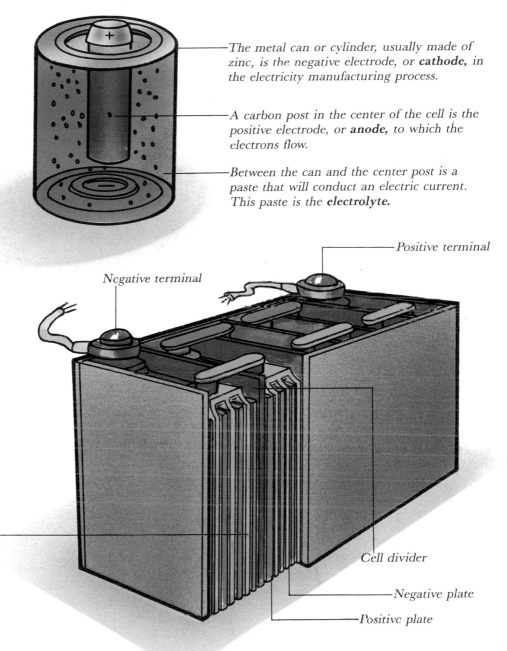

The metal can or cylinder, usually made of zinc, is the negative electrode, or **cathode,** in the electricity manufacturing process.

A carbon post in the center of the cell is the positive electrode, or **anode,** to which the electrons flow.

Between the can and the center post is a paste that will conduct an electric current. This paste is the **electrolyte.**

FLASHLIGHT BATTERY

The ordinary flashlight battery is a dry-cell battery. Modern technology has produced dry-cell batteries that can power a watch, a flashlight, or burglar and fire alarm systems for many years.

As a rule, the larger the battery the longer it will last. AA cells last longer than AAA cells, and D batteries last longer than C batteries.

Batteries are sized for their application and the rate of power consumption that is anticipated. The voltage output of a dry cell is 1½ volts per cell. To obtain more voltage, cells can be connected together in a series (see page 20); to obtain greater capacity, or longer-lasting output, cells can be connected in parallel (see page 20). This is often done inside a single housing, such as in a 9-volt battery.

Negative terminal

Positive terminal

The two sets of plates are interleaved together in each cell.

Cell divider

Negative plate

Positive plate

AUTOMOBILE BATTERY

Storage batteries, such as an automobile battery, contain an acid electrolyte. A cell consists of two sets of metal plates submerged in the solution. Most commonly, one set of plates is made of lead peroxide and the other is a porous lead casting called sponge lead. Insulating cell dividers separate the sets of plates for each cell. An automobile battery is a 12-volt battery that has six 2-volt cells connected in **series** (see page 20).

While the battery is in use, the active material of the plates interacts with the acid electrolyte to produce electricity. To charge the battery, the connections are reversed, which reverses the chemical reaction. Sometimes distilled water is added to the electrolyte to maintain the fluid level.

A device called a hydrometer can measure the strength of the acid solution. Eventually some of the lead material from the plates is lost to the bottom of the battery. The battery can no longer hold a charge, and it has reached the end of its useful life.

SWITCHES

Electric current can flow only if the circuit is complete. An open circuit—one that has a gap within it—prevents an electrical device from operating. By creating a gap intentionally, we can turn the device off and on. That is what a switch does.

Inside a switch are two or more terminals. When the switch is turned off, nothing connects the two terminals, and the circuit is interrupted just as if a wire had been removed from the device. When the switch is turned on, the two terminals are connected, the circuit is closed, and current can flow again.

This opening and closing action occurs very quickly to prevent electrical sparking between the contacts. The current tries to bridge the gap between the contacts as the gap narrows; the heavier the current flowing in the circuit, the more electric sparking can occur. Some switches operate with a tube of mercury, which conducts electricity. When the switch is on, the tube tilts, and mercury flows down between the two contacts, When the switch is off, the mercury flows away from the contacts, and the circuit is open. This design reduces the possibility of sparking and also makes the operation quiet.

TOGGLE SWITCH ▶

A U-shaped metal yoke provides the connection between the two terminals. When the switch is off, the yoke is held away from the terminals by a spring.

Pushing the switch lever to the "on" position compresses the spring, which snaps to the opposite side of the yoke. The spring then expands again, forcing the yoke against the terminals. Contact is made between the two terminals, and electricity flows.

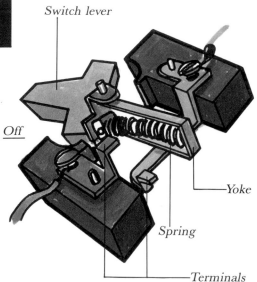

Switch lever

Off

Yoke

Spring

Terminals

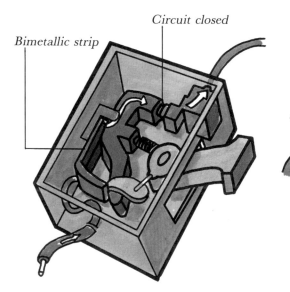

Bimetallic strip

Circuit closed

On

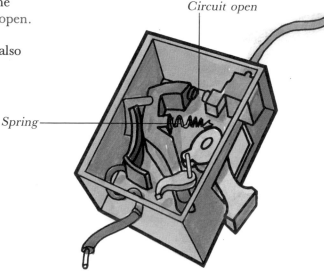

Circuit open

Spring

◀ CIRCUIT BREAKER

A circuit breaker is also a type of switch. It is designed to open the electrical circuit if the circuit is asked to carry more current than it can handle. Too many appliances plugged into the circuit, or a large appliance containing a large heating element plugged into the circuit, can cause an overload of the circuit.

Inside the circuit breaker is a latch consisting of a **bimetallic strip.** When this bimetallic strip is exposed to enough heat, it warps because of the different rates of expansion of the two metals. This warping opens the circuit. A built-in heating element generates this heat when the circuit is overloaded.

GENERATORS

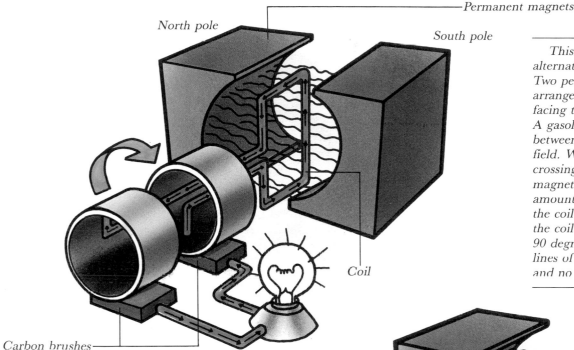

North pole

Permanent magnets

South pole

Coil

Carbon brushes

This generator produces alternating current (see page 20). Two permanent magnets are arranged with the north pole of one facing the south pole of the other. A gasoline engine rotates a coil between them, within the magnetic field. When the moving coil is crossing the lines of force of the magnetic field, the maximum amount of voltage is produced in the coil. This amount decreases as the coil rotates. When it has rotated 90 degrees, it is not crossing the lines of force of the magnetic field, and no current is produced.

A generator generates electricity. Small generators are found in many houses for use during power outages or to power outdoor equipment. Motor homes and boats use them to provide power for electric appliances. And, of course, very large generators provide the power that flows across transmission lines from power plants into our houses.

Magnetism makes the generator work. A magnetic field can cause electric current to flow in a wire that's moving in that field. If you wind a long conducting wire into a coil and move it through a strong magnetic field, an extremely large amount of voltage, or **electromotive force,** can be generated. The current produced by this voltage passes through conductors or wires to the point where the electricity will be used.

It continues to rotate another 90 degrees and once again crosses the magnetic field, but in the opposite direction. At this point the current is flowing the opposite way, producing alternating current.

The current is transferred away from the generator by means of two metal rings attached to the coil. These rings rub against pieces of carbon called brushes, which pick up the current.

23

ELECTROMAGNET

Any wire conducting electricity generates a circular field of magnetism around it, with the wire at the center. The greater the amount of current flow in the wire, the more magnetic force is generated around the wire. With a single wire, this is usually a very small amount of force, but it can be multiplied many times by wrapping the wire in coils around a soft iron core. This creates an electromagnet. It can be used to repel and attract, just as any other magnet can, but simply by turning the power off or on, the magnetism is turned off or on.

Electromagnets have many uses in home and industry. They are often used in conjunction with a switch to form a device called a **relay.** A very small amount of current can be used to turn the electromagnet on. The force of the generated magnetism can then open a large set of switch contacts carrying thousands of times more current than the circuit that turns on the magnet. The relay of a thermostatic control circuit for a central air conditioner works in this way. The thermostat controls only a low-voltage electromagnet; this in turn operates contacts that do the really big job of turning the air conditioner on.

Electromagnets can also be used to perform mechanical functions such as shifting gears in a washing machine. These magnetic coils are usually stronger than those used in most power relays. Electromagnets that create motion, or mechanical energy, from electrical energy are called **solenoids.**

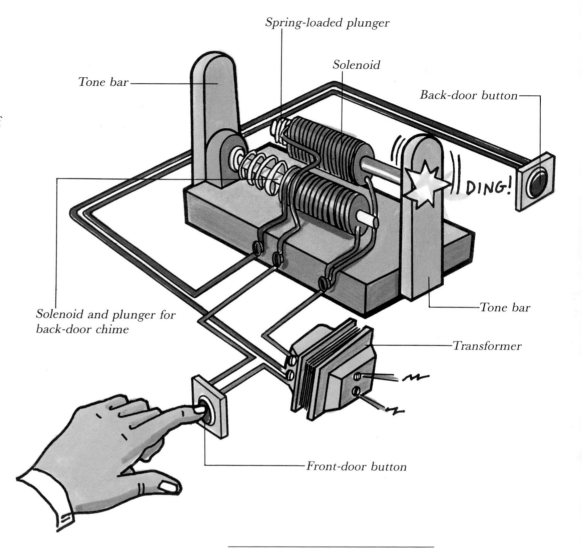

Spring-loaded plunger

Solenoid

Tone bar

Back-door button

DING!

Solenoid and plunger for back-door chime

Tone bar

Transformer

Front-door button

ELECTRIC DOOR CHIME

The door chime contains two solenoids, which use electrical energy to move a rod that rings the chime.

When you push the button, the circuit is closed, and power from the transformer flows through the coil, forming an electromagnet. An iron rod on a spring is then drawn inside the solenoid and hits the right-hand tone bar.

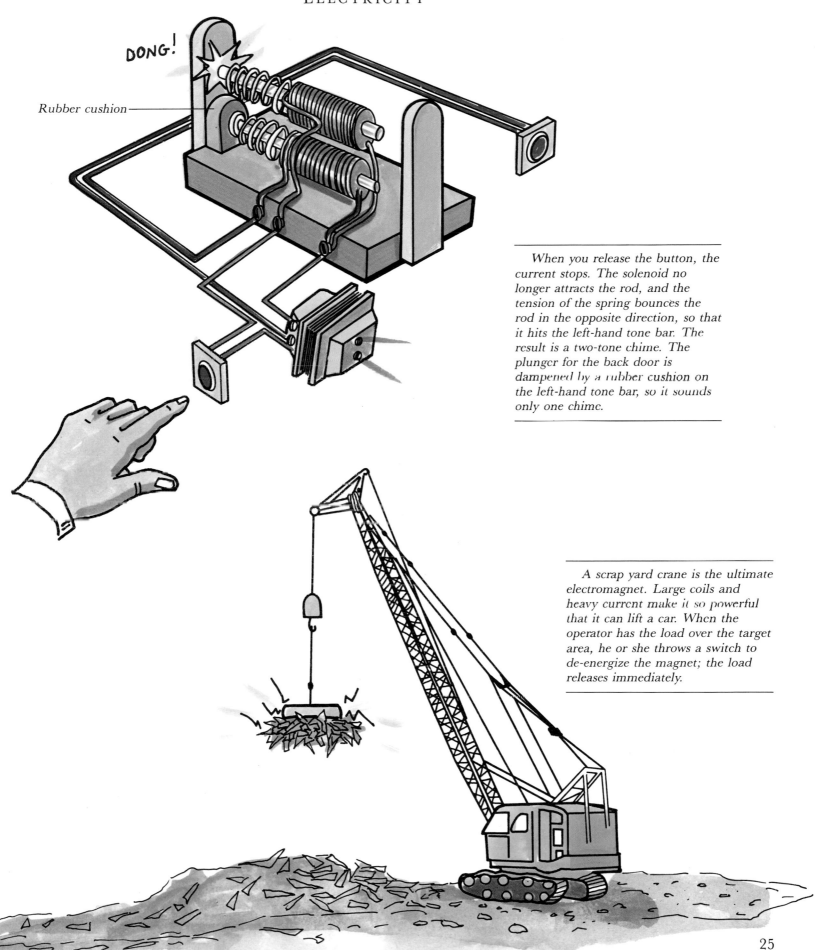

DONG!

Rubber cushion

When you release the button, the current stops. The solenoid no longer attracts the rod, and the tension of the spring bounces the rod in the opposite direction, so that it hits the left-hand tone bar. The result is a two-tone chime. The plunger for the back door is dampened by a rubber cushion on the left-hand tone bar, so it sounds only one chime.

A scrap yard crane is the ultimate electromagnet. Large coils and heavy current make it so powerful that it can lift a car. When the operator has the load over the target area, he or she throws a switch to de-energize the magnet; the load releases immediately.

MOTORS

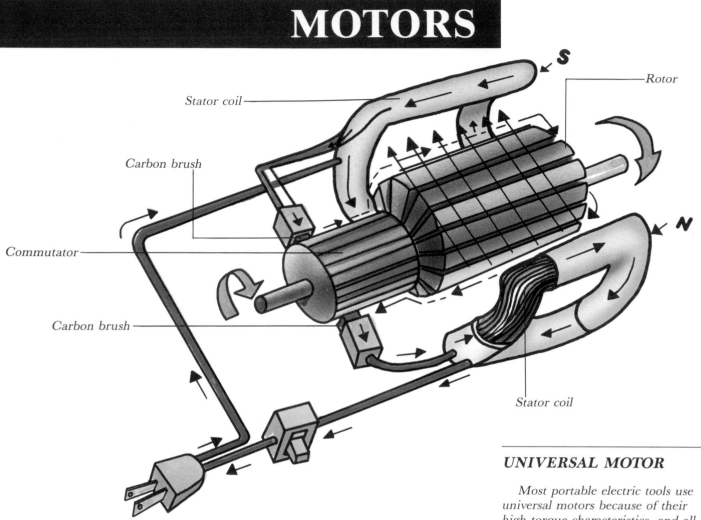

Stator coil

Rotor

Carbon brush

Commutator

Carbon brush

Stator coil

It has been estimated that the average household contains more than 40 electric motors. They power the refrigerator, the washer and dryer, the vacuum cleaner, the record player, the pump that brings in water, and many, many other appliances.

A motor is nothing more than a device that converts electrical current into rotary motion. The basis of the motor is that opposite magnetic poles attract each other and like poles repel each other. A motor sets up a stationary magnetic field and another magnetic field that is free to rotate within it. The rotating magnetic field rotates in an attempt to line up its poles with the opposite poles of the stationary field.

There are two basic types of motors—the universal motor and the induction motor. The universal motor is so called because it will operate on either AC (alternating current) or DC (direct current). It has strong **torque** (twisting force) from the time that power is applied. Induction motors, on the other hand, have low starting torque and operate on alternating current only. Their strong points are low maintenance and smooth operation. Every motor is an **electromagnetic** device (see page 24), but the ways they accomplish their work differ greatly.

UNIVERSAL MOTOR

Most portable electric tools use universal motors because of their high torque characteristics, and all cordless tools use them because they will operate on the direct-current power provided by rechargeable batteries.

The universal motor has an **armature,** or **rotor,** which is free to rotate within a magnetic field. The magnetic field is generated by two **stator** (stationary) coils consisting of many turns of fine wire. The stator coils are on either side of the rotor.

The rotor is wound with coils of wire. Each coil is connected to two opposite copper bars in the **commutator,** which is a cylindrical arrangement of copper bars on one end of the rotor. Two carbon brushes rub against a pair of opposite bars of the commutator. These brushes are pieces of carbon that are soft enough to make good contact with the bars. They are connected by wire to the power source.

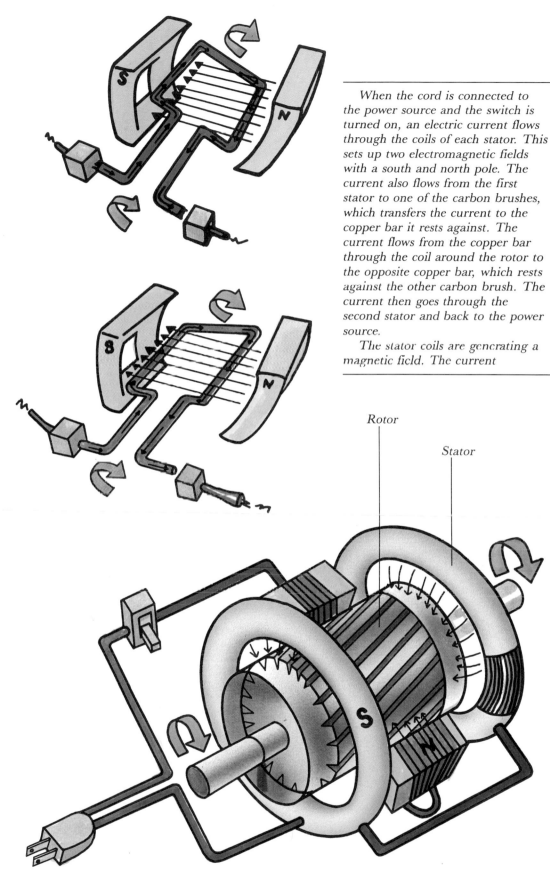

Rotor

Stator

When the cord is connected to the power source and the switch is turned on, an electric current flows through the coils of each stator. This sets up two electromagnetic fields with a south and north pole. The current also flows from the first stator to one of the carbon brushes, which transfers the current to the copper bar it rests against. The current flows from the copper bar through the coil around the rotor to the opposite copper bar, which rests against the other carbon brush. The current then goes through the second stator and back to the power source.

The stator coils are generating a magnetic field. The current traveling through the commutator bars and the rotor are generating an opposing magnetic field, which repels the field generated by the stator. The rotor tries to line itself up with the stator's magnetic field by rotating slightly. But by doing so, the carbon brushes in the commutator are now in contact with two different copper bars. The current is removed from the original coils and is applied to a different set of coils. Once again, the two magnetic fields are in opposition. The result is that the rotor continues to rotate in a futile attempt to line itself up with the magnetic current of the stator.

INDUCTION MOTOR

The motor used in most major appliances is an induction motor, which operates only on alternating current (see page 20).

The stator in an induction motor contains many coils of fine wires, much as the stator in a universal motor does. The rotor, however, has no coils or electrical connections, no brushes, and no commutator.

When power is supplied to the stator coils, a strong magnetic field is created. This magnetic field induces an electric current in the rotor. (The induction of this current gives the induction motor its name.) The electric current in the rotor sets up a magnetic field in opposition to that of the stator. The rotor turns in order to align the magnetic fields. By that time, however, the alternating current has alternated, reversing the polarity of the stator coil's magnetic field. With a 60-cycle alternating current, this happens 120 times per second, so the rotor continues to rotate.

TRANSISTORS

In the middle of the 20th century, television sets had tiny screens and huge cabinets; "portable" radios were about the size of two shoe boxes and weighed ten pounds or more; calculators that could perform elementary mathematical functions were about the size of one shoe box. Computers—the few that existed at the time—filled huge rooms the size of an entire floor of an office building.

Today, we have portable radios so small we can wear them while jogging. We can wear a television receiver on our wrist. Calculators that perform functions beyond the imagination of the mathematicians of the 1940s are carried in shirt pockets; they could be even smaller, but the buttons would be too small for our fingers. The computers that were used in writing this book are smaller than many portable radios of the '50s, and they outperform in every way—speed, capacity, and durability—the pioneer computers that filled a 30 × 50-foot room. All of this was made possible by the development of the **transistor**—and its later evolution into the integrated circuit or **microchip** (see page 30).

The end function of most electronic devices is to amplify a tiny signal into a larger one that we can hear or see. For example, the radio amplifies tiny impulses in the air into signals powerful enough to drive a loudspeaker. Until the transistor was invented, this was accomplished with vacuum tubes. They consume a great deal of power, since they depend on heated filaments. They generate a great deal of heat, they are relatively large in size, and they have a relatively short life. The transistor uses little power and generates almost no heat, and some varieties can last indefinitely.

Transistors are made of semiconductors, and they amplify electric signals because of the behavior of semiconductors. Semiconductor materials will conduct electricity under one set of conditions but will act as an insulator under another set of conditions. This is primarily due to their atomic composition; their atoms have electrons that can be shared by neighboring atoms. The best-known semiconductors are silicon and germanium. Semiconductor materials are usually crystalline and, in their pure form, behave like an insulator. They are made into semiconductors by adding an impurity in a process called "doping." This impurity disturbs the arrangement of electrons, which gives the material the ability to conduct electricity.

If the material is doped with one type of impurity, more free electrons are created. An electric current will cause the electrons, which have a negative charge, to flow toward a positive terminal or away from a negative terminal. This is an **n-type** semiconductor.

Another type of impurity will create vacancies for electrons to fill. These vacancies are called "holes," and they carry a positive charge. This is a **p-type** semiconductor, in which the current is carried by the positively charged holes. The position of a hole is changed when it is filled with a free electron; another hole is created in a nearby atom. In this sense, the holes move in one direction as the electrons move in another.

A transistor is a sandwich made of these two types of semiconductors. It can be made of a p-type layer sandwiched between two n-type layers, or it can be the opposite. The two constructions work in similar ways.

In a transistor with a p-type layer between two n-type layers, one of the n-type layers is an emitter (of electrons); the p-type layer is the base; the other n-type layer is the collector. A small current fed between the emitter and the base causes the electrons in the emitter to travel toward the holes in the base.

The key feature is the size of the base, which must be thin enough for electrons to move through it from the emitter to the collector. This results in a much larger current in the collector, since it collects many more electrons as they move from the emitter across the two junctions into the collector.

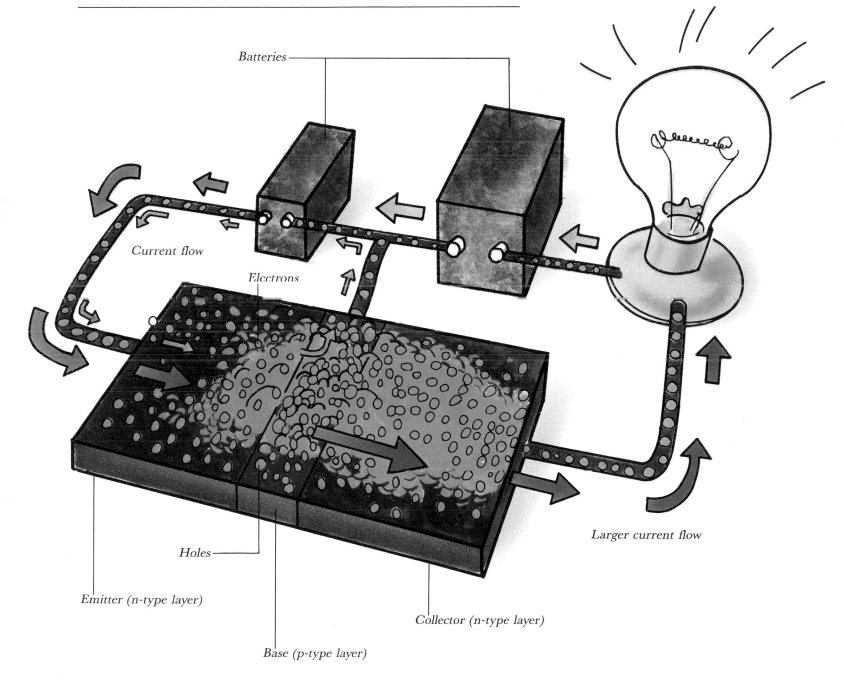

Batteries

Current flow

Electrons

Holes

Emitter (n-type layer)

Base (p-type layer)

Collector (n-type layer)

Larger current flow

MICROCHIP

The microchip, or integrated circuit, not only allows for smaller electronic products, but also makes those products cheaper and dramatically more reliable. The minuscule size of a microchip actually improves its performance, because all its working parts are so close together. It can behave like a small computer, enabling us to have telephones that redial the last number called, automobile ignition and fuel systems that are constantly computer-controlled, and personal computers that are more powerful than the room-size computers of a generation ago.

Microchips are very tiny assemblages of electronic components on slices of silicon. They vary in design according to the task each microchip is meant to perform. They can contain tens of thousands of transistors, along with **capacitors** to receive and store electricity, **resistors** to govern the current or voltage through the system, and **diodes** to allow the current to flow in only one direction.

All these electronic components are manufactured into the microchip directly. In the same way that a semiconductor is produced (see page 32), a piece of silicon is treated with certain impurities in specific regions. These regions form the individual components. A layer of metal is set on the surface of the chip to connect the components.

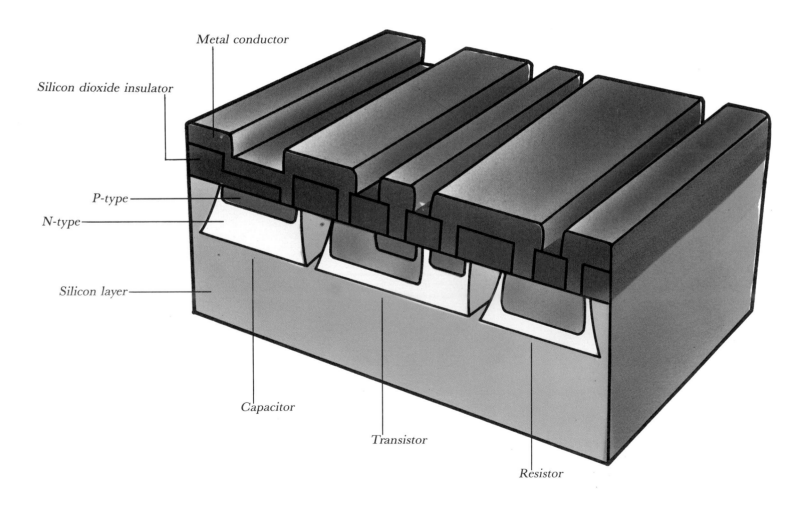

Metal conductor

Silicon dioxide insulator

P-type

N-type

Silicon layer

Capacitor

Transistor

Resistor

QUARTZ WATCH

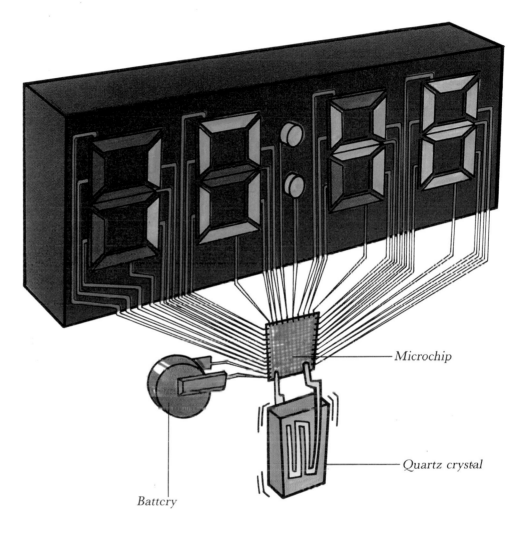

Certain crystals generate an electric current when subjected to stress and vibrate or change their thickness when exposed to an electric current. This property is known as **piezoelectricity.** Quartz is such a crystal; quartz crystals vibrate at a particular frequency when electricity is passed through them. Because its oscillations, or variations from one extreme to the other, are so regular, a quartz crystal can be used in making extremely accurate watches and clocks. Furthermore, such watches never need winding, because they are powered by batteries.

The faster the oscillation of the quartz crystal, the more accurate the watch or clock. Manmade crystals are used today in order to get a crystal that vibrates at the most desirable frequency. The first quartz watches oscillated at about 8,000 vibrations per second. They were accurate within seconds per month and sold for more than $2,000. Today, most quartz watches oscillate at 32,768 vibrations per second, and for less than $30 you can purchase one that may be accurate to within seconds per year.

Microchip

Quartz crystal

Battery

A quartz crystal watch with a digital display has no mechanical components. A bar-shaped sliver of quartz receives an electric current from a battery, which causes the quartz to vibrate at its natural frequency of 32,768 pulses per second. The quartz crystal is connected by an electrode to a microchip, which counts these vibrations and sends a signal to the digital display to advance the time one second for every 32,768 pulses. Either a liquid crystal display (LCD; see page 155) or a light-emitting diode (LED; see page 59) can be used for the digital display.

Some watches with hands rather than digital displays are governed by a quartz crystal. A microchip converts the quartz's vibrations to the gradations of the dial and operates a motor that drives gears to move the hands.

SUPERCONDUCTORS

A superconductor is a material that loses all resistance to the flow of electrical current once the material has been cooled to a certain temperature. The potential advantage of such a property is tremendous for any type of use in which the heat of resistance to electricity is enough to significantly reduce the device's efficiency. This is true of almost every device that deals with large currents. Motors and generators could operate without the tremendous efficiency losses they now suffer. A single power line could carry the total output of an entire electric power plant. Superconducting magnets could power all forms of transportation, levitating trains and automobiles above their path in a silky-smooth, frictionless self-propelling glide. And superconducting components could have a greater effect on electronics than that of the transistor and microchip.

What stands in the way of this dream world is the fact that superconducting materials have to be cooled to extremely low temperatures in order to become superconductors. The temperature at which the change occurs, which is different for different materials, is called the transition temperature.

For a long period of time, superconductivity could only be accomplished by cooling the material with liquid helium at temperatures approaching absolute zero (0° Kelvin, or –459.69°F). Recently, scientists have discovered new ceramic compounds whose transition temperatures are higher—around 90° Kelvin—so that they can be cooled to that temperature with liquid nitrogen, which is much cheaper than liquid helium.

In a conductor such as copper, an electric current is produced when electrons flow through the material. The structure of the material is in the form of a crystal lattice. Some of the electrons scatter around because they encounter imperfections in the lattice structure. The movement of the electrons causes heat, which makes the lattice structure vibrate. This in turn hinders the movement of the electrons.

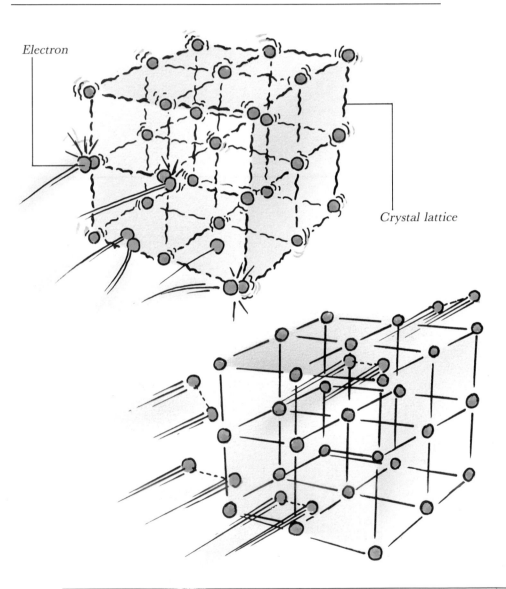

Electron

Crystal lattice

At supercold temperatures, these effects disappear. The current theory as to why that happens is the BCS theory, named for the three physicists—Bardeen, Cooper, and Schrieffer—who proposed it. The theory holds that at the transition temperature, for reasons that defy precise explanation, the electrons form tightly bound pairs with a very large combined diameter. These pairs can sweep through the lattice unhindered, and all resistance to current flow vanishes.

SUPERCONDUCTING MAGLEV

Magnetically levitated trains (see **Maglev,** page 182) are suspended above guideways by electromagnets that act to repel each other. Superconducting maglevs, which are still in the experimental stage, use liquid helium at a temperature close to absolute zero to create superconducting electromagnetic coils. Because of the lack of electrical resistance in the superconductors, the train would use fuel only to cool the liquid helium and overcome the friction of the air. Existing maglevs run on a cushion of air less than an inch deep; superconducting maglevs could maintain a six-inch distance from the guideway with greater stability.

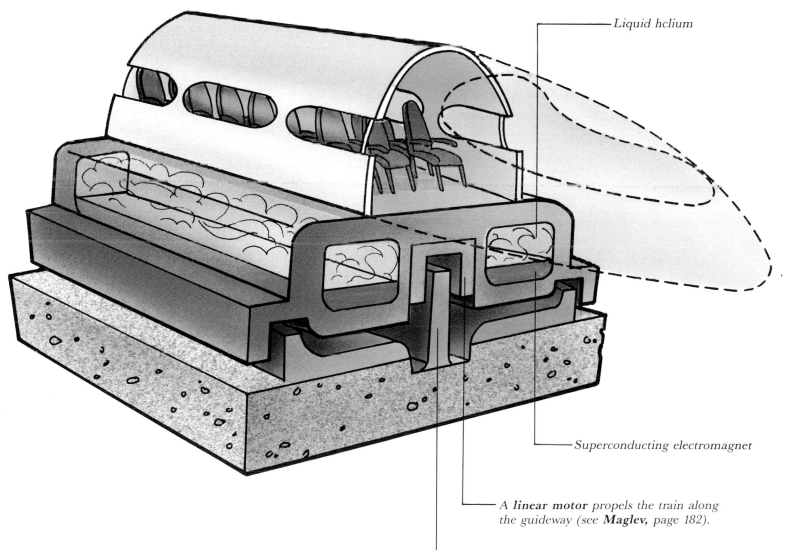

Liquid helium

Superconducting electromagnet

A **linear motor** propels the train along the guideway (see **Maglev,** page 182).

Aluminum guideway

LEVERS

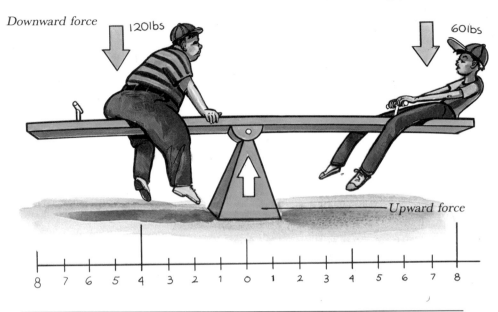

Downward force 120lbs *Downward force* 60lbs

Upward force

8 7 6 5 4 3 2 1 0 1 2 3 4 5 6 7 8

When you ride a teeter-totter, the rigid plank that you sit on is actually a lever with its **fulcrum,** or balance point, in the center. The teeter-totter, like any lever, has three forces acting on it at different points along its length. The two forces at the ends act in one direction (down), and the third force at the fulcrum acts in the opposite direction (up). The force at one end of the lever might be much greater than the force at the other. By positioning the riders properly in relation to the fulcrum, a light load at one end can balance a heavy load at the other end. This increase in force for the light load is the result of the **mechanical advantage** that the lever provides (see page 35).

TEETER-TOTTER

To balance the teeter-totter, the heavier person must be closer to the center and the lighter person farther out. A lever is balanced when the weight times the distance to the fulcrum is the same on each side. If a 120-pound child rides the teeter-totter with a 60-pound child, the 60-pound child sits twice as far from the fulcrum as the 120-pound child. By means of the teeter-totter, the lighter child is lifting twice his or her own weight but must move twice as far as the heavier child to do so.

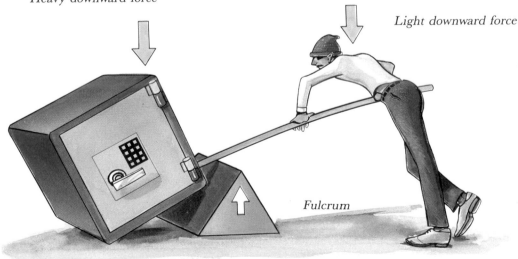

Heavy downward force

Light downward force

Fulcrum

The fulcrum has to be sturdy enough to bear the sum of both the light and heavy forces.

CROWBAR

A crowbar is a lever used to move heavy weights. The person applies force at the far end of the crowbar, while the point or working end is wedged under the heavy object to be moved. This places the fulcrum quite close to the working end. If the long section is five feet and the short section is one foot, a 150-pound person can lift 750 pounds.

WEDGES

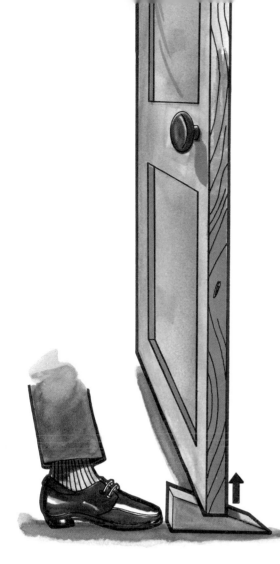

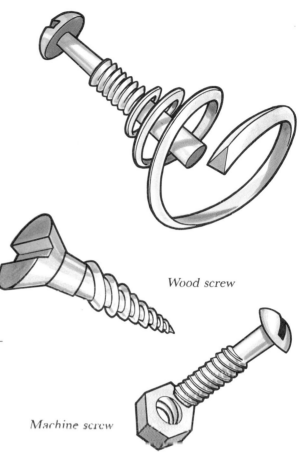

DOORSTOP

A door that closes automatically can be held open with a wedge. By nudging the short side of the doorstop lightly with your toe, you can exert enough pressure to keep the door open. Propping it open with your foot requires much more effort. The light pressure of your foot on the short side of the doorstop is magnified by the long, sloping side of the doorstop to a greater force that is exerted upward on the door and downward on the floor, creating a large frictional resistance.

Wood screw

Machine screw

LOG SPLITTER

To pry a log apart would require a great deal of effort. By pounding a wedge into a crack in the log, the force of the hammer is magnified outward, and the log splits apart. However, friction hurts, rather than helps; you need to hammer harder to overcome it.

B y using a wedge, a force can be magnified to a much greater force. The effort applied to the short side of the wedge is small compared with the force it exerts out from the long, sloping side. This magnification of force is called **mechanical advantage,** which enables you to do more work with a tool than you could without it. The amount that a wedge can magnify the force is the length of the long, sloping side divided by the height of the short side (although friction can reduce that force considerably). If the long, sloping side is two inches long and the short side is one inch high, the force the wedge exerts would be twice the force that is applied to it, if there were no friction. Wedges can be used to lift heavy weights, to split logs, and to plow the soil.

SCREWS

The screw is merely a long wedge wrapped around a rod so that it forms a spiral. The thread of the screw is the long surface of the wedge; the shaft of the screw is the shorter one. When you turn the screw with a screwdriver, you are effectively forcing the wedge forward. That twisting force, or **torque,** is magnified by the wrapped wedge. The magnified force pulls the screw down.

A wood screw has a tapered shaft in order to let the screw force the wood fibers apart. The magnified force pulls the screw into the wood.

A machine screw is not tapered; the magnified force from the thread pulls the nut and screw together to provide a clamping force.

WHEELS AND AXLES

For thousands of years, people moved things by lifting or dragging them. Even as recently as the seventeenth century, native Americans could move only what they could carry themselves or put on horseback or have a horse drag on a stretcher; they learned of that extraordinary invention, the wheel, from the European settlers.

Wheeled vehicles are effective because rolling friction is much lower than sliding friction. (See **Friction,** page 7.) Round sticks used as rollers under heavy objects were the forerunners of the wheel; the rollers, of course, had to be moved from behind the load to the front. The Egyptian pyramids were built out of immensely heavy stone blocks that were moved in this manner.

Eventually a way was found to eliminate the perpetual recirculating of the round sticks. Each end of the stick was whittled down and inserted into a drilled hole in a block, which was then attached to a platform. This drilled block was a **bearing.** It was a short step from this device to disc-shaped wheels and spoked wheels. Finally, ball bearings or roller bearings reduced the friction to almost negligible levels.

*The invention of the **bearing** made it possible to have a roller that could roll over the ground while still remaining attached to a platform.*

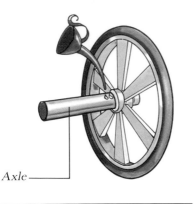

Axle

*The typical wheel still has a hub rotating on a shaft called the **axle.** Several types of bearings can be used. With a "plain bearing," the hub rotates directly on a fixed axle. The sliding action requires a lubricant to reduce friction.*

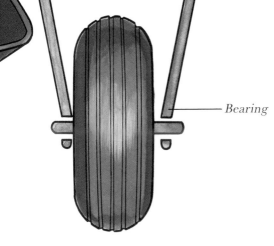

Bearing

Bearing

*A **ball bearing** provides a secondary rolling action between the axle and the hub of the wheel. This eliminates friction almost entirely.*

UNICYCLE

The unicycle consists of a single spoked wheel mounted between two bearings, which are attached to the seat structure. Pedals are attached to either side of the axle. Every time the rider's foot pedals one revolution, the wheel turns one revolution. To stay in balance, the rider leans forward or backward until the unicycle is about to fall and then pedals quickly to straighten up.

SPRINGS

Most springs are made of a special kind of steel that will bend or twist under pressure rather than break and, just as important, will return to its original shape once the pressure is released. Certain plastics have been designed to work as springs and, of course, rubber has springlike qualities as well.

The elasticity of springs make them useful for storing and furnishing energy, absorbing shock, and holding items in position. They are used to return pressed telephone buttons to their original position, as paper clips to hold a number of sheets of paper neatly, to hold the tumblers in a lock against the key (see **Locks**, page 152), and to close doors.

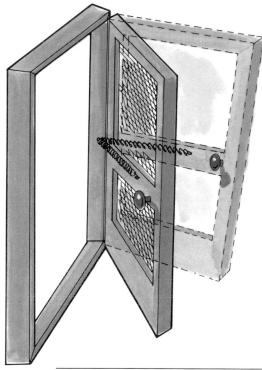

SCREEN DOOR

A screen door is held shut by a long coil spring. When you push open the door, the spring uncoils and extends. You feel resistance to the movement of the door, but it is easily overcome. When the door is released, the spring returns to its original length, pulling the door shut.

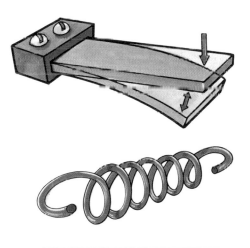

Some leaf springs are simply flat strips of metal that are fixed at one end and free to flex at the other. They look just like miniature diving boards projected or cantilevered out so as to permit the free end to move. Others are fastened at both ends and flex when force is applied to their centers.

Coil springs are made of wire wound into a helical shape. When the coil is extended or compressed, the twisted wire resists. When the force is released, the coil returns to its original length.

AUTOMOBILE SUSPENSION SYSTEM

A leaf spring attached to the chassis is often included in automobile suspension systems. A length of spring steel is shaped into an arc. The ends of the arc are attached to the chassis; the center is fixed to the wheel axle. The leaf spring permits the wheel axle to move up and down with small bumps in the road without jarring everybody in the car. To stiffen the spring, several leaves clamped together are often used. If these springs are of different lengths, a progressive spring—one that's soft when it starts to flex, but stiffens as it flexes farther—is possible.

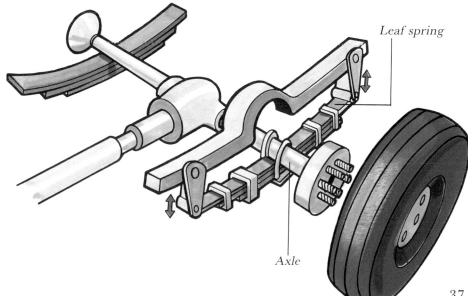

Leaf spring

Axle

CAMS AND CRANKS

A **crank** consists of an arm attached at right angles to a shaft. A second, short shaft acts as a handle, which transfers force to or from the shaft through the crank. A handcrank acts as a lever that rotates around its fulcrum. The circular motion of the arm and shaft can convert rotational movement into back-and-forth (**reciprocating**) movement. It can also convert reciprocating movement to rotational movement.

WISHING WELL

The crank handle moves in a circular direction, winding the rope around a cylinder and raising or lowering the bucket in an up-and-down, or reciprocating, movement. Less force is needed to turn the handle than would be used to lift the rope with your arms. This is because the distance between the handle and the axis is greater than the distance between the coiled rope and the axis. In turning the crank, your arm is traveling a long distance, but exerting greater force on the cylinder with the rope around it. The light force exerted in turning the handle times the distance between the handle and the axis is equal to the force (weight) of the bucket times the distance of the coiled rope from the axis. This is the mechanical advantage of the crank.

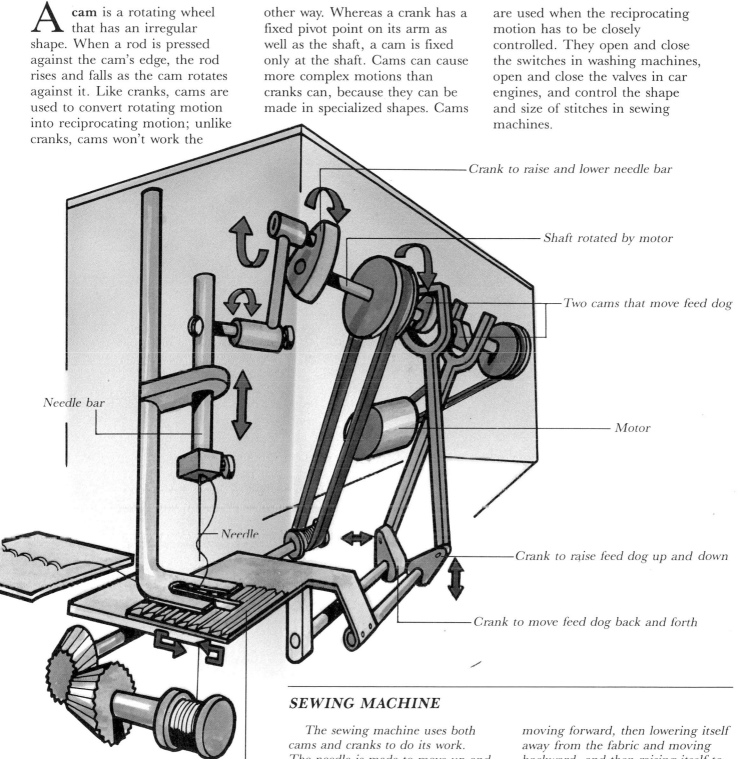

A cam is a rotating wheel that has an irregular shape. When a rod is pressed against the cam's edge, the rod rises and falls as the cam rotates against it. Like cranks, cams are used to convert rotating motion into reciprocating motion; unlike cranks, cams won't work the other way. Whereas a crank has a fixed pivot point on its arm as well as the shaft, a cam is fixed only at the shaft. Cams can cause more complex motions than cranks can, because they can be made in specialized shapes. Cams are used when the reciprocating motion has to be closely controlled. They open and close the switches in washing machines, open and close the valves in car engines, and control the shape and size of stitches in sewing machines.

Crank to raise and lower needle bar

Shaft rotated by motor

Two cams that move feed dog

Motor

Crank to raise feed dog up and down

Crank to move feed dog back and forth

Needle bar

Needle

Feed dog

SEWING MACHINE

The sewing machine uses both cams and cranks to do its work. The needle is made to move up and down by a crank connected to the shaft of a motor. As the motor shaft rotates, the crank drives a connecting rod, which pushes and pulls the needle in its up-and-down motion.

The feed dog is the toothed plate that moves the fabric. It does this by moving forward, then lowering itself away from the fabric and moving backward, and then raising itself to move forward again. These motions are controlled by two cams. One moves a rod that is connected to a crank that moves the feed dog up and down; the other moves a rod connected to another crank that moves the feed dog back and forth.

GEARS, CHAINS, AND BELTS

The function of gears, chains, and belts is to enable one rotating shaft to turn another, often at a different speed and possibly in a different direction. In a clock, the works should turn the hour hand around the dial only once every 12 hours, but the minute hand should turn once per hour. Proper gearing allows both to operate from the same rotating shaft. An automobile engine might run most efficiently at 2,500 revolutions per minute, but if those revolutions of the engine were directly connected to the wheel axle, the car would be going about 340 miles per hour. To reduce the speed of the wheel axle to a more practical level, one or more pairs of gears intervene between the engine and the wheel axle.

At the same time that gears, chains, and belts change the speed of rotation of different shafts, they change the force the shafts can exert. The geared-down rear wheels of an automobile can exert a much greater rotary force, or **torque,** than the engine can directly.

GEAR

A gear is a disc rotating on a shaft. The disc has teeth cut into its outer edge, which mesh with mating teeth on an adjacent gear. One gear drives the other. The teeth on the driving gear push the teeth on the driven gear, causing it to rotate in the opposite direction. If the two gears are the same size, the driven gear will rotate at the same speed as the driver. But if the driving gear has 12 teeth and the driven gear has 24 teeth, the driven gear will rotate only half as fast as the driver. In this case, the larger gear can exert a greater force than the smaller one.

BICYCLE CHAIN

Chain drives are similar to gears in that teeth are cut in the circumference of the discs, but the two discs are set apart from each other and a chain connects them. The discs of a chain drive are called **sprockets;** the chain is made up of a number of round rods linked together so that they exactly match the distance between sprocket teeth. The two sprockets rotate in the same direction.

A bicycle uses a chain drive to transfer rotation from the pedal shaft to the rear wheel, which has on it sprockets of different sizes. In high gear, the chain is moved to a small sprocket, so that the rear wheel moves faster than the pedals. In low gear, for going uphill, the chain is moved to a large sprocket. The rider pedals faster, but with less force. The rear wheel moves slowly, but with great force.

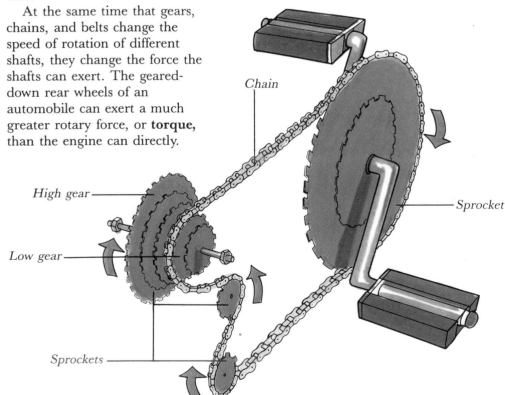

Chain

High gear

Low gear

Sprockets

Sprocket

BELT DRIVE

Belt drives are similar in operation to chain drives, but they substitute a flexible belt for the chain and smooth wheels for the sprockets. Because of slippage between the belt and the wheels, the ratio of speed between one wheel and the other is not as precise as in a chain drive. Belt drives are used when the speed ratio need not be exact. They are also used when the noise level produced by a gear or chain drive is unacceptable.

ESCALATOR

An escalator moves by a large chain drive. Each stair tread is actually a small four-wheeled platform. The front two wheels are on long extensions; the back wheels are close to the stair tread. The front wheels run on one track, and the back wheels run on another. Each tread is linked to the others with a heavy chain that loops around sprocket wheels at the top and bottom of the escalator. The treads make a continuous loop running up the escalator, around a wheel at the top, and down under the escalator to the bottom again.

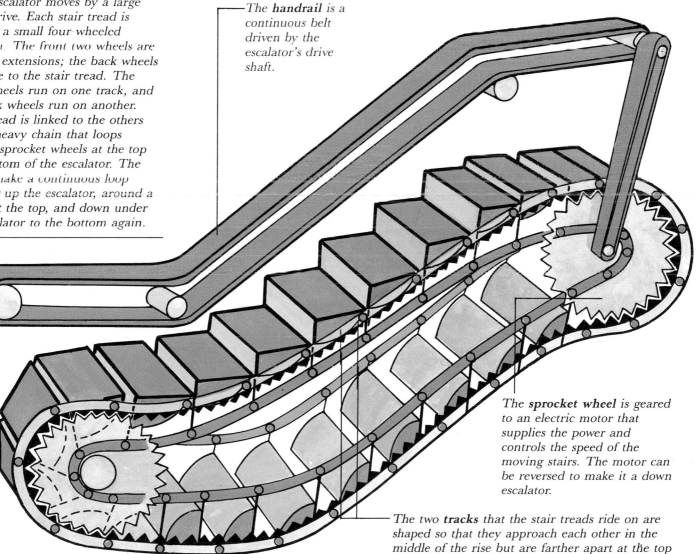

The **handrail** is a continuous belt driven by the escalator's drive shaft.

The **sprocket wheel** is geared to an electric motor that supplies the power and controls the speed of the moving stairs. The motor can be reversed to make it a down escalator.

The two **tracks** that the stair treads ride on are shaped so that they approach each other in the middle of the rise but are farther apart at the top and bottom. This configuration causes the stairs to flatten out at the top and bottom of the escalator.

PULLEYS

A simple pulley is a rope guided around a wheel with a groove in its circumference to hold the rope in place. This type of pulley, when used to lift something attached to one end of the rope, changes the direction of the force used to lift the load (from upward to downward).

Adding a second pulley, so that one is attached to the load and the other to a fixed support, cuts in half the effort it takes to lift the load, but makes you pull twice as far.

OPENING A WINDOW

A double-hung window has two sashes that slide past each other to allow you to open and close the window. Since the sashes can be quite heavy, each side of the sash has a counterweight tied to a rope or chain that rides over a pulley at the upper end of the window. The combined weight of the two counterweights for each sash is approximately equal to the weight of the sash. Because the two weights are nearly equal, the effort it takes to raise or lower the sash is only that of overcoming friction.

Pulley

Rope

Counterweight

Elevator car

Ropes

Counterweight

ELEVATOR

Since an elevator weighs thousands of pounds and a great deal of effort is required to lift it, a pulley system is used to lighten the work of the motor. Multiple pulleys are used so that the strain can be spread among more than one. These pulleys hold a number of wire ropes that attach to the elevator on one end and a counterweight on the other. The motor does not lift the entire weight of the elevator—only the difference in weight between the counterweight and the elevator.

WHIRLING WHEELS

"A body in motion tends to remain in motion; a body at rest tends to remain at rest." This simplification of Sir Isaac Newton's law defines what we know now as **inertia.** Every object has inertia; it takes effort to get an object moving, and once it is moving, it takes effort to change its speed.

The amount of inertia an object has is closely associated with its mass or weight. The greater the weight the greater the inertia, all other things being equal. For example, a small ice cube sliding across the floor will stop much sooner than a large one sliding at the same speed. A heavy cast-iron wheel will continue to spin freely much longer than a light plastic one. However, it takes more pushing to get the heavy wheel going.

Inertia of objects that move in a straight line is strictly dependent on the object's mass or weight. Inertia of rotating bodies, such as wheels, depends not only on the amount of the mass, but also on its distribution. A wheel that has its mass concentrated in the rim has much greater inertia than a wheel with its mass distributed throughout its radius. The farther away from the pivot point the mass is, the greater the inertia.

GYROSCOPE

A gyroscope is a weighted flywheel mounted in special mounting rings called **gimbals,** which permit the axis of the flywheel to take any position. When the flywheel is made to spin at high speed, it will maintain exactly the same position in space no matter how you twist or turn the outer mounting rings. Gyrocompasses make use of gyroscopes. The axis is set so that one end points north, and the wheel is set in motion. As the gyroscope travels across the Earth, the gyroscope keeps the compass pointing north. Since the Earth is a heavy mass turning on its axis, it also acts as a gigantic gyroscope. That is why the Earth's axis always points in the same direction.

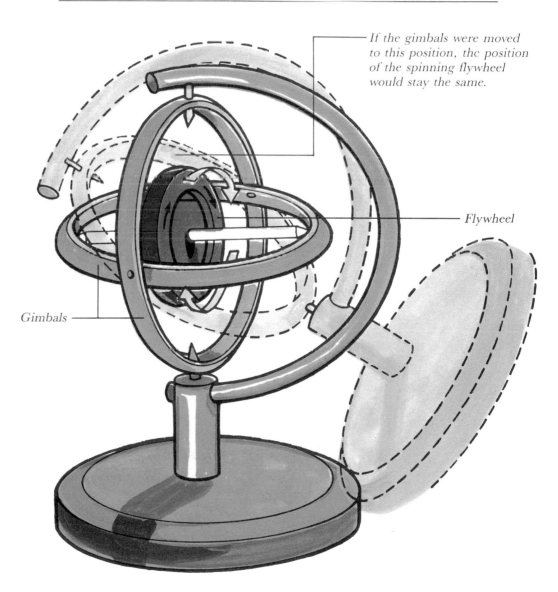

If the gimbals were moved to this position, the position of the spinning flywheel would stay the same.

Flywheel

Gimbals

HYDRAULICS & PNEUMATICS

Air, water, and oil can be as useful a working part of a machine as the metal parts—gears, linkages, wheels, and springs. If the fluid in the machine is oil or water, the system is said to be a **hydraulic** system; if the fluid is air, the system is a **pneumatic** system.

Whether air, oil, or water is used, a pump is needed to pressurize the fluid. The pressurized fluid is then directed through pipes or hoses to a cylinder and piston arrangement. When the fluid enters the cylinder, it pushes the piston out ahead of it and the outer end of the piston does whatever task it has been designed for.

Valves are used to regulate the flow of fluid. Valves are like faucets in that they control the amount of fluid sent to the cylinder, but whereas faucets are always opened or shut by hand, valves are often controlled by electric solenoids (see **Electromagnet,** page 24) so that the push of a button or the pull on a lever is all that is required to operate the valve. Other valves, called check valves, act as "turnstiles" to permit fluid to flow in only one direction.

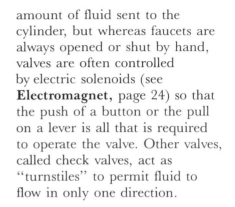

BICYCLE PUMP

This hand-powered pump forces air into a storage tank or tire. On the downstroke of the piston, the air already in the cylinder is forced through the outlet check valve, while the inlet check valve is held closed by the internal pressure. When the pump handle is raised, air is sucked into the pump cylinder through the inlet check valve, and the outlet check valve is held shut.

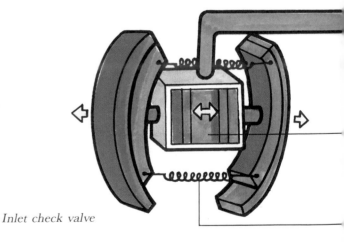

Inlet check valve

Outlet check valve

AIR HAMMER

An air hammer is supplied with a constant flow of high-pressure air. Air is admitted when the air hammer's trigger is pulled. The air is guided by a rocker valve into the cylinder, where it forces the piston against the chisel. The chisel then moves downward to do its work. The air leaves through the exhaust hole, which is exposed when the piston passes it. This causes a drop in air pressure inside the cylinder, so that the incoming high pressure rocks the valve and opens the opposite air passage. Now the pressurized air is directed under the piston, forcing it up so that it is ready for the next downward stroke.

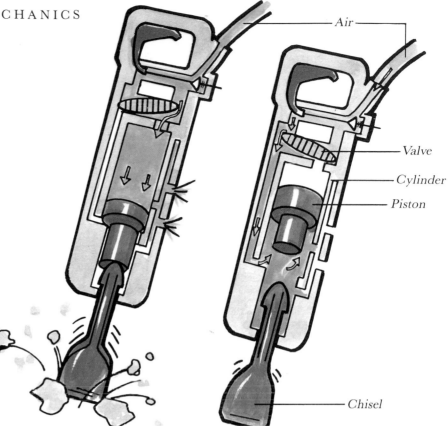

Air

Valve

Cylinder

Piston

Chisel

The brake fluid is forced out of the cylinder into the brake lines and out toward the wheels.

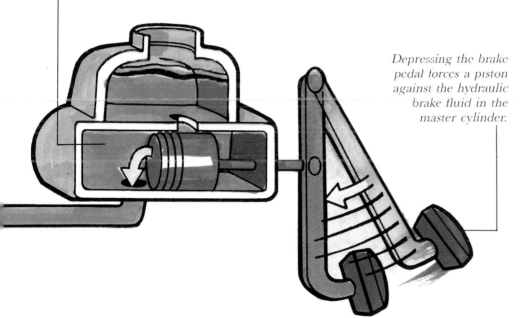

Depressing the brake pedal forces a piston against the hydraulic brake fluid in the master cylinder.

Slave cylinders at each of the wheels have their own pistons; the pressure of the hydraulic fluid causes the pistons to move outward, forcing the brake shoes against the drums.

Return springs pull the shoes away from the drums when the brake pedal is released.

HYDRAULIC BRAKES

The brake system in a car is based on the fact that fluids cannot be compressed. In addition, pressure exerted on a fluid in a closed container is spread out equally in all directions.

Near the brake pedal is a master cylinder with a piston in it; the master cylinder is connected by a pipe system—the brake lines—to slave cylinders, also with pistons, at each of the wheels. The whole system is filled with hydraulic brake fluid. Pressure exerted on the fluid in the master cylinder travels, by means of the brake fluid, all the way to each of the wheels. At the wheels, the pistons in the slave cylinders are forced outward, pressing the brake shoes against the drums in the wheels. The pressure exerted by your foot on the brake pedal is transmitted through the hydraulic fluid equally to all four wheels. The meandering pipe system is no disadvantage; hydraulic power can be transmitted long distances and around corners.

WAVES

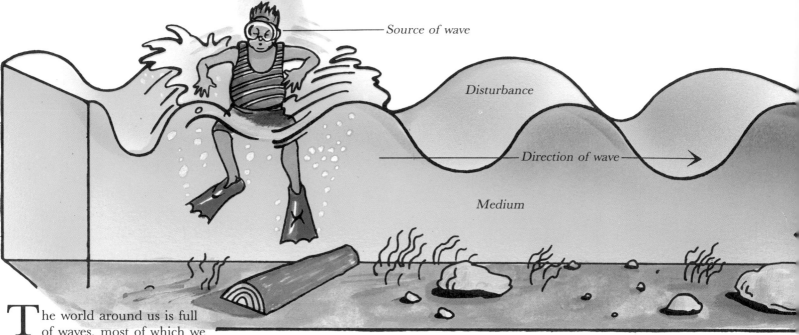

Source of wave

Disturbance

Direction of wave

Medium

The world around us is full of waves, most of which we can neither see nor hear. Sound and light are both waves, but they are of very different types.

Sound waves (see page 48) consist of vibrations traveling in a material such as air. Since the material itself vibrates, sound waves are called **mechanical waves.** Water waves and earth tremors are also mechanical waves.

Light waves, on the other hand, do not need a material to travel through. Their vibrations are movements of electronic and magnetic fields that interact with each other as the wave moves through space. Light waves are the visible part of a whole family of waves called **electromagnetic waves** (see page 50). Other examples of electromagnetic waves are radio waves, X rays, microwaves, and gamma rays.

Electrical waves are variations in electron charge that travel along an electrical conductor such as a wire.

Despite these differences in their nature, waves of all kinds behave in similar ways.

A wave is a chain of disturbances that propagates in a medium. The medium can be a liquid, a gas, a solid, or space. To produce a wave, energy is required. The energy comes from a source, and it travels along with the wave as the wave radiates from the source.

Although the wave itself moves long distances, the medium in which the wave is traveling moves very little in comparison. An individual disturbance passes its energy along to its neighbors, causing that area of the medium to move.

How fast the wave travels depends on the kind of wave it is— mechanical, electromagnetic, or electrical—and on the kind of medium that it is traveling in.

*The direction of the disturbance in this wave is perpendicular to the direction of the traveling wave. That makes it a **transverse** wave. Sound waves (see page 48) are **longitudinal** waves, meaning that the disturbance is in the same direction as the wave's travel.*

*The **amplitude** of the wave is the magnitude of the disturbance in the medium.*

*The **wavelength** is the distance between corresponding parts of the wave disturbances.*

WAVES

How often a disturbance is generated, or how often it moves past a fixed point, is called the **frequency** of the wave. The frequency is commonly expressed in **hertz.** One hertz is one cycle per second.

Wavelength and frequency are related: Longer waves have lower frequencies, and shorter waves have higher frequencies.

Low frequency

High frequency

A narrow group of waves going in the same direction is called a **ray.**

A collection of parallel rays is a **beam.**

When waves travel in the same direction, side by side, varying in amplitude and frequency by the same amount at the same time, they are in **phase,** and their corresponding parts form a **wave front.**

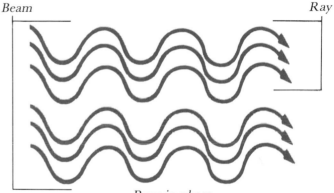

Beam

Ray

Rays in phase

Rays not in phase

When a ray or beam travels between media that have different properties, the direction of the ray changes. It can be **reflected,** or sent back, from the surface of the new medium.

The ray can also be **refracted,** or bent, as it goes into the new medium.

An obstacle in the path of a ray causes the ray to bend slightly or change direction as the ray hits the edges of the object. This is called **diffraction.**

Every medium, or change of medium, offers some hindrance to the wave as it travels through. Some of its energy is dissipated and some might be reflected back toward the source. This characteristic is known as **impedance.** How much impedance is present depends on the nature of the medium and the power of the wave.

47

SOUND

If a tree falls in the middle of a forest in which there are no people, is there a sound?

Sound pressure waves, or just sound waves, are pressure fluctuations that travel through the air or other materials. The human ear can detect changes in air pressure, as anyone knows whose eardrum has popped when going up or down in an airplane. When air pressure fluctuates repeatedly, between about 20 and 20,000 times per second, the brain interprets the rapid pops as sound. The number of times per second that the air pressure fluctuates is the **frequency** of the sound wave. For many people, fluctuations outside a range of 80 to 8,000 times per second are inaudible. Some animals, birds, and fish can detect much faster and slower fluctuations in pressure.

When the tree fell, it created vibrations in the air of many frequencies, not all of them within the range of human hearing. Pressure fluctuations with frequencies below the audible range of humans are called **infrasonic,** from the Latin word for "below." Infrasonic waves that have frequencies of a few cycles per second (hertz) or less and wavelengths several miles long can be detected by instruments such as **seismographs** (see page 264). They are sometimes felt as earth tremors and changes in atmospheric and underwater pressure. The falling tree may have caused nearby trees to shake from infrasonic vibrations.

Pressure fluctuations with frequencies above the human range of hearing are called **ultrasonic,** from the Latin word for "beyond." Ultrasonic waves have frequencies as high as millions of hertz (megahertz) and have wavelengths as short as those of visible light. They can be produced and detected by bats, ocean mammals, and insects. Some animals in the forest may have detected ultrasonic vibrations from the falling tree.

Sound waves can be propagated, not only in the air, which is a gas, but in liquids and solids as well. By putting your ear to a railroad track, you can hear an oncoming train through the track before your ear can hear the sound through the air. The sound is traveling faster through the track than through the air.

The velocity of sound depends primarily upon the density and temperature of the material in which it is traveling. In air at a temperature of 70°F, the speed of sound is 1,130 feet per second. Fresh water at 77°F carries sound at 4,897 feet per second. The speed of sound in the earth depends on whether it is traveling through loosely packed sand or solid rock. In metal, a longitudinal sound wave travels about four times as fast as in water.

A person at the edge of the forest with an ear to the ground may have detected earth tremors from the falling tree, whereas the vibrations that went through the air weakened and died before they reached the other ear.

Was there a sound when the tree fell in the forest, or not?

A SOUND PRESSURE WAVE

Striking the drum causes the drum head to vibrate. When the drum head moves outward, the air molecules next to the drum are pushed away, pressing them against molecules on the other side of them. This creates an area where the molecules are denser and the air pressure is increased. The compressed air continues to move away.

*When the drum head draws inward, it leaves a space with fewer molecules and a lower air pressure. This area is called a **rarefaction,** because the air molecules are rarer. Surrounding air molecules rush in to fill the space the drum head used to occupy.*

As the drum head moves back and forth, areas of higher and lower air pressure—compressions alternating with rarefactions—are propagated. This is a sound pressure wave. When the wave strikes the eardrum, it vibrates back and forth at the same rate as the originating drum.

*The disturbances in a sound wave are not perpendicular to the wave, as they are in a water wave. The medium is disturbed in the same direction as the wave. Such a wave is called a **longitudinal** wave, as opposed to a transverse wave (see page 46).*

The air molecules near the drum do not travel all the way to the ear. They move back and forth only infinitesimal distances, a small fraction of a millimeter. It is the sound pressure wave that travels to your ear.

Compression

Rarefaction

MEGAPHONE

A megaphone not only gives direction to the sound by preventing the sound from going off to the side, it also pulls more sound out of the vocal tract. When the traveling sound wave produced by your voice encounters an abrupt change in the medium it is traveling in—such as the change from the narrow air column in your vocal tract to the wide open air—some of its energy is **impeded** from making the transition. The impeded energy is reflected back into your mouth. Sound waves bunch up and block subsequent sound from coming out.

A tapered megaphone acts as an impedance-matching transformer for sound waves. The small end matches the impedance of the vocal tract; at the wide end, the impedance is more closely matched to that of the open air. The tapered shape makes the transition more gradual and helps the sound waves continue ahead. To be most effective, the length of the tapered megaphone should be longer than the sound wavelengths. Because the lowest frequency of the human voice has a sound wave several feet long, a cheerleader's megaphone should be several feet in length.

ELECTROMAGNETIC WAVES

Unlike sound waves, which are disturbances transmitted through a medium such as air, electromagnetic waves need no medium in which to propagate. They can travel through a void. They create interacting electric and magnetic fields, which vary in strengths along the wave.

The visible part of the family of electromagnetic waves is composed of light waves. Other examples that are not visible are radio waves, microwaves, X rays, and gamma rays. This entire group of waves is called the **electromagnetic spectrum,** which consists of waves of a wide range of frequencies.

At one end are waves of low frequency and long wavelength, such as electrical and radio waves. They are 10,000,000,000,000,000 times longer than the shortest and highest-frequency waves yet discovered—cosmic rays—at the other end of the spectrum. The waves of highest frequency have the highest energy and the shortest wavelengths. These waves are so short that they are measured by a unit called an angstrom (Å), which is the diameter of one hydrogen atom. Much longer waves, such as radio waves, are usually described not by their length but by their frequency expressed in **hertz** (cycles per second).

ELECTROMAGNETIC WAVE

A moving electric charge creates a changing electric field. This induces a changing magnetic field perpendicular to the electric field, which in turn generates another changing electric field farther away, and so forth. The magnetic field is perpendicular to the electric field as well as perpendicular to the direction of the wave. These self-propagating, changing electric and magnetic fields, pulling each other along, constitute an electromagnetic wave.

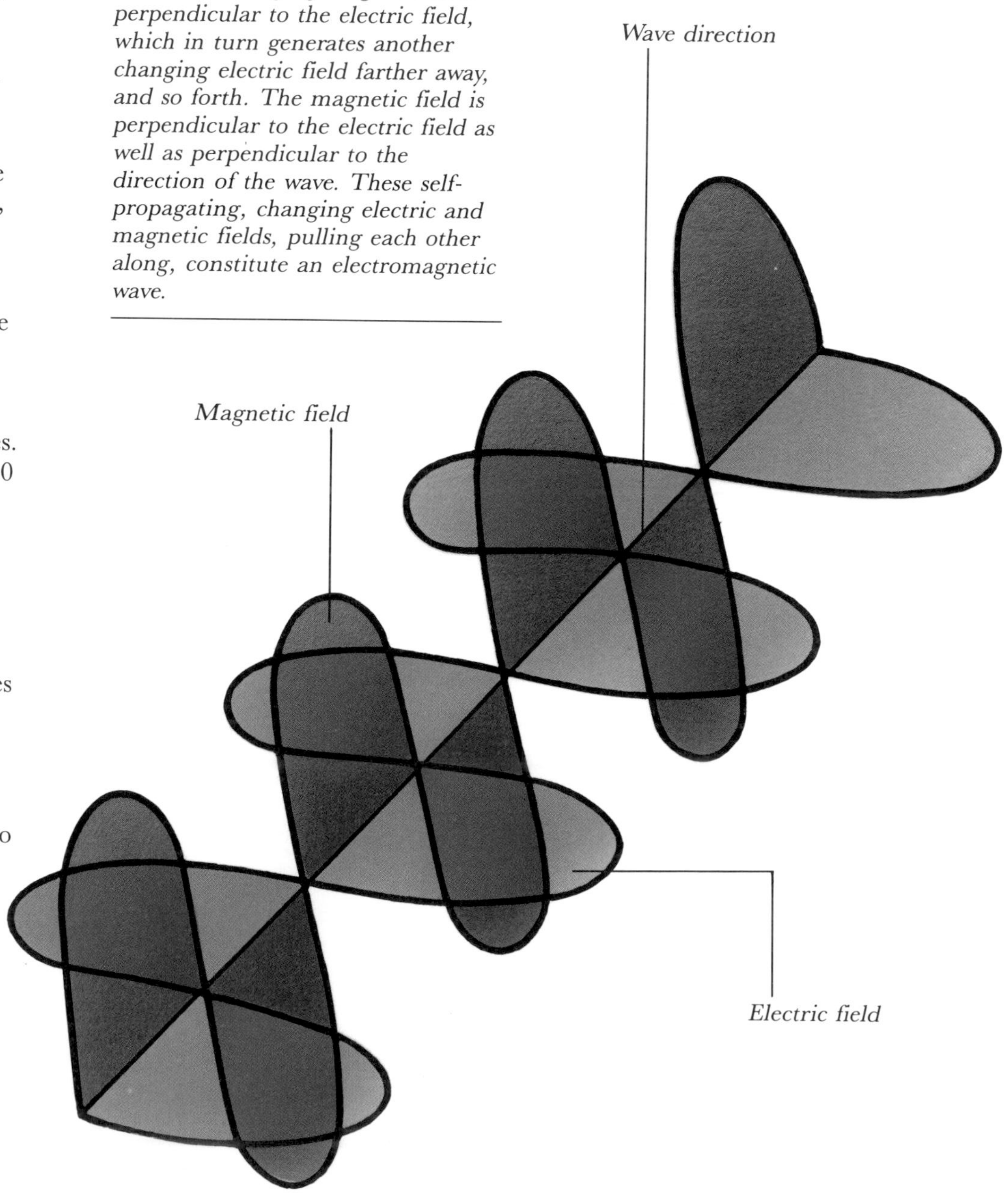

Wave direction

Magnetic field

Electric field

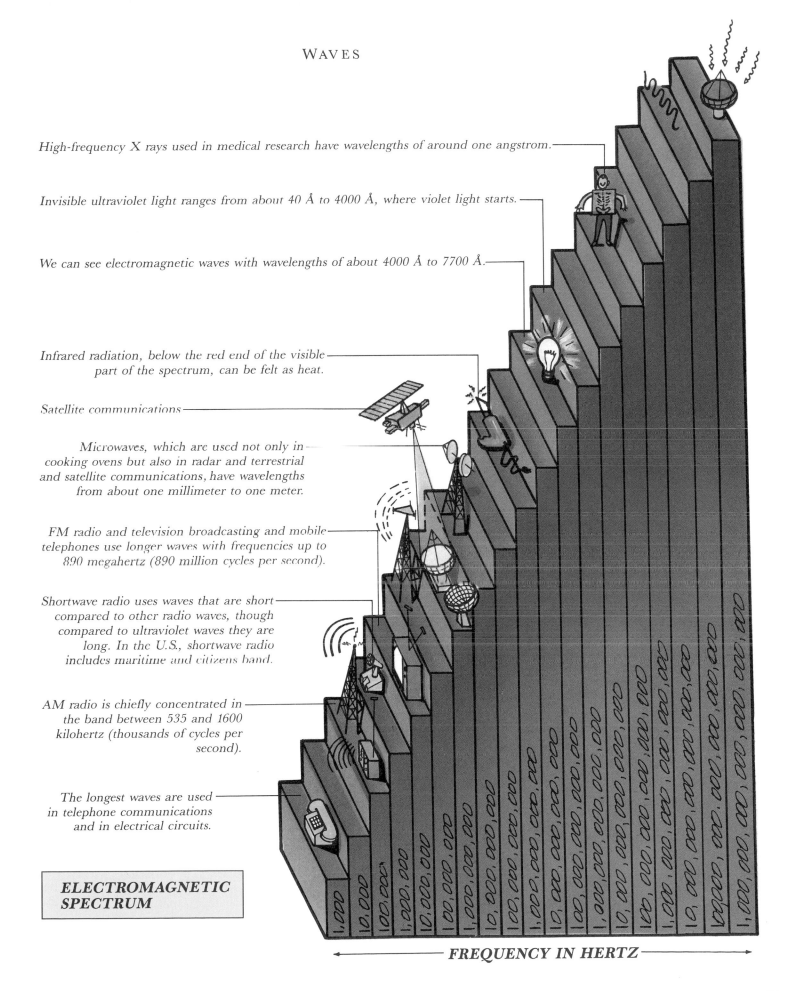

High-frequency X rays used in medical research have wavelengths of around one angstrom.

Invisible ultraviolet light ranges from about 40 Å to 4000 Å, where violet light starts.

We can see electromagnetic waves with wavelengths of about 4000 Å to 7700 Å.

Infrared radiation, below the red end of the visible part of the spectrum, can be felt as heat.

Satellite communications

Microwaves, which are used not only in cooking ovens but also in radar and terrestrial and satellite communications, have wavelengths from about one millimeter to one meter.

FM radio and television broadcasting and mobile telephones use longer waves with frequencies up to 890 megahertz (890 million cycles per second).

Shortwave radio uses waves that are short compared to other radio waves, though compared to ultraviolet waves they are long. In the U.S., shortwave radio includes maritime and citizens band.

AM radio is chiefly concentrated in the band between 535 and 1600 kilohertz (thousands of cycles per second).

The longest waves are used in telephone communications and in electrical circuits.

ELECTROMAGNETIC SPECTRUM

← FREQUENCY IN HERTZ →

RADIO WAVES AND MICROWAVES

HOW A RADIO TRANSMITTER AND RECEIVER WORKS

The energy in waves can be used to carry information. Such a wave is called a **signal.** The information is contained in the changes in the wave. A regularly undulating wave would be meaningless; only a wave that changes irregularly carries information. The simplest change consists of turning the signal on and off; this is the principle of Morse Code, which uses long and short signals.

To send a message on a wave, the wave's frequency or amplitude is altered in ways that represent the information. Radio waves are electromagnetic waves (see page 50). They range from those with frequencies of a few kilohertz (thousands of cycles per second) and wavelengths of hundreds of miles to those with frequencies of 300 gigahertz (billions of cycles per second) and wavelengths of perhaps 40 to 150 feet. This portion of the electromagnetic spectrum is called the radio spectrum. The electric and magnetic fields of radio waves can be modulated, or changed, so as to carry information over great distances at the speed of light.

Although they consist of less than one billionth of the electromagnetic spectrum, radio waves are used for a great many forms of wireless communication. AM and FM radio broadcasting, television, telephone calls by microwave radio, radar, satellite communications, military

communications, citizens band radio—even the small radio transmitters that open and close garage doors—use radio waves. Many medical, industrial, and scientific applications also use radio waves.

All these uses could easily interfere with each other if they attempted to use the same frequency of wave for their communications. To minimize this possibility, the frequencies in the radio spectrum are assigned by international and national agencies. The users of radio waves compete fiercely for allocations of the spectrum, which is viewed as a limited natural resource. However, continuing technological advances permit more of the radio spectrum to be used and allow those frequencies to be used more effectively.

Microwaves are very short radio waves—up to three centimeters—with frequencies from about 300 megahertz to 300 gigahertz. They extend into the infrared range.

1. A microphone changes sound vibrations into an electrical current.

2. The current varies with the variations in sound; it is called an **audio-frequency signal.**

3. The radio station is assigned a particular wave frequency for its use in broadcasting. This is a current of a much higher frequency, and it does not vary. This current is called the **carrier** current.

4. So that the carrier current can carry the audio-frequency signal, it is **modulated**—or changed—to resemble the audio-frequency signal (see page 100). Either the frequency or the amplitude of the carrier current can be modulated.

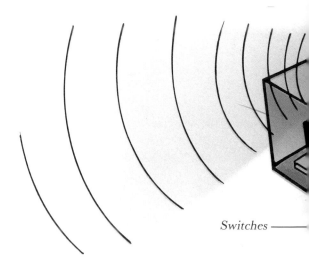

Switches

6. The current is sent to an antenna, which generates radio waves to correspond with the current and broadcasts them in all directions.

8. The signal is **demodulated**; that is, the audio-frequency signals are removed from the carrier signal.

7. Considerably weakened over distance, these waves meet a radio receiver aerial. They induce in the aerial a varying radio frequency current, which is then amplified.

9. The audio-frequency signals are used to drive a loudspeaker that converts them to sound waves.

5. The modulated carrier current is amplified so that it will be strong enough to send a long distance.

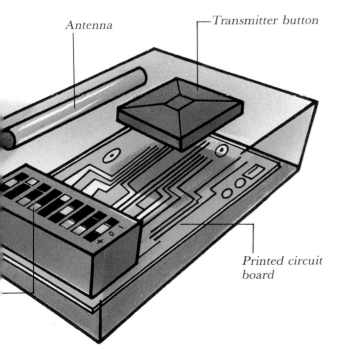

Antenna

Transmitter button

Printed circuit board

GARAGE DOOR OPENER

With radio waves, you can open and close your garage door without getting out of your car. These systems consist of a battery-operated, low-powered radio transmitter and, in the garage, a receiver tuned to the same frequency as the transmitter. Pressing a button on the transmitter broadcasts a signal a distance of about 100 feet. The receiver detects it and turns on an electric motor that opens or closes the garage door.

To avoid having neighbors' garage door transmitters or even CB radios accidentally opening your garage door, the signal is encoded. The transmitter sends a series of pulses of two or three different radio frequencies. In one widely used system, the code is set by flipping each of nine small switches to one of three positions, up, down, or centered. These positions represent three different radio frequencies. A corresponding row of switches on the receiver is set the same way, and the receiver is then tuned to act only on a signal in the private code. This system will allow more than 19,000 combinations, or individual codes.

LIGHT

White light

Refraction

Spectrum

Visible light is that very narrow portion of the electromagnetic spectrum from about 100 million megahertz (millions of cycles per second) to 1,000 million megahertz that the receptors in our eyes respond to. By custom, light is usually described in terms of its wavelength rather than its frequency. Visible light in a vacuum has wavelengths between about 4,000 and 7,500 angstroms. One angstrom equals one hundred millionths of a centimeter, or the diameter of a hydrogen atom.

Light exhibits all the properties of electromagnetic waves. It radiates from a source in a straight line. It can be reflected. In fact, we see more reflected light than we do light from a direct source. Reading this page, you are seeing light reflected from the paper.

Light can be refracted, or bent, when traveling from one transparent medium, such as air, to another, such as water. An oar in the water looks as if it bends at the water line.

Light can also be diffracted, refracted, absorbed, and transmitted, depending upon the material it shines upon (see **Waves,** page 46).

No form of matter or energy can travel as fast as light. In a vacuum, its speed is 186,000 miles per second. In air, the speed of light is just slightly less. In water, light travels 75 percent as fast as in air. In a transparent medium such as glass, light travels about 124,000 miles per second.

Most of the light we see comes from the sun. Light is also generated on Earth when the atoms of certain materials are excited to higher-than-normal energy levels. They throw off this added energy in the form of **photons,** which are bundles of radiant energy. This is called **luminescence.**

Luminescent materials give off light without much heat. **Incandescent** materials give off both light and heat when they are heated to the point at which they glow.

Some substances, called **phosphors,** emit light when they are excited by ultraviolet rays or electron beams, thus making light with little heat. This is called **fluorescence;** not only fluorescent lights, but video display tubes, use this principle.

Certain solid semiconductor materials (see page 28) can be made to luminesce without heat by passing an electric current through them. This effect, called **electroluminescence,** is used in light-emitting diodes (LEDs).

And some organisms—fireflies, for example—emit light through bioluminescence, generated by the energy of chemical reactions in their bodies.

SEEING THE COLOR SPECTRUM WITH A PRISM

White light actually contains the colors red, orange, yellow, green, blue, indigo, and violet. Each color has a different wavelength. This spectrum of colors can be seen by shining a beam of sunlight through a prism—a triangle of polished clear glass or some other transparent material.

As white light passes through the glass, it is slowed by the density of the glass. Each of the different wavelengths that comprise white light is slowed and refracted a slightly different amount. (See **Waves,** page 46.) Red light, with the longest wavelength, is slowed the least, and therefore is refracted the least. Violet light, with the shortest wavelength, is slowed the most and refracted the most. This spreading of white light is called **dispersion.**

Raindrops in the air act as miniature prisms when sunlight shines through them. They disperse the white sunlight into a spectrum of colors and reflect them into the air. Standing at just the right place between the sun and the raindrops, you will see a rainbow.

LASER

The radiated white light from an electric light bulb is composed of different wavelengths, emitted in short bursts at different times in different directions and with the waves largely out of phase with each other. It is light that is disorganized, or **incoherent.**

Laser light, however, is highly organized, or **coherent.** It is composed of waves whose wavelengths and frequency hardly differ at all. Moreover, the waves move forward parallel with each other and in phase with each other, like soldiers on parade marching forward in step.

Coherent light can be produced from many different kinds of lasers—gas, liquid, crystal, semiconductor, or chemical. Each produces a laser light with unique characteristics that are suited to different purposes. The emitted light can be continuous or pulsed, low power or high power, visible or invisible.

One kind of semiconductor laser may be used by communications systems to transmit information. It emits a beam of infrared light pulses so pure—nearly a single frequency—and so coherent that it can carry billions of bits (on/off cycles of light) per second over threads of glass. That would be equivalent to sending the entire text of the *Encyclopaedia Britannica,* error-free, in less than a half-second.

Lasers are used in making **holograms** (see page 72), and the high coherence of laser light allows it to be used for extremely precise scientific measurements. Because a laser beam is parallel,

making a bright straight line over a long distance, it is useful in surveying.

Some lasers emit highly intense beams, which can be precisely focused and used for cutting or welding—even for surgery (see page 294).

When an electric lamp is switched on, an electric current flows through the filament, passing its energy to the filament's atoms and heating it so hot that the filament glows. It glows because as soon as the atoms are energized, they spontaneously emit their excess energy as radiating **photons**— bundles of radiant energy.

Atoms with high energy

Energy source

Laser light also glows from the emission of photons. But in a laser, the emissions are stimulated and coordinated. This gives the light its coherent quality.

Energy—electrical, chemical, radio, or light—is pumped into a specially prepared material. The energy of the atoms in this material is increased until there are more with high energy than those with normal energy. The atoms do not emit their excess energy right away. The additional energy raises them to a higher, unstable energy level.

LASER

Photon

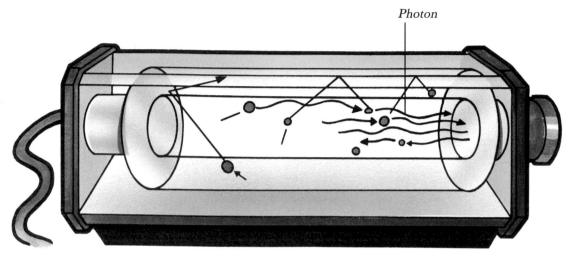

When one of them does drop back to a lower energy level, it radiates the extra energy as a bundle of energy (a photon). This photon has a characteristic wavelength determined by the nature of the laser material.

The emitted photon collides with another high-energy atom before it has had a chance to drop back to a lower energy level. This collision stimulates the atom to release its extra energy as a photon with the same wavelength. Together, the emitted photons stimulate more excited atoms to release more photons.

Intense beam of coherent light

Mirror

Mirror

Photons reflected by mirror

In a laser, the material is enclosed in a chamber with parallel mirrors at each end. The photons are reflected back and forth, stimulating even more excited atoms to emit even more photons of the same wavelength and in phase. Only those photons with wavelengths and phases that fit precisely a multiple number of times between the mirrors will be reflected. Photons with other wavelengths, phases, and directions will be dampened. The avalanche of photons with the same phase, wavelength, and direction continues to build up.

A small number of the selected photons leave the chamber, through a partially reflecting mirror, as an extremely intense continuous beam or series of bursts of coherent light. This process gives the laser its name, which is an acronym for Light Amplification by Stimulated Emission of Radiation.

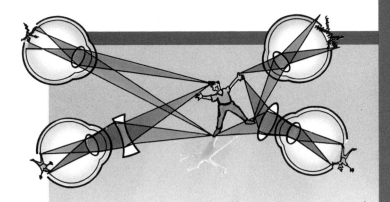

Sight and Sound

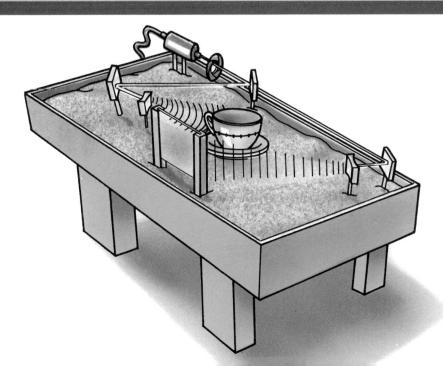

LIGHT BULBS

Romantic though candlelight and gaslight may be, the development of light by electricity has transformed our indoor world into a brighter environment. The electric light bulb is a relatively recent invention; Thomas A. Edison developed the first commercially practical incandescent bulb in 1879.

TUNGSTEN BULB

The tungsten bulb is a globe of clear or frosted glass sealed to a brass socket. Inside, a glass stem holds up a wire frame with a zigzag-shaped or coiled wire filament of tungsten. When electricity flows through the filament, it heats up. It gets so hot that it turns red and then white. This property is called **incandescence.** The main disadvantage of a tungsten bulb is the amount of heat it gives off in order to keep the filament at white heat. Oxygen in the air would cause the filament to burn out right away. To prevent this, the air is removed from the globe of glass and replaced with other gases, such as nitrogen.

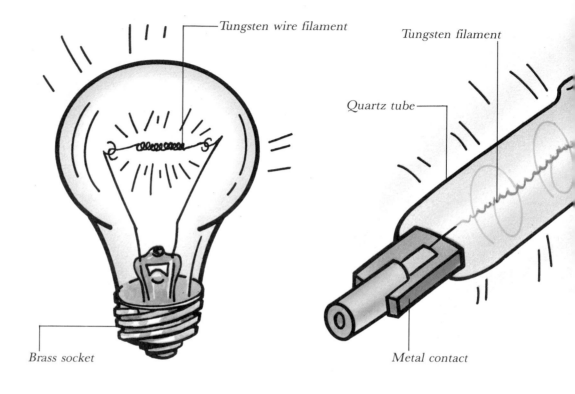

Tungsten wire filament

Tungsten filament

Quartz tube

Brass socket

Metal contact

FLUORESCENT TUBE

The fluorescent tube is more efficient than the tungsten bulb. It not only gives off much less heat, but the same light output can be had with lower electric bills. The fluorescent tube is filled with mercury vapor. This allows electricity to flow between the metal contact rings at either end of the tube. The mercury vapor produces invisible ultraviolet light when current flows through it. The ultraviolet rays strike a phosphorescent coating on the inside of the tube, which then glows brightly. The original fluorescent tubes produced a cold blue light; later variations in the phosphors produce warmer colors.

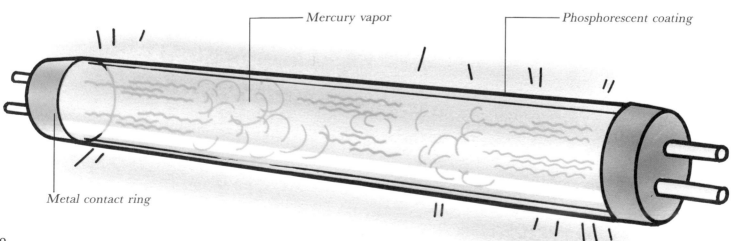

Mercury vapor

Phosphorescent coating

Metal contact ring

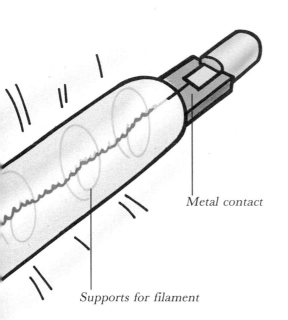

Metal contact

Supports for filament

HALOGEN LIGHTS

Halogen lights are a special form of incandescent light. A tiny amount of iodine or bromine is added to the gas filling. The halogen light operates at higher temperatures than a tungsten bulb does, and the pressure of the gas filling inside is higher. For this reason, a quartz tube, which is sturdier than glass, encloses the filament. Halogen lights are often used as street lights or automobile headlights.

ELECTROLUMINESCENCE

Certain materials called **phosphors** glow when exposed to alternating electric voltage. Electroluminescence depends on this property. A coating of metal so thin that it is transparent is sprayed on a glass panel. A layer of electroluminescent phosphor is placed on the metal; a thin sheet of metal foil completes the sandwich. When alternating current is connected to the two metal layers, the metal conducts the electricity and the phosphor between them glows.

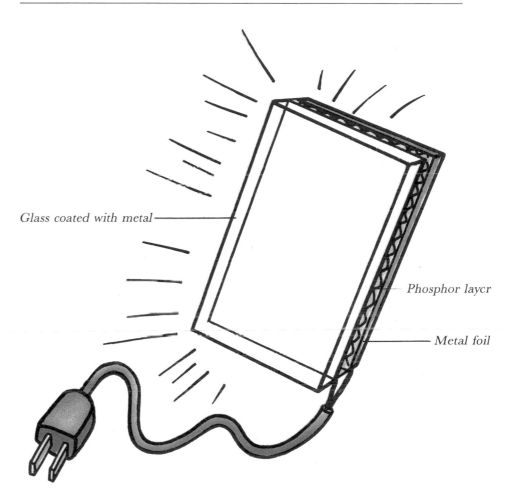

Glass coated with metal

Phosphor layer

Metal foil

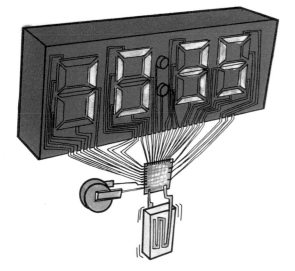

LEDs

Light-emitting diodes (LEDs) are special types of transistors (see page 28). LEDs are crystals of gallium arsenide with special deposited impurities that enable them to turn electric energy into light. Because of their compatibility with transistor circuits and their low power drain, they are often used in electronic circuits as signal lamps. They do not provide enough light to be used in home lighting systems.

MIRRORS

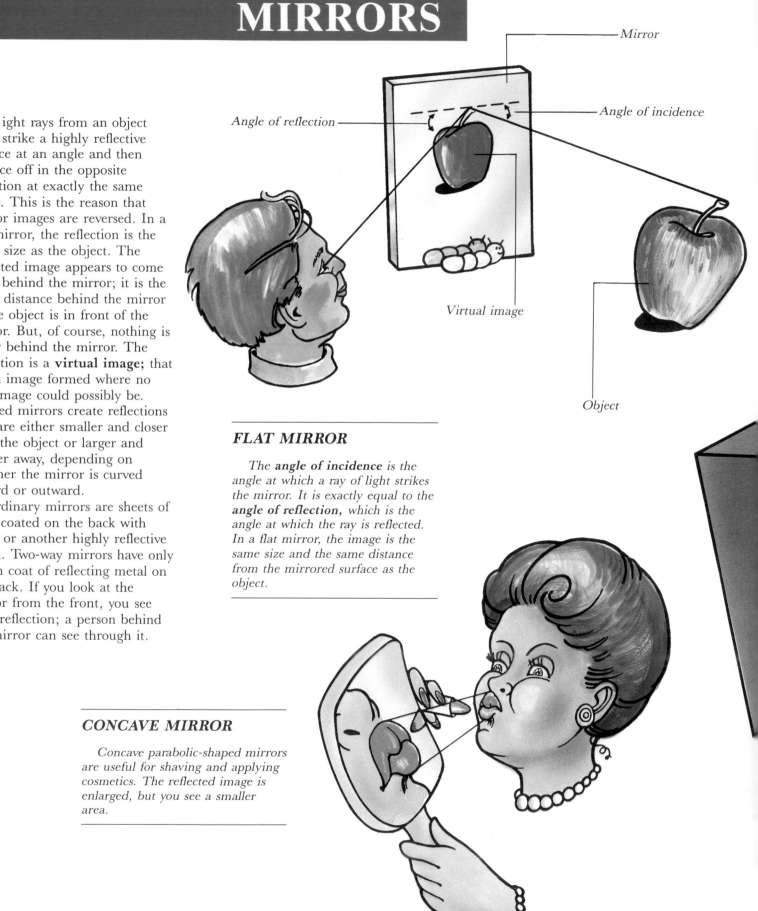

Light rays from an object strike a highly reflective surface at an angle and then bounce off in the opposite direction at exactly the same angle. This is the reason that mirror images are reversed. In a flat mirror, the reflection is the same size as the object. The reflected image appears to come from behind the mirror; it is the same distance behind the mirror as the object is in front of the mirror. But, of course, nothing is really behind the mirror. The reflection is a **virtual image;** that is, an image formed where no real image could possibly be. Curved mirrors create reflections that are either smaller and closer than the object or larger and farther away, depending on whether the mirror is curved inward or outward.

Ordinary mirrors are sheets of glass coated on the back with silver or another highly reflective metal. Two-way mirrors have only a thin coat of reflecting metal on the back. If you look at the mirror from the front, you see your reflection; a person behind the mirror can see through it.

Mirror

Angle of reflection

Angle of incidence

Virtual image

Object

FLAT MIRROR

*The **angle of incidence** is the angle at which a ray of light strikes the mirror. It is exactly equal to the **angle of reflection,** which is the angle at which the ray is reflected. In a flat mirror, the image is the same size and the same distance from the mirrored surface as the object.*

CONCAVE MIRROR

Concave parabolic-shaped mirrors are useful for shaving and applying cosmetics. The reflected image is enlarged, but you see a smaller area.

CONVEX MIRROR

Convex mirrors, which are curved outward, reduce the size of the reflected image. This permits a relatively small mirror to show a much larger field of view than it would if it were flat—a real convenience for the passenger-side mirror on a car.

Partially silvered coating

Darkened room

TWO-WAY MIRROR

Two-way mirrors allow persons behind the mirror to see without themselves being seen. They are sometimes used in store security systems. If you look at the mirror from a well-lit room at the side without the coating, you see your reflection. If you look at the mirror from a darker room at the coated side, you are not dazzled by the reflecting metal and can actually see through the glass. Although the light rays reflect on one side of the mirror, they penetrate the thinly coated side to reach the darkened room.

EYEGLASSES

To see an object sharp and clear, the light rays coming from that object have to be focused precisely on the retina in the back of the eye. The eye's lens focus these light rays on the retina. The image on the retina is, naturally, smaller than the object you see and is upside-down. This image is changed into electrical pulses that travel along the optic nerve to the brain. It interprets this information so that you see the object in its true position and size.

The function of eyeglasses is to improve the image on the retina of the eye when the eye is not working properly. Their purpose is not to magnify the image, although that often occurs, but to change the paths of the light waves coming from the object you see so that the eye's lens can focus the image precisely on the retina.

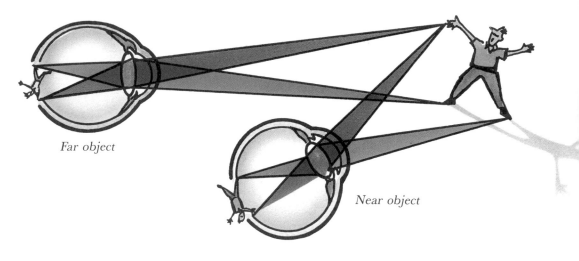

Far object

Near object

When a person with normal vision looks at a distant object, the eye lens, which is somewhat elastic, takes the shape of a narrow lens that brings the light waves coming through the cornea together to form an image precisely on the retina. The brain perceives the image as sharp and clear (in focus).

When a person with normal vision views a close object, the eye lens bulges into a thicker shape. The light waves are traveling a shorter distance, and the thicker shape bends the light waves so that the image on the retina is in focus. Since the image is larger than an image from a distant object, the brain interprets the object as being near. Normal eyes are able to adjust themselves to see objects in focus as close as about 10 inches and as far away as many miles.

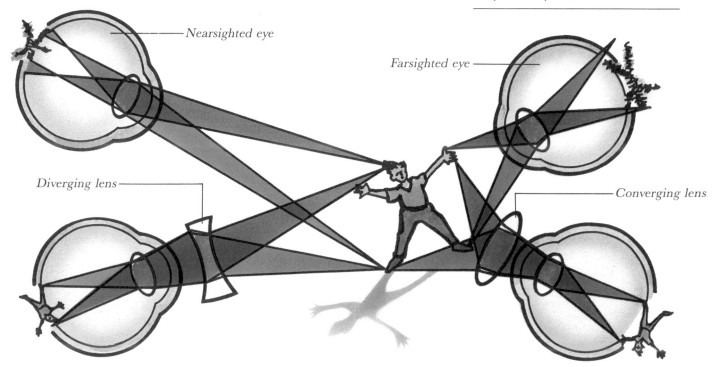

Nearsighted eye

Farsighted eye

Diverging lens

Converging lens

NEARSIGHTEDNESS

If a person's eye is a little thicker than normal, the eye may be able to see very near objects quite well. In fact, the images may be larger and more detailed than a normal eye can see at this near point—the person has a kind of built-in magnifying glass. For distant objects, however, the eye lens may not be able to change shape enough to focus the object's image onto the retina. The light waves converge in front of the retina, and the brain perceives a fuzzy image. This eye is called **myopic,** or nearsighted.

The cure for this is to place an eyeglass made of a **diverging (concave) lens** in front of the eye. It bends the practically parallel rays from the distant object outward, so that they appear to be coming from nearby. The cornea and the eye lens are able to focus them to converge on the retina.

FARSIGHTEDNESS

If a person's eye is a little thinner than normal, the eye may be able to see far objects very well. But the eye lens may not be able to change shape enough to focus the images of near objects onto the retina. The place where the image would be in focus is behind the retina, and the brain perceives a fuzzy image. The eye is called **hyperopic,** or farsighted.

To see near objects better, the farsighted eye needs a **converging (convex)** eyeglass. The light rays from the near object are bent inward so that they appear to come from a more distant point. The cornea and the eye lens can focus the image onto the retina.

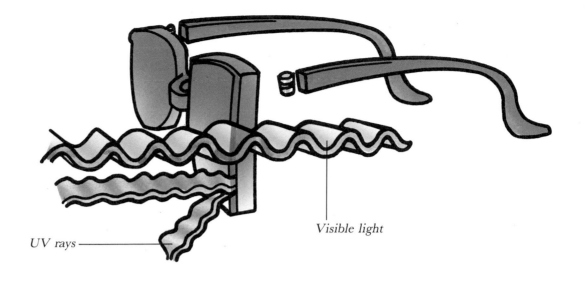

UV rays

Visible light

SUNGLASSES

Some sunglasses are designed to block specific light frequencies such as ultraviolet (UV) rays. Normally, the cornea and the eye lens absorb these invisible rays to protect the retina, but in so doing they may suffer harm. To reduce this possibility, thin film coatings are added to the sunglass lenses during manufacture. The coatings selectively reflect all but one or two percent of the extremely short wave lengths of ultraviolet rays and let through the relatively longer waves of visible light. Some UV glasses are also tinted to reflect blue light. Because blue light waves are shorter than other visible rays, the eye lens focuses them slightly in front of the retina. By blocking the blue waves and letting through those rays that focus more directly on the retina, objects appear to be sharper, although less bluish.

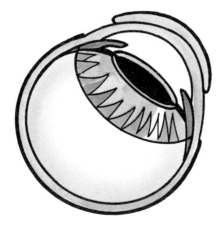

CONTACT LENSES

Contact lenses are specially shaped discs of transparent glass or plastic that are worn in contact with the cornea. Actually, they ride on a layer of tears on the surface of the eye. They correct vision in essentially the same manner as eyeglasses, with the difference that the eye sees the image at the same angle as it would see the object without the lens. The size of the image on the retina, therefore, is not distorted.

TELESCOPES

A simple telescope has two lenses. The **objective** is the lens at the large end of the telescope, which you direct at the object; the **eyepiece** is at the end you look through. The two elements are separated by a relatively long tube blackened on the inside, made in two sections so that one section nests inside the other. This permits the telescope to be focused. One section slides in or out of the other to change the distance between the objective and the eyepiece until a distant object is made to appear clear and sharp.

The eyepiece is a magnifying glass used to enlarge the image. Simple eyepieces are made in the form of a single convex lens. More expensive and sophisticated eyepieces, used in most telescopes and binoculars made today, have two to four glass lenses, or **elements,** making up their eyepieces.

A convex lens inverts the image before the image reaches the eyepiece. To correct this, most telescopes use a third lens system about halfway down the tube, which inverts the image again. The brightness and clarity of the original lens system is decreased, but that sacrifice is outweighed by the advantage of having the image right side up.

Telescoping barrels

Objective

Eyepiece

To create an erect image, an additional lens is added in the center section of the telescope.

Eyepiece

*The **objective** is a curved mirror, which creates the image.*

ASTRONOMICAL TELESCOPE

A lens large enough to serve as an objective for an astronomical telescope, used for photographing stars is difficult to make. An astronomical telescope usually has a concave mirror instead of a lens as its objective. The mirror serves exactly the same function as a lens—it creates an image just in front of the eyepiece.

The image is reflected off a small flat mirror to remove the eyepiece and observer from the path of light reaching the mirror.

BINOCULARS

Binoculars are two telescopes mounted side by side, with **prisms** between the eyepieces and the objectives. A ray of light entering a prism through the diagonal face exactly perpendicular to the surface is not refracted, as it would be if it entered at another angle (see **Prism,** page 54). Instead, it travels through to the slanting side and is totally reflected. It bounces between the slanting surfaces and back out the diagonal face. For this total reflection to happen, the prism has to be cut so that the front and back surfaces are in the shape of right triangles. This is a standard 90-degree prism. Each side of the binoculars has two such prisms.

The prisms offer two advantages for the binoculars. They turn the image right-side up with virtually no loss in light, and they increase the distance between the left and right tubes so as to exaggerate the stereoscopic effect. When properly focused, the images from each tube superimpose and produce a true three-dimensional image.

A single focusing control telescopes both tubes simultaneously.

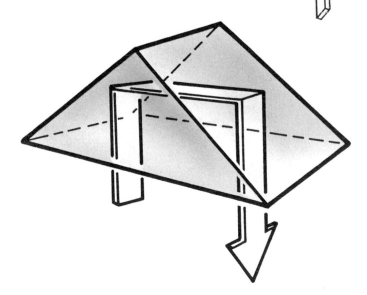

A prism is simply a triangularly cut piece of clear glass. Light rays that enter perpendicular to the surface of the diagonal face of the prism are totally reflected back in the same direction, but reversed left and right, as in a mirror.

Two prisms separate the objectives and erect the image. One prism reverses the image from top to bottom, so that it is right-side up; the other reverses the image from left to right. Light from the viewed scene makes four right-angle turns before reaching the eyepiece.

SINGLE-LENS REFLEX CAMERA

A camera uses a lens system to focus the light rays that come off any subject onto a sheet of film, which is exposed to the light for a brief instant. The light makes chemical changes in the coating on the film, which, when developed, produces an image in black and white or in color (see **Film,** page 68).

To do this, the photographer adjusts the camera in three ways. The lens, which is in front of the film, is positioned so that the image on the film will be in focus—crisp and clear, rather than blurry. A **focusing ring** on the camera moves the lens toward and away from the film.

The camera also needs to be adjusted to let in just the right amount of light to properly expose the film. The **diaphragm** or **aperture** is a hole that can be varied in size to let in more or less light. The opening is measured in **f-stops.**

The third adjustment is the **shutter speed.** The shutter keeps all light away from the film until the moment the photograph is taken. At that point it opens for a fraction of a second, allowing the light to reach the film. How long the shutter should be open is determined by how fast the subject is moving. At a slow shutter speed, a moving object will show up in the picture as a blur.

Since a slow shutter speed also admits more light to the film, the f-stop setting you choose and the shutter speed you choose are related to each other.

Many cameras have two lenses, one that you look through and one that actually makes the picture. A single-lens reflex camera has only one. You see through the viewfinder the same image that comes through the lens, which will later make the picture. The image is reflected up to the viewfinder by a movable mirror. When the shutter is released, the mirror quickly flips up and out of the way so that the image can be focused on the film.

Leaves of the diaphragm

f/3.5   f/5.6   f/8   f/11   f/16

DIAPHRAGM

*The diaphragm is built into the lens system. It consists of a series of thin, overlapping leaves of metal that can reduce the open hole to a fraction of an inch. The **f-stop** refers to the size of that hole. The largest hole, which is determined by the diameter of your lens, has the lowest number. Each succeeding higher f-stop number will let in half as much light as the previous number.*

In a modern single-lens reflex camera, the diaphragm stays at the widest open position, even after you have set the f-stop, until the release button is pressed. This feature allows you to see the subject through the lens with the greatest amount of light.

FOCAL-PLANE SHUTTER

Most single-lens reflex cameras use a focal-plane shutter. It consists of curtains of flexible material with an adjustable gap, rather like a partly open theater curtain. The curtains move across the film at a constant speed; the width of the gap determines the exposure time. You can set the shutter to open the gap wide for long exposures of 1/8 of a second or more; or you can set it to narrow the gap for exposures of 1/500 of a second or less.

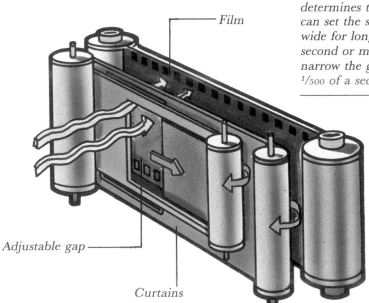

Film

Adjustable gap

Curtains

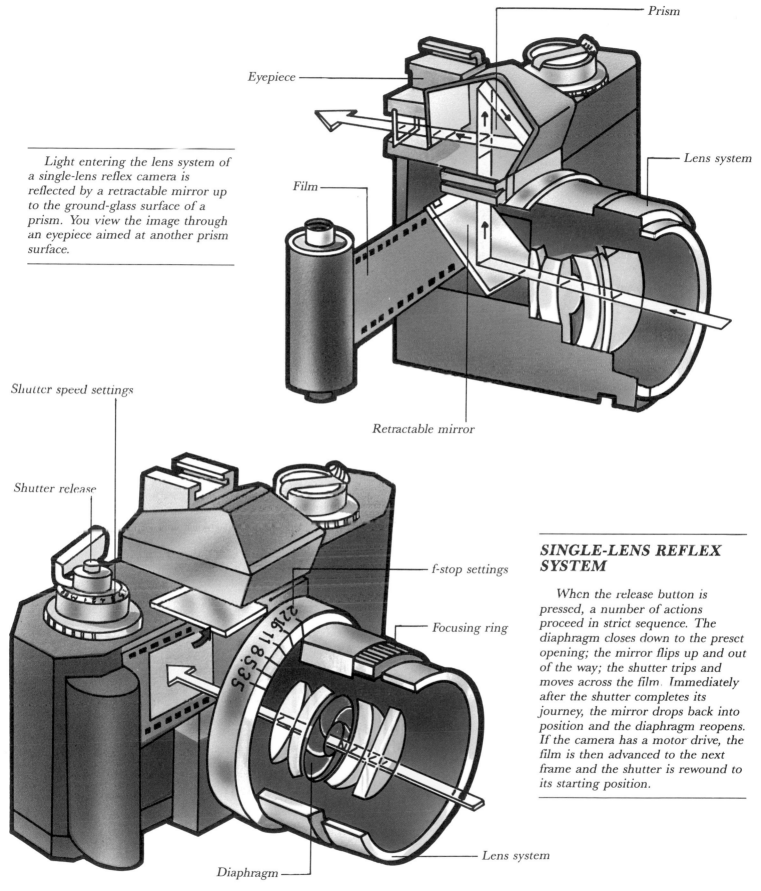

Light entering the lens system of a single-lens reflex camera is reflected by a retractable mirror up to the ground-glass surface of a prism. You view the image through an eyepiece aimed at another prism surface.

Prism

Eyepiece

Lens system

Film

Retractable mirror

Shutter speed settings

Shutter release

f-stop settings

Focusing ring

SINGLE-LENS REFLEX SYSTEM

When the release button is pressed, a number of actions proceed in strict sequence. The diaphragm closes down to the preset opening; the mirror flips up and out of the way; the shutter trips and moves across the film. Immediately after the shutter completes its journey, the mirror drops back into position and the diaphragm reopens. If the camera has a motor drive, the film is then advanced to the next frame and the shutter is rewound to its starting position.

Diaphragm

Lens system

FILM

The camera film that will eventually produce a picture consists of a chemical coating on the surface of a thin transparent plactic base. The chemical coating is exposed to light for a brief instant of time. This forms a potential image (or **latent** image) on the film. The image is made visible when it is developed later in special chemical baths.

The chemical coating, or **emulsion,** is made up of crystals of silver halide in suspension in a layer of gelatin. When light strikes the silver halide crystals, a few atoms of silver separate out from the compound. The latent image has now been formed. Areas hit heavily by light have relatively large numbers of freed silver atoms. Areas touched only slightly by light have few or no silver atoms separated from the silver halide.

So few silver atoms in the emulsion have been separated that the change is invisible to the eye and even to the most sensitive instruments. However, when the film is placed in the developing solution, these isolated atoms of silver collect much more silver around them, and the clumps become plainly visible. When the unexposed silver halide crystals have been washed away, you can take the film out into daylight without fear of ruining it by further exposure.

The film is now a **negative,** with clumps of black where the bright areas in the original scene were, and clear areas where the original scene was dark. Positive prints can be made from the negative by directing light through it onto light-sensitive paper.

BLACK-AND-WHITE FILM

Black-and-white film is made up of three layers.

*The **emulsion** contains the silver halide crystals mixed in and held by a layer of gelatin.*

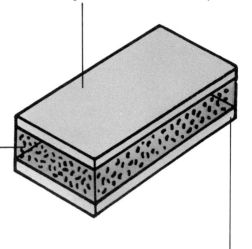

A **scratch-resistant coating** protects the emulsion layer.

*The **base** is a supporting layer of transparent plastic.*

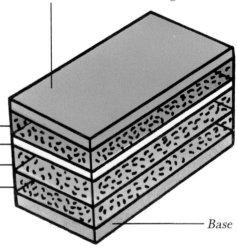

Scratch-resistant coating

Blue-sensitive layer
Yellow filter
Green-sensitive layer
Red-sensitive layer

Base

COLOR FILM

In color film, the emulsion is actually made up of three separate sensitized layers with filtering dyes to control the transmission of red, green, and blue light. A silver image is formed in each layer corresponding to the red, green, and blue light of the original scene. This image is made visible in the development process, which creates red, green, and blue areas in the film instead of black ones.

Transparencies, or color slides, are color-for-color duplicates of the original scene. They must be viewed by transmitted light—the light must pass through and be filtered by the transparent colors in the film. You can project transparencies directly onto a screen to show a lifelike rendition of what the camera originally saw.

COLOR NEGATIVES

Negative color films are also available. They produce color negatives in which the colors in the original scene are replaced by their complements—red appears as cyan on the film, blue appears as yellow, and green appears as magenta. The color negative is used to make a print on sensitized white paper that also has three layers of emulsion, just as the film does. The paper, when developed, is a positive color print of the original scene that you can look at without projection.

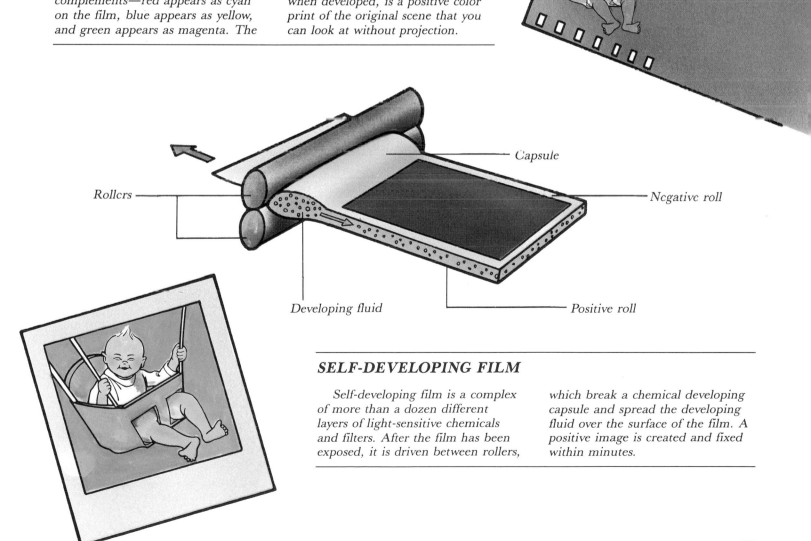

Capsule

Rollers

Negative roll

Developing fluid

Positive roll

SELF-DEVELOPING FILM

Self-developing film is a complex of more than a dozen different layers of light-sensitive chemicals and filters. After the film has been exposed, it is driven between rollers, which break a chemical developing capsule and spread the developing fluid over the surface of the film. A positive image is created and fixed within minutes.

MOVIE CAMERAS AND PROJECTORS

Almost every child has played with a pack of cards showing a series of pictures in which the subjects differ only a slight amount from picture to picture. When you flip through the pack quickly, the appearance of movement is extraordinarily convincing. The eye is fooled into thinking it sees continuous motion when the cards flip by at some speed.

Motion pictures fool the eye in the same way. Movie cameras and projectors use long strips of film on which individual frames have been photographed 1/24 of a second apart. The film is pulled through the camera or projector by means of sprocket holes cut into the edges of the film. Slowing the motor to shoot fewer than 24 frames per second and projecting the film at 24 frames per second produces time-lapse photography; speeding the motor produces slow-motion.

In most other respects, the movie camera is like a still camera. Light from the subject passes through a lens system, often a zoom lens to give the cinematographer control over how much of the subject appears in the frame. The light passes through a revolving shutter, which opens briefly. After the frame has been exposed, the shutter closes and the film is advanced one frame. The shutter then opens again for the next frame exposure.

In amateur cameras, sound is recorded on a magnetic strip running along one edge of the film so that it will be in sync with the visual images being filmed.

In professional filmmaking, the sound is recorded on special tape recorders that keep the sound in sync with the pictures. The sound is added later to the final print of the movie.

MOVIE CAMERA

A movie camera focuses light from the subject on one of a series of frames on a long strip of film. Each frame of film is pulled down in front of the gate by an intermittent claw. While the claw returns for the next frame, the disk shutter opening allows light to reach the film.

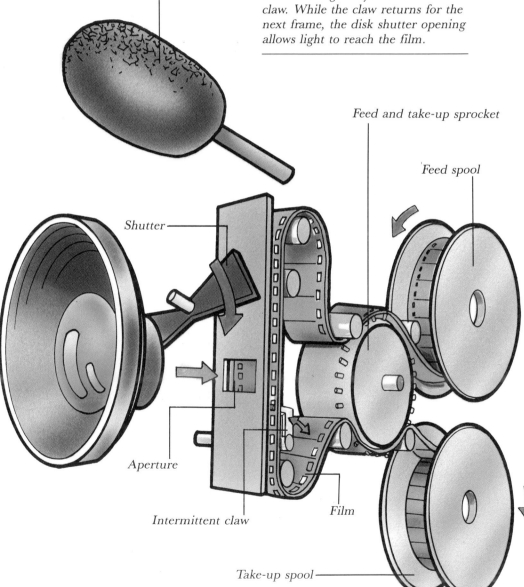

Microphone

Shutter

Aperture

Intermittent claw

Film

Feed and take-up sprocket

Feed spool

Take-up spool

Four sizes of film are now in current use.

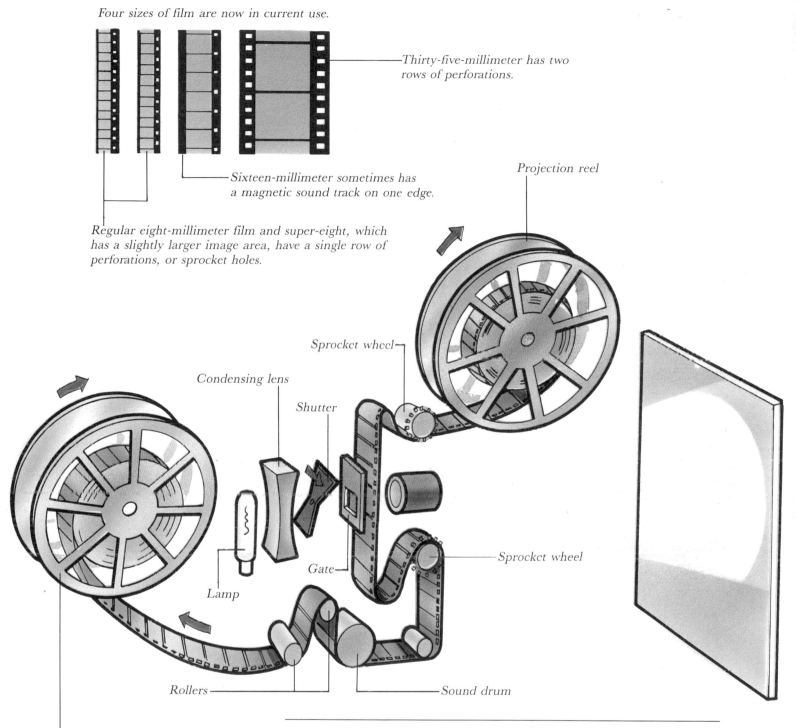

Thirty-five-millimeter has two rows of perforations.

Sixteen-millimeter sometimes has a magnetic sound track on one edge.

Regular eight-millimeter film and super-eight, which has a slightly larger image area, have a single row of perforations, or sprocket holes.

Projection reel

Sprocket wheel

Condensing lens

Shutter

Sprocket wheel

Lamp

Gate

Sound drum

Rollers

Take-up reel

MOVIE PROJECTOR

The film is pulled off the projection reel in a series of jerks that places one frame at a time in front of the lens. In between frames, a rotating shutter prevents light from shining through while the next frame is moved into position. Sprocket wheels ensure that each frame is correctly placed, and a take-up reel winds the film after it has passed through the projector. All this is accomplished with a single motor driving the various elements with gears and belts (see **Gears, Chains, and Belts,** page 40).

71

HOLOGRAMS

The image seems to float in space, solid and real, but when you try to grab it, your hand passes through it. It seems like magic, but it is a hologram. Like a photograph, a hologram is a recording of visual information—but much more.

When you photograph a teacup, you are recording the intensity of light waves reflected from the cup onto the film (see **Film,** page 68). A hologram is a recording not only of the intensity of the reflected light waves, but their wavelengths, amplitudes, and phases (see **Waves,** page 46). This information includes the distance and direction of all visible parts of the object—all the information that the waves contain. When the film is developed and "played back" by shining a light through it, the original light waves reflected from the teacup are reconstructed, and an image is projected that has depth and perspective—a seemingly solid image made of nothing but light.

A hologram captures information so completely that every part of the film plate holds a record of light from all parts of the object. If the plate is broken, each piece can reveal the entire teacup image, although the image will be fainter and fuzzier.

The first holograms were made in 1947, but they remained small and crude until the invention of the laser, which made available a source of intense coherent light. (See **Laser,** page 55). Coherent light is needed to make good holograms.

Holographic images can now be reconstructed, however, by shining white light instead of coherent light on the hologram, and techniques for creating color holograms have been developed. In addition, copy holograms can now be produced from a master negative. These developments have made it possible to use holograms as illustrations in books and magazines and on items such as credit cards.

MAKING A HOLOGRAM

A beam of coherent light from a laser is split so that one part—called the **reference beam**—*reflects off a mirror onto a film plate. Some of the light scattered from the object also falls onto the film plate. The other part of the beam is reflected off another mirror onto the teacup. This is the* **object beam.**

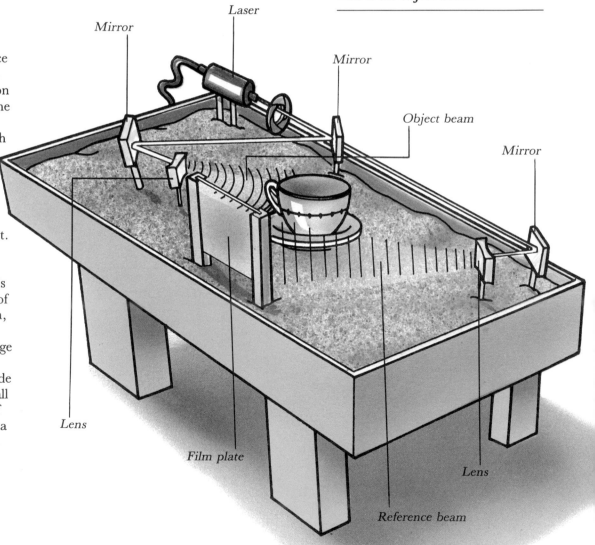

Laser

Mirror

Mirror

Object beam

Mirror

Lens

Film plate

Lens

Reference beam

The developed film shows no recognizable image. Instead, the plate records an invisible pattern of fine lines and whorls called **interference fringes,** which look similar to a fingerprint. They are caused by the interaction of the reference beam and the light reflected from the object. This pattern contains all the information about the object that the object beam sees.

Room light at the same angle as the reference beam

Reflection beam

Reference beam

VIEWING A HOLOGRAM

When rays of light are directed through the hologram plate along the same angle that the reference beam took, they are bent and scattered by the interference pattern so that they are identical with the light rays originally reflected from the object. In effect, the developed hologram is a set of instructions that tells a beam of light how to construct an infinite number of two-dimensional images of the same object.

Virtual image

Reference beam

HOLOGRAM

As you peer into the plate from the side opposite the beam, you see a **virtual image** of the teacup. (A virtual image is an image that appears to form where the object does not exist.)

The light shining on the hologram produces multiple two-dimensional images. Each image shows more of one side of the teacup and less of the other. Your two eyes are each seeing different two-dimensional images, just as they do when looking at a real teacup. Your brain interprets these two-dimensional images in a hologram as one round three-dimensional view. If you shift your head to the side, you see different rays of light from the plate, providing a different view of the teacup and the illusion of depth. If the hologram was done perfectly, with the best possible photographic film, a laser with precisely controllable frequency to reconstruct the image, and no vibration in the object or apparatus during the recording, the image will look remarkably solid and real.

73

STEREOPHONIC SOUND

Stereo sound is the blending of sounds recorded from slightly different directions into sounds that the ears interpret as coming from a solid, three-dimensional source. Although stereo sound is usually transmitted from two loudspeakers spaced apart, the effect is much more than just hearing sound coming from two directions. The sound has depth and appears to come from many directions.

The difference between stereo recordings and monophonic recordings is that, in stereo, more than one channel is used for recording and playback. In the recording studio, at least two microphones—usually many more—pick up the sounds. The signal from each microphone is sent separately to a magnetic tape recorder and recorded as a separate sound track.

When recording an orchestra, 24 sound tracks or more are recorded that run parallel to each other on a magnetic tape. The sounds from one musical instrument or group of instruments might be picked up by one of the microphones and recorded on one of the tracks. The signals from several microphones might be combined on one track. The tracks can even be recorded at different times in different studios.

The sound from all the various tracks are combined (this process is called **mixing**) and put onto a stereo tape that has two sound tracks, or channels. Each channel will come out of a different loudspeaker when played back.

If the sound engineer, during the mixing, were to put half the original sound tracks on the channel for the left speaker and the other half on the channel for the right speaker, the effect when played back would be that half the instruments would come out of one speaker and half out of the other. The orchestra would sound as if it were split into two sections. Instead, the engineer carefully allocates the amount of sound from the original 24 tracks so that the instruments appear to be spread out. Less sound in the left channel and more in the right channel will appear to "move" the instrument toward the right speaker. The instruments will sound as though they are each a different distance from the microphones, just as they are at a live concert.

With a monophonic, one-channel system, all sounds appear to come from the direction of the loudspeaker. Using several speakers does not produce a stereo effect, because the sound from each of the speakers is the same.

With a stereophonic system, the listener gets the impression that the musicians are spread out behind the loudspeakers as if on stage at a concert hall. To get the full effect of stereo, the two speakers must be placed precisely the same distance from the listener and the room acoustics must be just right.

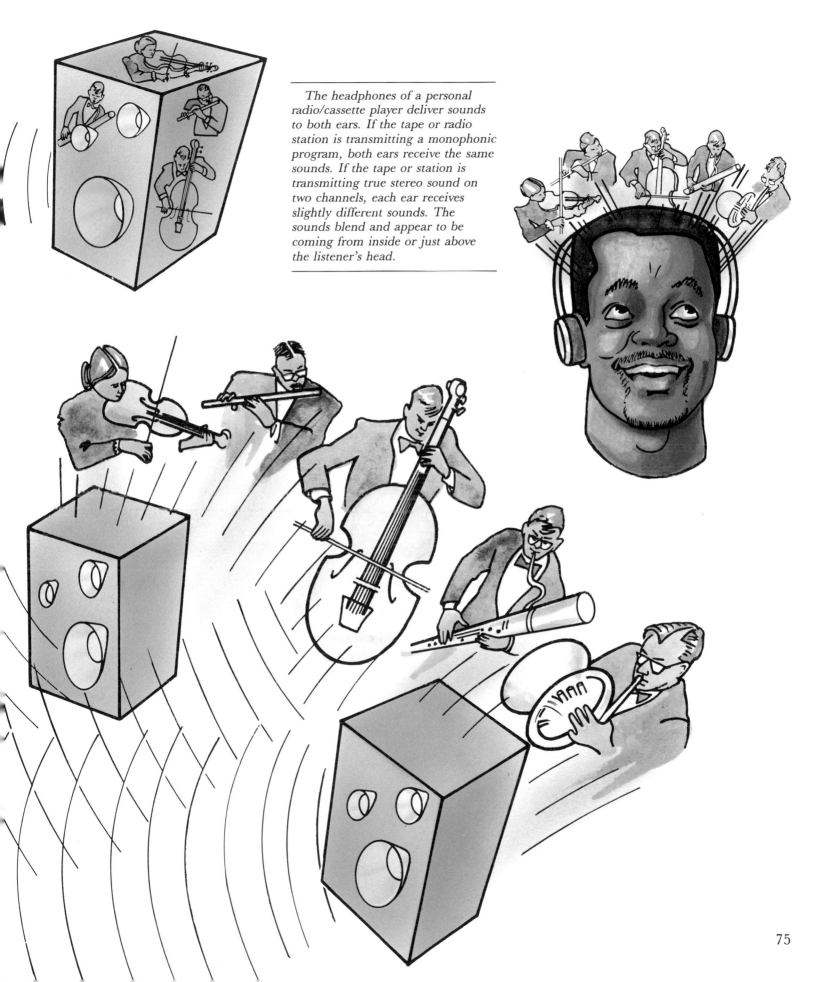

The headphones of a personal radio/cassette player deliver sounds to both ears. If the tape or radio station is transmitting a monophonic program, both ears receive the same sounds. If the tape or station is transmitting true stereo sound on two channels, each ear receives slightly different sounds. The sounds blend and appear to be coming from inside or just above the listener's head.

LOUDSPEAKER

A loudspeaker can also be a "softspeaker" when the original sound is soft. Either name would suggest what the device does—it converts the electrical energy emerging from the amplifier of a radio, for example, into mechanical energy. The mechanical energy is typically in the form of vibrations of a diaphragm. Anything vibrating creates sound waves in the surrounding air. The vibrations of the diaphragm, therefore, go into the surrounding air as sound—a more or less faithful reproduction of the sound originating in the radio station.

A common arrangement for a loudspeaker is a coil and a magnet in association with each other. The magnet produces a magnetic field. The coil, located in that field, is energized by the electrical signal coming from the amplifier. The result is that the coil develops a varying magnetic field of its own, so that the coil is at one time attracted to the magnet and at another time repelled by it. The coil therefore moves. Because the coil is attached to the diaphragm, the diaphragm also moves, generating the sound waves that you hear.

If a radiating surface such as a diaphragm is large, it works most efficiently at lower sound frequencies—the bass sounds. If the surface is small, it works best at high frequencies. For this reason, a loudspeaker is likely to contain at least two types of radiating surfaces, one large and one small. They go by the splendid names of woofer and tweeter. The woofer is most

efficient in reproducing the low-frequency, wooflike sounds, and the tweeter works best at reproducing the high-frequency, tweetlike ones. The combination gives good sound in both ranges.

*The source of the sound emerging from a loudspeaker is likely to be a **diaphragm.** In the absence of an electrical signal from the amplifier, meaning the absence of sound in the radio station, record, or tape, the diaphragm does not move.*

When an electrical signal comes from the amplifier, reflecting a sound at the source, the diaphragm vibrates, thus transmitting sound waves to the surrounding air.

*A small **tweeter** contributes the high-frequency sounds.*

*The **vent** helps to strengthen the bass tones.*

*The **terminals** connect the loudspeaker to the amplifier.*

*A loudspeaker designed to emphasize bass sounds contains a large **woofer** to reproduce low-frequency sounds.*

LP RECORD

What was a marvelous advance 40 years ago—the "long-playing" or LP record—may be in the dinosaur class in another decade or so because of the advent of the **compact disc** (see page 79) and the prospective arrival of **digital audio tape** (see page 78). The LP record got its name because it played so much longer than its predecessor, the record of similar size that ran more than twice as fast and therefore had less than half the content. That record ran at 78 revolutions per minute (RPM), whereas the LP turns at 33⅓ RPM. To play a 78, you needed a record changer that plopped one disc on top of the preceding one, at the same time withdrawing and then gently replacing the tone arm. With the 33⅓, you could listen for a long time to a single record.

An electric motor causes the turntable to turn at 33⅓ RPM. When the pickup arm is placed at the beginning of the record, a jewel-tipped needle, or stylus, in the pickup cartridge rests in a continuous concentric groove, which is about 27 thousandths of an inch wide. The needle riding in the groove vibrates according to variations in the groove, which represent variations in the original sound. The vibrations go to an electromechanical device in the pickup cartridge called a **transducer,** which converts the motions into an electrical signal. Because the signal is quite weak, it must go next to an amplifier to strengthen it, and then to a loudspeaker, which converts the electrical signal into vibrations that create sound waves in the surrounding air.

To get the sound onto the record in the first place, a cutting head—an electrical motor with a stylus—is mounted above a rotating table. On the table is an aluminum platter coated with plastic. The stylus of the cutting head digs into the surface, cutting a v-shaped, spiral groove. Varying electrical current in the cutting head causes the stylus to vibrate in ways that correspond with the original sound. These vibrations make the stylus cut an irregular groove with wiggles in it. Loud sounds make the stylus cut deeper and wider; high-pitched sounds make it wiggle back and forth more rapidly than low-pitched sounds. For a stereo record (see **Stereophonic sound,** page 74), the stylus vibrates in two directions, cutting different wiggles on each side of the groove corresponding to the signals from the two stereo tracks. The aluminum platter serves as a master record; what you put on your turntable is one of many copies of the master.

The turntable is turned by a belt running off an electric motor. The pickup arm contains, among other things, a cartridge that holds a stylus made of diamond or sapphire. The stylus runs in the record's groove, which varies in shape according to variations in the original sound.

The pickup cartridge converts the vibration of the stylus to an electrical signal. As the stylus vibrates in the walls of the groove, its motion is transferred to the cartridge, which can contain a moving magnet surrounded by fixed coils. The movement of the stylus causes the magnet to move, generating electrical signals in the coils. Sometimes the opposite arrangement is used—moving coils that generate signals in a fixed magnet. For a stereo recording, the right-hand signal is recorded in one wall of the groove, the left-hand signal in the other.

Stylus

Cartridge

Groove

Pickup arm

AUDIOTAPE RECORDING

The tape that plays music in an audiotape player is a thin plastic material coated with particles of a magnetic material, such as iron oxide. Music is recorded on the tape because of the principles of **electromagnetism** (see page 24). An electric current flowing in a coil of wire generates a magnetic field, and a magnetic field moving near a wire generates a current in the wire. In recording, the sound is converted to a changing electrical signal by a microphone, amplified, and fed into a coil that is part of the recording head on the tape recorder. The current causes a fluctuating magnetic field in the head. As the tape being recorded moves across the head, the arrangement of the iron oxide particles is changed according to the magnetic field.

When you play back a tape, the opposite occurs. The tape moves past a playback head, and the magnetic field of the tape induces an electric current in the coil.

Most tapes are based on analog recording, meaning that the voltage of the original sound is represented directly on the tape by the continuously varying orientation of the magnetic particles. Tape can also be recorded by digital recording. In digital audiotape (DAT), the sound is represented indirectly through magnetic particles oriented in one of two directions. The orientations represent "on" or "off" signals derived from an intense sampling of the original sound (see also **Compact disc player**). An analog tape reproduces, along with the sound you want to hear, all the noise and distortion that occur in the recorded signal. A digital tape is free of them. It therefore produces a cleaner recording.

If you are recording, the **erase head** removes any previous recording by rearranging the magnetic particles on the tape into a random pattern.

In the playback mode, only the **playback head** operates; it reads the magnetic pattern on the tape and generates the electric signal that drives the loudspeaker.

The **record head** then magnetizes the particles into a pattern that is an analog of (analogous to) the sound waves from the sound being recorded.

AUDIOTAPE DECK

A home cassette tape deck includes a motor and a system of gears and belts to drive the tape. In a machine that lets you either record sound or play it back, the path of the magnetic tape carries it past three kinds of electrical devices called heads. Recording and playback are governed by electronic controls, and a variety of other controls enables you to regulate the sound volume, run the tape rapidly forward or backward, and eject the cassette. A magnetic tape carrying an analog recording has its particles of iron oxide magnetized in degrees depending on the voltage in the input signal. This in turn varies continuously according to the strength of the original sound. A magnetic tape carrying a digital recording stores the signal as a sequence of highly magnetized and lightly magnetized areas. They represent the "on" and "off" signals that describe the large number of samples of the original sound.

COMPACT DISC PLAYER

The key to the compact disc is the difference between analog and digital recording. The same difference occurs between a speedometer of the dial type and a digital-reading speedometer. On a dial-type speedometer, the needle varies continuously with the fluctuation of the speed of the car. In analog sound recording, the sound being recorded is observed continuously, and the recording shows variations that are parallel with—or analogous to—the sound itself.

A digital-reading speedometer shows a number that reflects frequent samplings of measurements of the speed of the car. Digital recording of sound is similar—a large number of individual samples of the sound signal are taken, and the recording reflects those samples.

When music is being recorded, the variations in the intensity of the sound are transformed by the microphone into variations of voltage. In a stereophonic CD recording, the electrical signal from the microphone is sampled 44,100 times per second from each of the two stereo channels. Each sample winds up on the disc as either a pit or a flat space, reflecting the variations in voltage. The functioning side of the disc also bears a reflecting layer of aluminum and is protected by a coating of clear plastic.

When you put a CD in a player, a laser beam of low power reads along the spiral track from the inside to the outside of the disc as the disc rotates in the machine. A light-sensitive device called a **photodiode** receives the readings as on/off signals and converts them into continuously varying, or analog, electrical signals. At a pit, the beam is not reflected, and the photodiode produces no signal. At a flat space, the beam is reflected, and the photodiode produces a signal.

Digital recording has several advantages over analog recording. It's more accurate, because it precisely records, in CD recording, some 65,000 levels of voltage. It is impervious to the electrical noise or hiss that is inescapable in analog recording. It offers a better representation of the dynamic range of the music (the difference between the loudest and softest sounds). In addition to these advantages, the CD, unlike a conventional record or a magnetic tape, undergoes no wear because it is touched only by light.

When the laser beam enters a pit on the underside of the compact disc, it is not reflected. To the light-sensitive photodiode, this is an "off" signal, and the photodiode does not produce an electric signal. The absence of a signal represents silence in the original sound.

When the beam strikes a flat space, it is reflected to the photodiode, which produces an electric signal representing a minute part of the original sound.

Lenses focus the beam on the microscopic pits and flat spaces engraved on the disc. The beam reads them as digital code—a series of on/off signals.

The *photodiode* converts the signals to analog signals.

A *prism* deflects the "on" signals to the photodiode.

The *turntable-drive motor* causes the disc to rotate.

The *laser* produces a beam of light at low power.

STRINGED INSTRUMENTS

*The four strings are positioned in an arch around the curved **bridge,** so that the bow can touch only one or two strings at a time.*

A violin has a narrow waist so that the bow can rub against the end strings without touching the wood.

*By pressing the string against the **fingerboard,** the string's vibrating length is shortened, and it sounds a higher note.*

*The front and back surfaces of the box are coupled together by a **sound post,** so that both surfaces of the instrument contribute to the sound.*

*The strings are wound around pegs in a **pegbox.** By turning the pegs, the tautness—and therefore the pitch—of the strings is controlled.*

Violins, guitars, and other stringed instruments produce sound by the vibration of a taut string. A vibrating string alone does not make a sound that is very loud. To amplify the sound, the strings are connected to a sounding board that vibrates driven by the strings. They can be set in motion by plucking them with the fingers, by striking them, or by rubbing them.

The vibrating string produces a tone at its natural frequency, called the **pitch.** This pitch depends on the length of the string, its weight, and its tautness. A short, tight string will produce a higher tone than a longer, looser one. By pressing down on a string with your finger, the string is temporarily shortened, and it will sound a higher note than it does when its full length is vibrating.

The strings are made of metal, nylon, or animal gut and wound with metal. Stringed instruments are made so that the strings lie across a **bridge** that stands up from the sounding board or box. This bridge transfers the vibrations of the strings to the sounding box, which magnifies them and transmits them to the air. The shape of the box, which is usually made of wood, and the holes in the box greatly affect the type of sound that the instrument makes.

Making a violin or a guitar is an art. Good instruments are still made by hand, because each piece of wood has individual acoustic qualities and must be shaped and treated with care. The varnish used to seal and preserve the wood is particularly important.

VIOLIN

Violins are actually a family of instruments that are shaped similarly but are of different sizes. Four strings, each tuned to a different note, are stretched along a **fingerboard** and over the sounding box and bridge to a tailpiece. Two f-shaped holes in the box are the sound holes.

The violin's strings are usually sounded by a **bow,** which is a thin rod of wood with a horsehair string stretched tightly between the raised ends. To increase the friction when it is rubbed against the strings, the horsehair is coated with rosin.

The smallest member of this family of instruments is the **violin,** which is held under the chin to be played. The **viola,** which is also held under the chin, is slightly larger and, consequently, produces tones that are a little lower. The **violoncello,** or cello, is about twice as large as a violin and sounds an octave lower than a viola. The player sits with the cello between the knees. The **bass viol** is about as tall as an average man, and is played standing up.

Violin

Viola

Violoncello

Bass viol

Frets

GUITAR

Guitars have **frets,** or metal inserts, set into the fingerboards to indicate the precise placement for the fingers. Banjos, ukuleles, mandolins, and lutes are also fretted instruments.

A guitar usually has six strings that are mounted over a flat bridge. The player plucks or strums them with the fingernails or uses a piece of plastic or metal called a **pick.**

81

PIANO

Inside a piano is a large, sturdy metal frame holding neatly lined-up strings. The sounding board, usually a pale, unfinished-looking slab of wood, is mounted under the strings. The top of a grand piano can be propped open so that sound from the sounding board can reflect off the top and out to the audience. The sounding board is what determines the characteristics of the piano's sound, more so than the strings, the case, or even the delicate touch on the keys by a skilled performer.

Striking one of the keys activates a complex system of levers, leather strips, and wire and felt cushions, producing a sharp blow by a cushioned hammer on the strings. Each hammer strikes one, two, or three strings, all tuned to the same pitch to provide a rich, full sound for each note. The bass notes each have only one string, which is steel that is wound with copper to make the string heavier. The middle notes have two steel strings each. For the high notes, thinner strings are used, but each note has three strings. Standard pianos have 88 keys, so more than 200 strings may be stretched across the frame. In a grand piano, the strings are arranged in two banks crossing each other to save space.

Pressing the key also lifts a damping pad off the strings so that the strings will continue to vibrate as long as the key is held down. Some pianos have two pedals, operated by foot, that also control the dampers. The loud pedal lifts the dampers from all the strings, causing them to vibrate for greater resonance of tone. The soft pedal allows the hammer to hit only two of the strings for its note, giving a softer tone. On grand pianos and some upright pianos, a third pedal holds the dampers off only the notes being played.

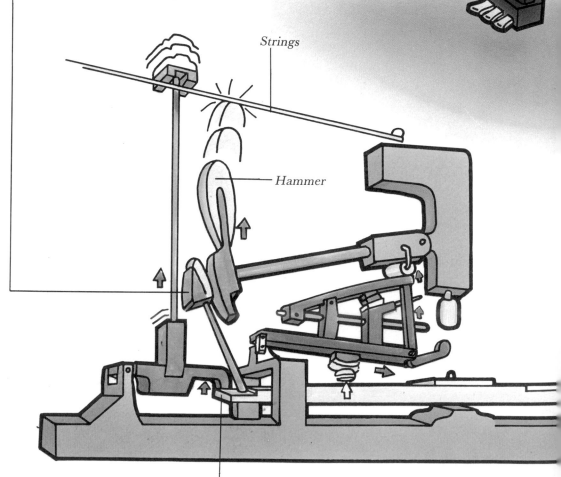

A check prevents the hammer from bouncing up to the string and hitting it again.

Strings

Hammer

The action of the key also raises the damper from the strings; as soon as the key is released, the damper falls back onto the strings.

HARP

Although the harp shares the piano's multiplicity of strings, each tuned to a different note, the strings are plucked by the player rather than struck with a hammer. The strings are stretched within a triangular frame and are attached at the bottom ends to a hollow box that is the sound resonator. To help the player find the right string in this bewildering array, some of the strings are colored.

Near its top, each string passes through a set of pegs on two rotatable discs. By depressing a pedal, the harpist can turn one or both of the discs. The effect is to shorten the vibrating length of the string, which raises its tone.

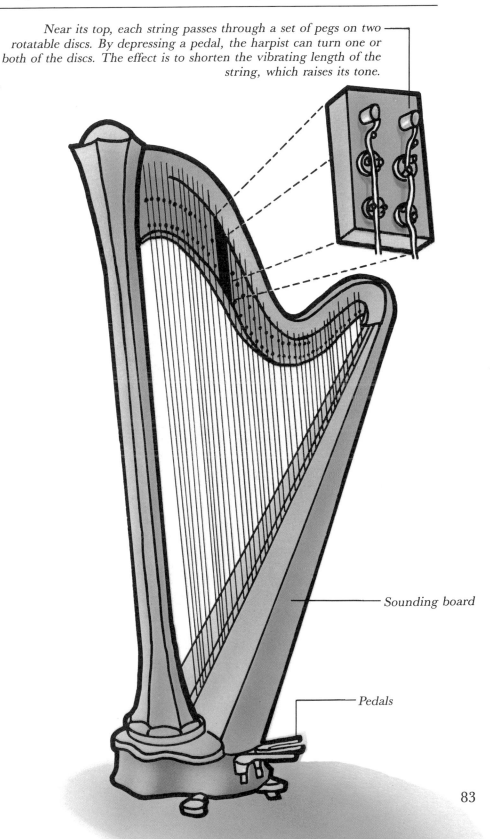

Sounding board

Pedals

HOW A PIANO KEY WORKS

Pressing the key activates a series of hinged levers that hurl the hammer against the string. The hammer immediately falls back so that the string can continue to vibrate.

WIND INSTRUMENTS

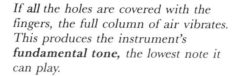

A large variety of musical instruments use the pressure of air to make a musical sound; these comprise the family of wind instruments. For most of them, you produce the air pressure by blowing into a metal or wooden tube in which the enclosed column of air is set vibrating.

The frequency or pitch of the sound is determined by the length of the column of air. A piccolo, which is only a few inches long, produces a high note; the 18-foot-long tuba booms out a bass note. The column of air does not have to be straight. Tubes that are longer than two or three feet are bent or coiled for easier handling.

To play different notes, the player can change the length of the column of air. Different instruments have different mechanisms for doing this.

If all the holes are covered with the fingers, the full column of air vibrates. This produces the instrument's **fundamental tone,** the lowest note it can play.

If a hole halfway along its length is opened, the vibrating column of air is shortened by half. This produces a note one octave higher.

RECORDER

The recorder is a simple **woodwind** instrument, with a mouthpiece in one end that you blow into. Holes along the side of the tube allow the player to change the note being played.

CLARINET ▶

The clarinet is a woodwind called a **reed** instrument; fixed to its mouthpiece is a single reed, which the player sets to vibrating by blowing past it. This in turn sets the column of air vibrating. Clarinets are made of lacquered wood with metal levers and caps to close the holes.

The saxophone, though it is made of brass, is classified with the woodwinds because it also has a single reed. It has a flared tube to intensify the sound (see **Megaphone,** page 49), which is curved upward.

Reed

Double reed

OBOE ▲

The oboe is the smallest of the family of **double-reed** woodwinds. The two reeds are held tightly between the player's lips and vibrate toward and away from each other when the player forces air between them. Larger members of this family include the English horn and the bassoon.

FLUTE

The flute is also considered a woodwind, although it is commonly made of silver or silver-plated brass or steel. The player blows across the mouthpiece on the side of the instrument. Padded caps operated with levers called **keys** open and close the holes in order to adjust the length of the air column.

The **trombone** has a slide that extends the tube out in front of the player.

BRASS INSTRUMENTS

The player's vibrating lip against a cup-shaped mouthpiece creates the pulsations of air pressure in the horn family of wind instruments. The player changes the tone by controlling the air pressure from the breath and the tightness of the lips. The length of the air column is changed, not with holes in the tube, but by changing the length of the tube.

In the **trumpet,** the length of the column of air is controlled by three valves, which may be opened to redirect the air through three bypasses of different lengths. These bypasses add length to the air column. Combinations of the three valves provide six different lengths of air column.

ELECTRONIC INSTRUMENTS

Musical instruments that depend on the vibration of a string or a column of air or the natural resonance of a physical object to make sound are called **acoustic** instruments. The invention of the vacuum tube and, later, the transistor opened the possibility of sound created entirely by electronic means. These developments encouraged scientists and musicians to reexamine exactly what a musical note is and why the sounds of one instrument differ from those from another.

A musical note has a number of distinctive characteristics. Sound is a pressure wave in the air which, when it reaches our ears, produces a matching vibration in our eardrums that we interpret as sound (see **Sound waves,** page 48). Ordinary noise contains sound waves of many frequencies. A musical note differs from noise in that it has a fundamental frequency, which gives the note a distinct **pitch.**

A note made up of just one frequency alone has a thin, whistlelike quality. Acoustic instruments produce notes with what is called **timbre** or **color;** a number of higher frequencies called **harmonics** are added to the fundamental frequency. Since the sounds of these higher frequencies are not as loud as the fundamental, they don't overwhelm the basic pitch that we hear. They do change the nature of the sounded note, and that is what makes the flute sound different from the piano, or the violin from the trumpet, even when the instruments are playing the same note.

An electric instrument creates notes from electric currents that are then sent to speakers to be turned into sound waves. The electric instrument shapes and controls each note according to a series of characteristics such as loudness, pitch, and duration. The notes become music when a number of them are sounded one after the other to make a melody, or when they are sounded at the same time to make a chord.

PRODUCING A NOTE ELECTRONICALLY

The first step in making a musical note electronically is an **audio oscillator,** *which is a device that changes an electric current to one that oscillates, or alternates, at a particular frequency. Four elementary waveforms are created by an audio oscillator.*

The **square wave** *sounds like the mixture of an oboe and a flute.*

The **triangle wave** *has a distinct flutelike sound.*

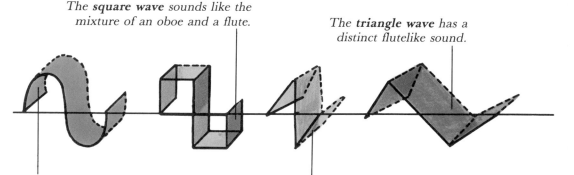

The **sine wave** *is the pure frequency of the note, and it has a thin sound like a whistle.*

The **sawtooth wave** *has a warmer and fuller sound, almost like a saxophone.*

Loudness is a measure of the extent of the sound pressure vibrating our eardrums. The volume control that makes the note louder or softer simply increases or decreases the amplitude of the electric wave.

Soft   Loud   Painful

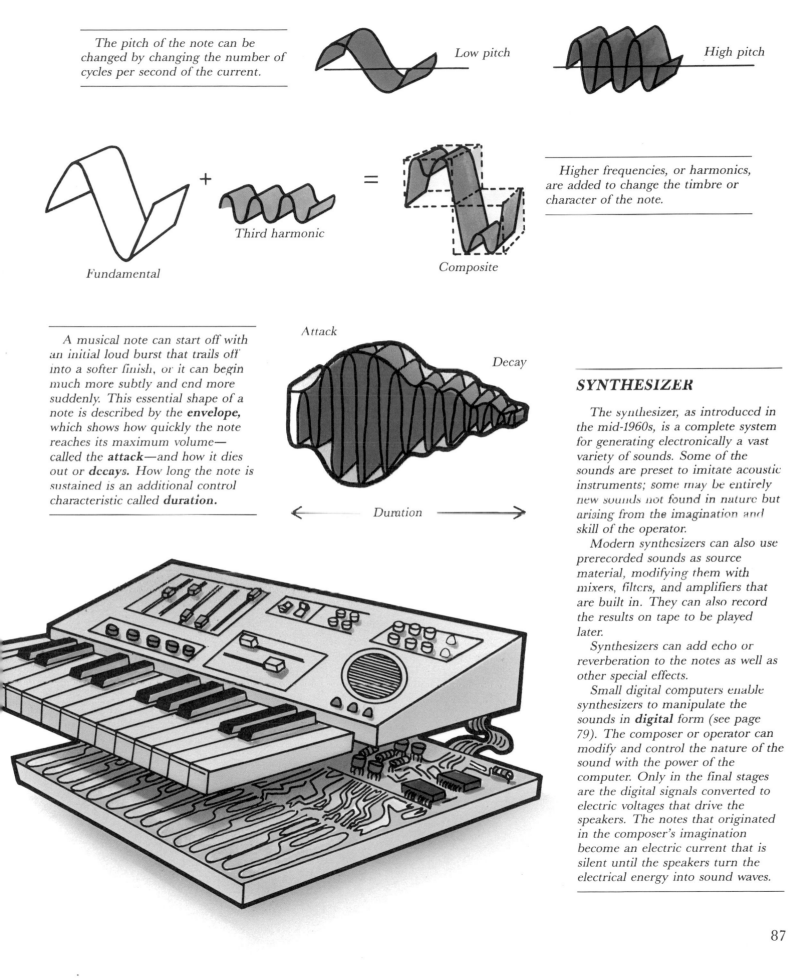

The pitch of the note can be changed by changing the number of cycles per second of the current.

Low pitch

High pitch

Fundamental

+

Third harmonic

=

Composite

Higher frequencies, or harmonics, are added to change the timbre or character of the note.

A musical note can start off with an initial loud burst that trails off into a softer finish, or it can begin much more subtly and end more suddenly. This essential shape of a note is described by the **envelope,** which shows how quickly the note reaches its maximum volume—called the **attack**—and how it dies out or **decays.** How long the note is sustained is an additional control characteristic called **duration.**

Attack

Decay

Duration

SYNTHESIZER

The synthesizer, as introduced in the mid-1960s, is a complete system for generating electronically a vast variety of sounds. Some of the sounds are preset to imitate acoustic instruments; some may be entirely new sounds not found in nature but arising from the imagination and skill of the operator.

Modern synthesizers can also use prerecorded sounds as source material, modifying them with mixers, filters, and amplifiers that are built in. They can also record the results on tape to be played later.

Synthesizers can add echo or reverberation to the notes as well as other special effects.

Small digital computers enable synthesizers to manipulate the sounds in **digital** form (see page 79). The composer or operator can modify and control the nature of the sound with the power of the computer. Only in the final stages are the digital signals converted to electric voltages that drive the speakers. The notes that originated in the composer's imagination become an electric current that is silent until the speakers turn the electrical energy into sound waves.

PIPE ORGAN

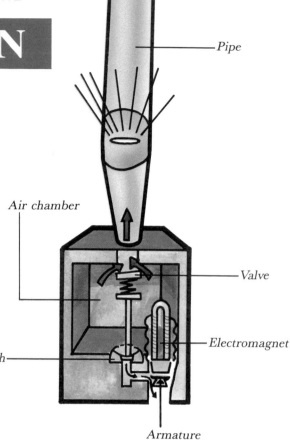

Pipe

Air chamber

Valve

Electromagnet

Pouch

Armature

Probably the largest of musical instruments, the pipe organ is a complex system. Like woodwinds and brasses, a pipe organ is a wind instrument, but the air comes from a pump. Air pressure is directed to the pipes; each note is sounded by its own pipe.

Differing designs of pipes produce the varied and dynamic sound of the pipe organ. Some of the pipes are **reed pipes,** each of which has a vibrating metal reed. The sound of the **flue pipes** comes from a vibrating column of air. In addition, organ pipes are shaped so as to produce a specific character of sound for each shape. Banks of pipes of similar design and shape, but of varying lengths, are called **stops.** Each stop is constructed to produce a particular type of sound.

Some organs have pipes that are six times taller than a man. Organs can have hundreds of pipes; the largest organs have more than 40,000 pipes. Each pipe organ is individually constructed to fit the building that will house it.

HOW THE AIR GETS TO THE PIPE

The **air chamber** holds air under pressure. When the player presses one of the keys, an electric current is sent to the **electromagnet** (see page 24), which pulls up the **armature.** This allows the air in the **pouch** to escape; the top membrane of the pouch, as it goes down, pulls open the **valve** and the pressurized air rushes into the pipe.

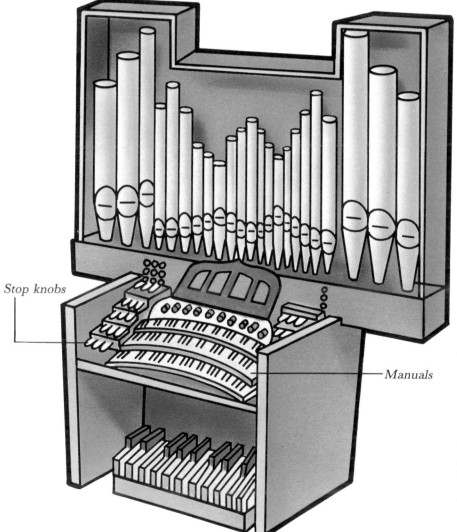

Stop knobs

Manuals

THE PLAYER'S CONTROLS

The player sits at a console with several keyboards, or **manuals,** and a set of large wooden keys that are played with the feet. **Stop knobs** at either side of the manuals direct the output to different banks of pipes.

Communications

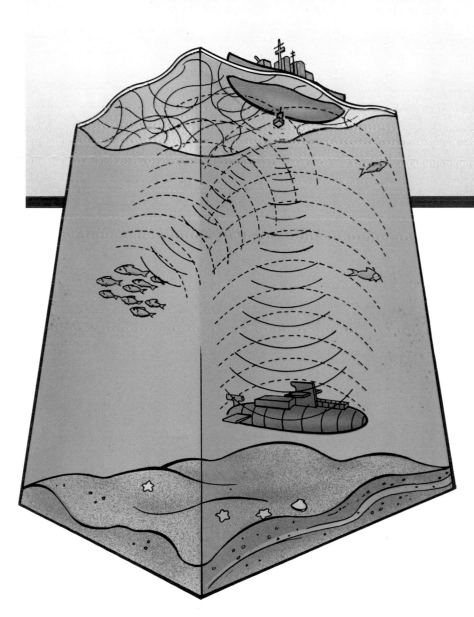

TELEPHONE

The only function of Alexander Graham Bell's original 1876 phone was to convert sound waves to electrical currents, and vice versa. A modern electronic telephone is a multipurpose instrument, full of integrated circuits and microprocessors, that not only performs the original function more efficiently, but also controls a worldwide telecommunications network at the touch of your fingers.

A modern phone can be programmed to store telephone numbers, forward calls to another phone, block or screen calls from certain phones, automatically redial numbers, or perform many other services. With the tone keys of a push-button phone, you can call up distant computers and either input information or request information.

2. It also contains a speaker, which converts the electric current of the incoming call into sound waves that you can hear.

4. This current closes a relay indicating that your phone is off the hook and requesting service. An electronic switching system connects your telephone line to the switching system's equipment and sends a steady humming signal—the dial tone—back to your phone, informing you that it is ready to handle a call.

*1. The handset contains a tiny **electret microphone** (see page 103), which picks up the sound waves of your voice and converts them to the electric current that goes through the telephone lines.*

3. When you pick up a telephone handset, the contacts of a switch hook in the base of the telephone close (they were held open by the weight of the handset). This forms a direct-current circuit through telephone wires between your phone and batteries in the local central office.

5. When you dial by pushing the buttons, the humming signal is removed, and your phone transmits audible tones to the switching system. The first three digits of the seven digits you dial identify the central office exchange of the telephone you are calling. The last four specify the phone line in that exchange. For long-distance calls, you must first dial the digit 1 and then three digits indicating the city or area you are calling.

6. These signals provide the routing information that the network needs to set up a connection. This information is sent over digital data channels—which are separate from the voice circuits you will use for your conversation—to long-distance switching systems and to the local exchange of the phone you are calling.

7. The switching machines quickly link together, out of thousands of possible alternative routes, the best available voice circuits between your local exchange and the exchange you are calling. If all circuits are being used, a busy signal (a rapid on/off tone) is sent to your phone.

9. When the person you are calling lifts the receiver, the plunger springs up and closes a switch in the base of the telephone, shutting off the ringing current and permitting a direct current to flow in the circuit from the local office. The connection is completed. The elapsed time from the end of dialing to ringing the called phone is typically less than two seconds.

By punching in tones from your phone, you have programmed the vast international network to connect your phone to another of the more than 600 million telephones in the world.

8. When the links are established, the called exchange transmits a 20-hertz signal to the called phone, causing its bell to ring or its tone alerter to sound.

HOW A PUSH-BUTTON PHONE WORKS

By pressing a button on the telephone, a pair of musical tones is generated and sent to the central office, which uses the tones to set up phone connections. The tones are pairs of very precise frequencies so that the detector in the central office can distinguish the signal from voice sounds and noise. It will accept the signal only if both tones are present simultaneously.

Pressing the 5 button will send tones of 1336 hertz and 770 hertz. Pressing the 9 button will send tones of 1477 hertz and 852 hertz.

Rotary dial phones send sequences of electrical pulses when the dial rotates. These are being supplanted by push-button phones because they are faster and less prone to error. In addition, the tone signals can be used to input and to request data from computers.

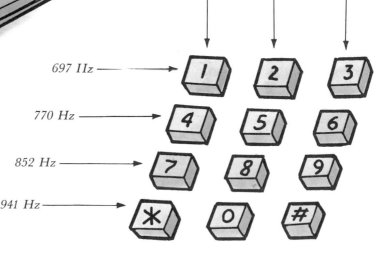

CORDLESS TELEPHONES

A cordless telephone is a portable extension phone capable of operating up to 1,500 feet from its base phone, which is wired into the telephone network. Calls are carried on two frequencies broadcast by the portable handset and the base phone. The two frequencies allow you to talk and listen at the same time. Otherwise, you would be able to either talk or listen, as with CB radios (see page 98).

Like a CB radio, the channels available for use by a cordless phone must be shared. Some cordless phones and their base phones are set at the factory to operate on one of ten channels allocated by the FCC in the range of 46 to 49 megahertz. If interference is encountered, they must be returned and reset to another channel. Some phones have a switch that will allow you to change channels. Most new cordless phones automatically scan the ten channels until they find one that is not being used.

If neighbors use a cordless phone on the same channel that you are using, you will hear their conversation and they will hear yours. To ensure privacy, future cordless phones and their base phones will have paired microprocessor chips that will scramble and unscramble the signals between them to prevent eavesdropping.

It is possible for someone to use a cordless phone to pick up the dial tone broadcast by your base phone and make calls through your phone line that will be charged to you. To prevent this pirating, a digital security system has been incorporated into some new cordless phone systems. Every time the portable handset is returned to the base phone cradle, a new security code is automatically chosen. When the handset signals the base phone for a dial tone, it also sends this coded signal. If the codes don't match, the base phone won't send a dial tone. Some cordless phone systems store up to a million different codes.

Portable handset

Base phone

The base phone is powered from the electrical wiring in your home; the portable handset is powered by nickel-cadmium batteries that are recharged when the handset is in the cradle of the base phone. Each has a miniature, low-powered FM radio transmitter and receiver and a telescoping antenna.

PHONE ANSWERING MACHINE

An answering machine automatically answers the phone with a recorded announcement, records callers' messages, and plays the messages back on demand. Some have built-in telephones; some are separate machines that are connected to your phone line.

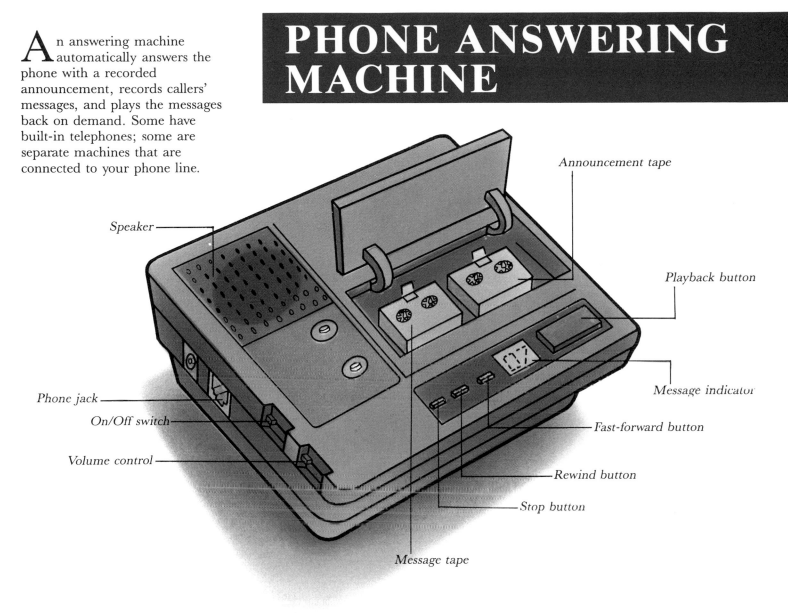

Speaker

Announcement tape

Playback button

Phone jack

Message indicator

On/Off switch

Fast-forward button

Volume control

Rewind button

Stop button

Message tape

The announcement—what you want the machine to say to the caller—is usually recorded on a magnetic tape microcassette via a microphone in the base of the answering machine. The machine automatically inserts a beep after the announcement is recorded. Some machines come with a microchip that has a prerecorded synthesized voice announcement.

The machine can be set to allow a number of rings (typically two to six) before it answers an incoming call. Once an activating "on" button has been pressed, the same signal sent to the telephone that causes it to ring (see page 90) is also routed to the answering machine.

When it answers, the caller will hear the greeting announcement followed by a beep. The purpose of the beep is to inform the caller that the machine is ready to take the message. After the beep, the machine will record the caller's message on a microcassette tape. To cut short long-winded messages, the machine can be set to stop recording after a few minutes.

When the caller hangs up, or is silent for a considerable period of time, such as ten seconds, the machine assumes the call is finished and resets itself to answer the next call. It indicates that a message has been received by activating a blinking light-emitting diode. Some machines have an electronic call-counter display.

To hear the messages, you rewind the tape and play the messages back consecutively.

When you're away, you can listen to messages by calling your phone from any push-button telephone. You can identify yourself by pressing a security code number on the keypad. Some machines use a pocket-size tone beeper that you hold close to the mouthpiece. The machine will then play back the recorded messages and set itself to receive more messages.

FAX MACHINE

Letters, handwritten notes, photographs, and printed text of all kinds can be sent over the telephone lines in less than a minute per page by **facsimile,** or fax.

When you put a document into a fax machine and dial the number of the receiving machine, the two machines automatically adjust themselves to send and receive at the same rate. Then the sending machine scans the document and converts what it sees to electric signals that can be sent over ordinary telephone lines. The receiving fax machine decodes the signals and prints out a duplicate of the original document.

There are more than two million facsimile machines in the United States and millions more in Europe and Japan. Government agencies and businesses use them for routine written communications among offices here and abroad. You don't have to own a facsimile machine to send a document. Nearly 20,000 public fax stations in shopping malls, hotels, office buildings, and airports allow you to fax a document to any other machine for a few dollars a page.

One type of fax machine can be programmed to store several documents in its memory and to transmit them to a hundred or more destinations at specified times. Businesses use this feature to send messages from a central office to multiple sales offices. And by this means direct mail advertisements (sometimes called junk mail) can be faxed to potential customers.

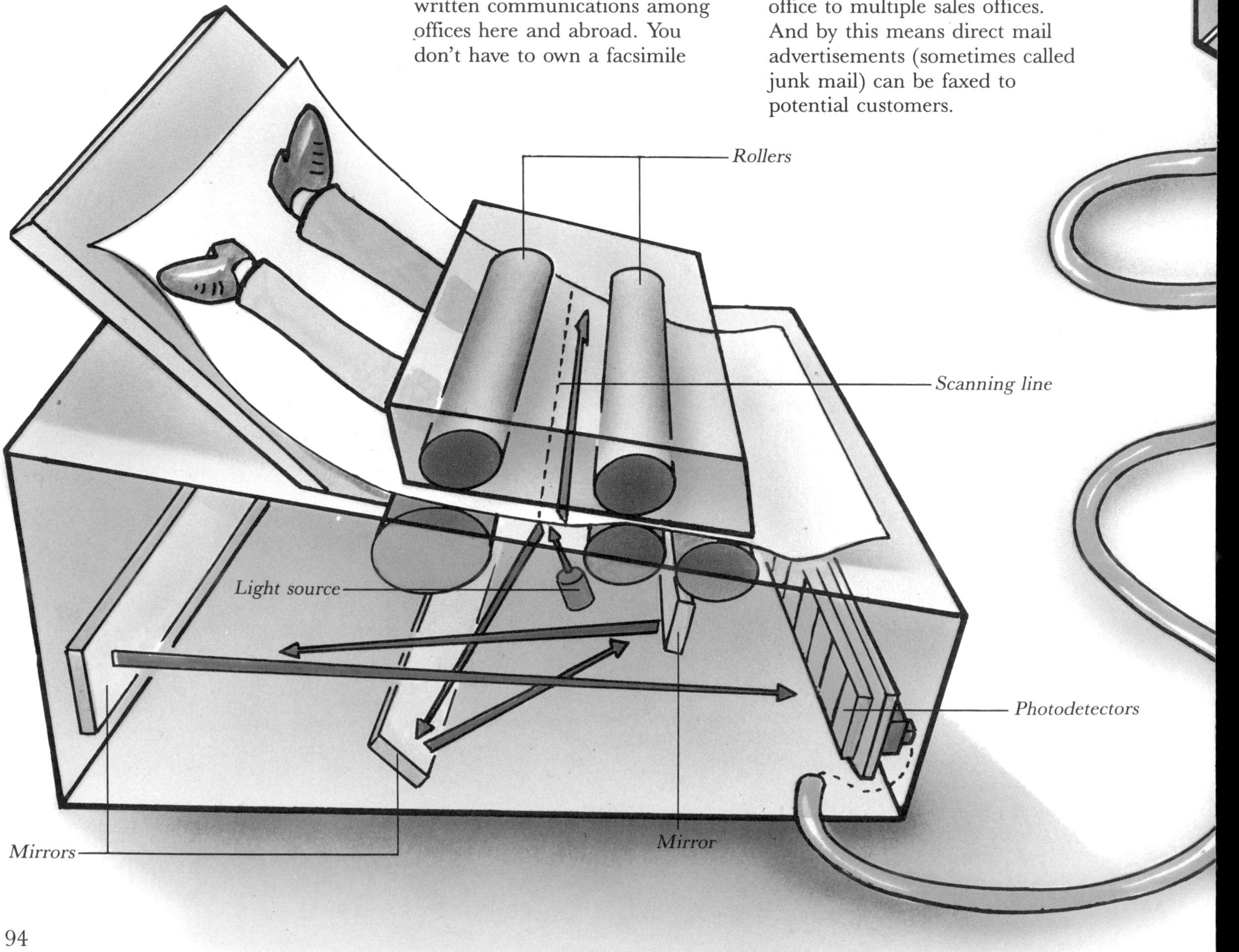

Rollers

Scanning line

Light source

Photodetectors

Mirrors

Mirror

In a facsimile machine, a scanner converts the image on a page into digital signals. A **modem** circuit (see page 163) converts the digital signals into analog tones compatible with telephone lines. When receiving signals, the modem does the opposite—converts the analog signals from the phone lines to digital signals. A printer recreates the original image.

As the page is drawn through the machine by rollers, it is illuminated by a fluorescent lamp. The image on the page is focused by mirrors and lenses onto a bank of light-sensitive devices called photodetectors. Electronic circuits scan the photodetectors, converting the image's light and dark areas, line by line, into currents that represent them. These currents are digitally encoded and fed to the modem.

Another method uses a line of 200 tiny light-emitting diodes (LEDs) as light sources. The light from each one is picked up by a matching line of optical fibers leading to light detectors. Each detector supplies a signal—either on or off—that instructs the printer at the receiving fax to print (or not print) a dot. The pattern of dots recreates the image.

The image can be printed in a number of ways. In some machines, special heat-sensitive paper is passed over a line of wire tips in a print head. The tips are heated to darken the paper wherever a signal is received, thus reproducing the dot pattern seen by the scanner of the sending machine. Some advanced machines use laser printers, which are more expensive but quicker and produce better-looking copy.

CELLULAR TELEPHONE

Communication between people who are on the move was recognized as possible and desirable soon after the invention of wireless radio. The first major use of the new technology was in ships at sea as an aid to navigation and safety. Then mobile radiotelephony, as it is now called, spread rapidly even though it lacked privacy; anyone with a radio tuned to the same frequency could listen in. Because the number of channels was limited, the circuits were often busy, requiring a wait of several minutes for an idle channel. Circuits were susceptible to static and interference. This was true on land, at sea, or in the air.

Today, cellular mobile radio provides quick connections, privacy, and sound quality as good as that of the wired telephone service in your home or office. Cellular telephone systems use low-powered FM transmitters (see page 52) in small hexagonal geographic areas called cells. The power of each transmitter is low enough that channel frequencies used in one cell can be reused in others a short distance away without interfering with one another. With cells as small as one mile in radius, hundreds of thousands of vehicles can be served in a single metropolitan area.

Public telephone switching office

Transmitter

Mobile telephone control center

Cell site
transceiver

When you telephone a vehicle from your home or office, the digits you dial are routed via the telephone network to a cellular telephone switching machine, which sends them over wire lines to all cell sites in the area. Each cell's transceiver broadcasts the signal on a radio channel to which all idle vehicle telephones are tuned. The vehicle phone you are calling replies with a signal telling where it is. The cellular phone switch then knows to which cell site to route the call. The cell site computer allocates an available radio channel for private use, sends a signal that tunes the vehicle phone to it, and then sends a ringing signal. When the driver picks up the phone, your call is connected. If the vehicle begins to move out of range of the cell transceiver while the call is in progress, the system senses the weakening signal and automatically transfers the call to the next cell station. It assigns a new radio channel and retunes the vehicle's radiotelephone to the new frequency instantly, without interrupting your conversation.

To make a call from a vehicle, just dial the number and press the send key. The digits and your mobile phone identification number are broadcast over a common radio channel. The cell site transceiver receives the information and passes it to a computer that assigns an available voice channel. The cell site transceiver sends out a signal that tunes your vehicle radio to it. The computer sends the dialing information by wire to an electronic switching machine that sets up a voice circuit in the telephone network just as in a normal call. The process usually takes less than ten seconds.

CB RADIO

Originally, CB radios were most used by truckers. Now many private automobiles have CBs for sociable chitchat with other drivers, for information on road conditions, and for use in emergencies. CB radios are different from cellular mobile telephones (see page 96) in that they are independent of the telephone system. Signals are broadcast through the air on radio frequencies allocated by the Federal Communications Commission for citizens-band radio, radio dispatch systems, and paging services.

Citizens band is an amplitude-modulated (AM) broadcast radio system (see page 100) that was created for noncommercial uses. A mobile unit consists of a microphone, a combination radio transmitter and receiver called a **transceiver,** and an antenna. It is powered by the vehicle's battery. CB radios can be in fixed stations; these operate through transformers on the power from electric utility companies.

CB users have a vocabulary of their own with phrases such as *smokey bear* for "police," *picture taker* for "radar patrol car," *pumpkin* for "flat tire," *good buddy* for "friend," *ten-four* for "message received," and *Keep the bugs off the glass and the bears off your tail* for "goodbye."

The transceiver broadcasts in all directions, and anyone tuned to the channel can listen in. You can either talk or listen—not both at the same time. Some CB radios have an automatic scanning circuit that sequentially switches the receiver from one channel to the next so that you can listen to all the channels and select the one you want to use. There are also scanners used only for monitoring emergency police and fire channels.

A squelch control quiets background noise and faint distant conversations.

The transceiver provides 40 communications channels, which you select with a switch. A more expensive CB radio called single sideband concentrates transmission on a very narrow band of frequencies. It has 120 channels and, with 12 watts of power, will transmit considerably farther, perhaps up to 100 miles, but can be received only by another single sideband radio.

An inexpensive vehicle transceiver with four watts of power will transmit signals up to about ten miles. The distance depends upon the length and position of the antenna, weather conditions, and whether you are in a city or out on the open road.

BEEPERS

People carrying beepers are actually carrying small battery-operated pocket receivers. Radio paging systems send one-way "beep" signals or short messages from one or more powerful radio transmitters to beepers within a radius of 30 or 40 miles. Although many beepers share the same radio channel, each reacts only to signals that are addressed to its unique call number, ignoring those addressed to other receivers. For two-way communication, the paged person must go to a telephone and call the pager. Some receivers can display the telephone number or name of the person who is calling.

To get in touch with someone who has a beeper, it's usually necessary to phone the controller at the central radio station and request that an alerting signal be broadcast. Some systems permit the beepers to be dialed directly without the intervention of a central operator. There are also nationwide paging systems that can reach virtually anywhere in the continental United States by broadcasting the paging signal from a satellite.

RADIO RECEIVER-AMPLIFIER

Wherever you go, indoors or outdoors, the air is filled with electrical signals from radio and television transmitters. (See **Radio waves and microwaves,** page 52). Yet without a receiver to collect them and an amplifier to increase their strength, they remain unheard.

An antenna is the first key component of a radio receiver. The signals in the air strike the antenna and transfer energy to it, resulting in weak electrical impulses in the antenna. These impulses travel to the tuner, which is the part you adjust when you turn the station knob on your set. When you set the tuner to a certain number, it shuts out all the signals except the ones for the frequency of the station you want to hear. From the tuner, the signals, still too weak for you to hear them, go to an amplifier, or more likely a series of them combined into what is called an integrated amplifier. Its job is to increase the strength of the signal.

In most modern equipment, this task is carried out by tiny transistors (see page 28). A transistor acts like a valve for electricity, letting changes in a very small electric current control the flow of a much larger current. Each transistor in an amplifier puts out a current that is like the current coming into it but larger. Amplification usually proceeds in several connected stages. A measure of what the system accomplishes is the amplifier's gain. Gain is the ratio of the current leaving the device to the current entering it. In an amplifier with several stages, the gain will range from 10,000 to 100,000.

Similar principles apply when you are listening to a record, a compact disc, or a tape. In those cases you need no antenna or tuner, but the signals from the record are still too weak to hear. The amplifier brings them into hearing range.

Carrier wave

Sound signal

Carrier wave

Sound signal

The tuner knob on a receiver enables you to choose the particular **carrier signal,** from the many that fill the air, for the radio station you want to listen to. That signal has to be at a high frequency—a large number of cycles per second from peak to peak—to transmit electromagnetic waves efficiently and with a minimum of noise.

The music or speech being broadcast, however, is at relatively low frequencies.

The carrier wave is **modulated** by the original sound signal. If the radio station is an AM station, the amplitude, or size, of the carrier wave is modulated to be the same size as the sound wave of the voice or music. The result is amplitude modulation—the basis of AM broadcasting.

If the radio station is an FM station, the frequency of the carrier wave is modulated to the same frequency of the original sound. The result is frequency modulation—FM broadcasting. Television sound also uses frequency modulation.

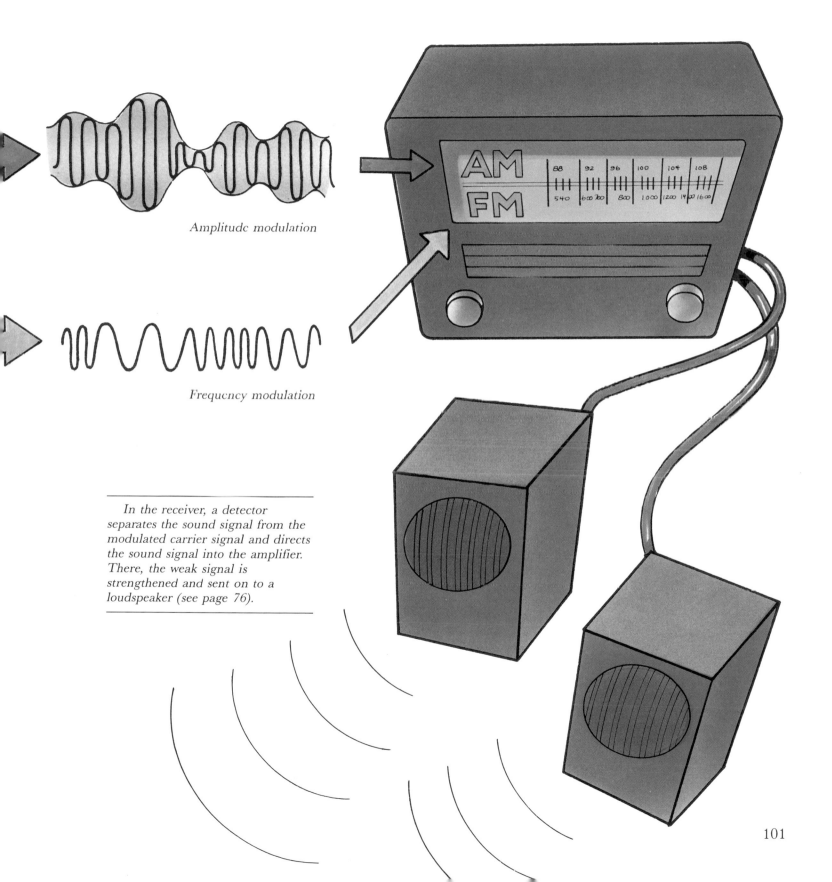

Amplitude modulation

Frequency modulation

In the receiver, a detector separates the sound signal from the modulated carrier signal and directs the sound signal into the amplifier. There, the weak signal is strengthened and sent on to a loudspeaker (see page 76).

MICROPHONES

Telephones, radios, records, movies, intercom systems—in fact, all devices that save sounds or send sounds long distances—depend upon microphones to work. Microphones change sound waves (see page 48) into electrical energy, which can be amplified and sent through electrical wires.

Diaphragm

Carbon granules

CARBON MICROPHONE

The simplest type of microphone is the carbon microphone. Relatively inexpensive to produce and not easily damaged, carbon microphones are in wide use in situations where high fidelity is not necessary. Carbon conducts electricity, although not as well as some metals. A small cylinder containing carbon granules is located behind a thin metal diaphragm. The sound waves cause the diaphragm to move back and forth, which causes the mass of carbon granules to alternately compress and expand. When the granules are closer together, they conduct electricity better than they do when farther apart. In this way, the electrical resistance of the carbon granules varies as the sound pressure waves vary. Since the microphone is connected to an electric circuit, the variations in current through the circuit represents the variations in the sound waves. This electric signal can be sent through wires. For many uses, the signal from a carbon microphone is strong enough so that the signal does not need to be amplified.

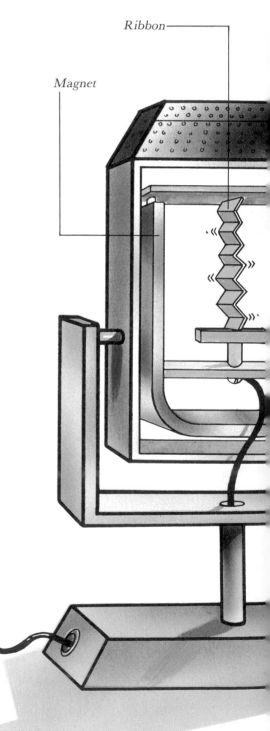

Ribbon

Magnet

RIBBON MICROPHONE

Much higher fidelity can be obtained with a **dynamic microphone,** of which the ribbon microphone is one type. A thin strip of metal—the ribbon—is suspended between the poles of a magnet. Sound waves strike the ribbon and set it vibrating within the magnetic field, which generates an electric current in the ribbon (see **Electromagnetism,** page 24). This electric current varies according to the vibration of the ribbon and forms the electric signal. The signal from a ribbon microphone is so weak that it needs to be amplified.

Another type of dynamic microphone is the moving-coil microphone, which is similar in design except that a coil of copper wire, which is attached to a diaphragm, takes the place of the ribbon. Movement of the diaphragm causes the copper coil to move within the magnetic field.

ELECTRET MICROPHONE

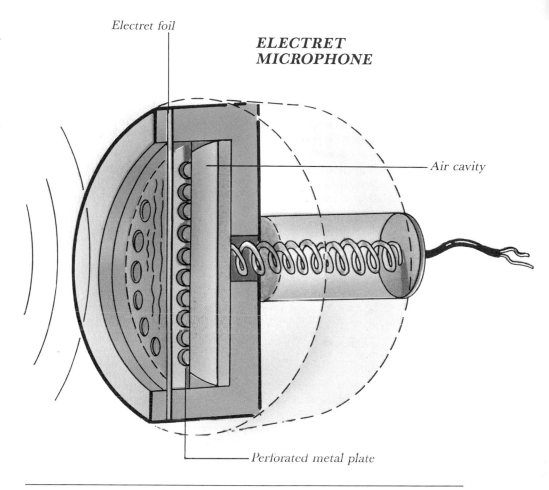

Electret foil

Air cavity

Perforated metal plate

CONDENSER MICROPHONE

A condenser microphone has a thin metal diaphragm in front of a metal plate and electric wires connecting the two. When an electric current passes through the wires, both the diaphragm and the metal plate become charged with electricity. Movement of the diaphragm in response to the sound waves changes the distance between the diaphragm and plate. This change in distance modifies the electric current in ways that correspond to the sound waves.

One form of condenser microphone is the **electret microphone.** An electret is a material that retains a permanent electric polarization, so that one end of it is positively charged and the other end is negatively charged. The electret microphone consists of an electret foil—a thin plastic membrane with an even thinner layer of metal evaporated onto it—stretched over a metal plate. The plate is perforated and touches the foil only at certain points, leaving shallow pockets of air that permit the foil to move back and forth. The foil has a permanent charge on it, which creates an electric field between the foil and the plate. Sound waves hitting the foil cause it to vibrate, changing the electric field and generating a small current that fluctuates in proportion to the changing sound pressure waves.

Electret microphones are rugged, highly sensitive, and capable of recording high-fidelity sound. They are frequently used in telephones, hearing aids, and lapel microphones.

TELEVISION

It is hard to realize now, when almost every household has a color television set, that until some 45 years ago there was no commercial television to speak of, and that until some 30 years ago color TV was a rarity. By the time color TV came in, black-and-white TV sets were so widespread in the U.S. that the Federal Communications Commission ruled that the color system adopted in the country would have to be compatible with black-and-white TV.

Like a motion picture camera, a television camera takes a number of still pictures each second, so many that when they are presented in sequence the viewer sees the impression of motion. But a television camera does not take a picture you can see on film; it converts the picture into electronic signals. Each picture is broken down into hundreds of horizontal lines, and each line is further broken down into hundreds of tiny individual picture elements called **pixels.** Each pixel is converted to an electrical signal that represents the pixel's color and brightness. These signals are transmitted sequentially, one line at a time, but at such a rapid speed that your home television screen seems to be completely lit up continuously. In the United States, the television camera takes 30 pictures per second, and each picture is made up of 525 horizontal lines. Each line contains some 435 pixels.

TELEVISION SET

In color television, the picture is separated into three images, one red, one blue, and one green. Your television set's picture tube contains three electron guns, one for each color. The guns scan across the screen, emitting beams. The television screen is coated with dots of a phosphorescent substance. These phosphor dots are in groups of three, each dot in a group being sensitive to one of the three colors. Between the dots and the electron guns is a mask that makes sure each beam is hitting a dot of the correct color. When the beam hits, the dot fluoresces, or lights up, in that color. Adjustments in the strength of the three beams regulate the relative brightness of each color. The effect is that the three superimposed images look like one picture in color. A black-and-white television set has only one electron beam, sending the same picture in monochrome.

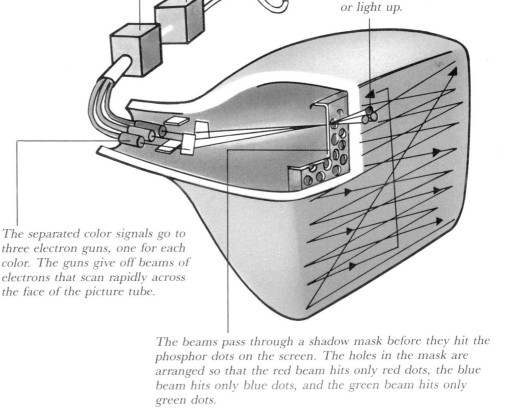

In your TV set, the signals go through a decoder that recreates the red, green, and blue signals picked up by the TV camera in the studio.

The electron beam hits phosphor dots on the screen, and they fluoresce, or light up.

The separated color signals go to three electron guns, one for each color. The guns give off beams of electrons that scan rapidly across the face of the picture tube.

The beams pass through a shadow mask before they hit the phosphor dots on the screen. The holes in the mask are arranged so that the red beam hits only red dots, the blue beam hits only blue dots, and the green beam hits only green dots.

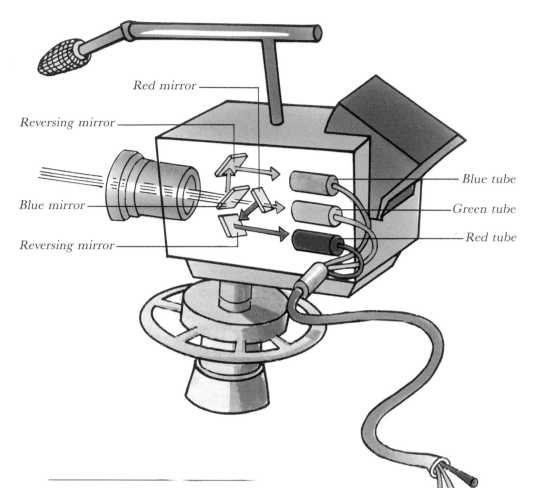

Red mirror

Reversing mirror

Blue mirror

Reversing mirror

Blue tube

Green tube

Red tube

TELEVISION CAMERA

In one kind of television camera, color-selective, or **dichroic,** mirrors separate the light entering the lens into three images: one red, one green, and one blue. The signals from each of the corresponding color tubes go to a color mixer, which produces a "luminance signal" indicating the brightness of each part of the scene being scanned by the three tubes. This luminance signal provides the information for black-and-white TV sets. Then the three color signals are combined into a "chrominance signal," representing the amount of each color in each part of the picture. The sound is added, and the combined signals go out to your TV set.

VIDEO GAMES

You can defeat evil sorcerers in magical kingdoms, play golf or football, or zap alien spaceships right in your living room with a video game system connected to your TV set. The games are really computer programs recorded on videotape.

Most video games systems consist of a base unit, controls such as a keyboard or joystick, and a variety of game cartridges that plug into the base unit. The base unit contains a small computer and a miniature TV transmitter. Instructions for the game are in a computer program stored in the cartridge. A microprocessor in the base unit converts those instructions to the display of color graphics and sound effects you see on your TV screen.

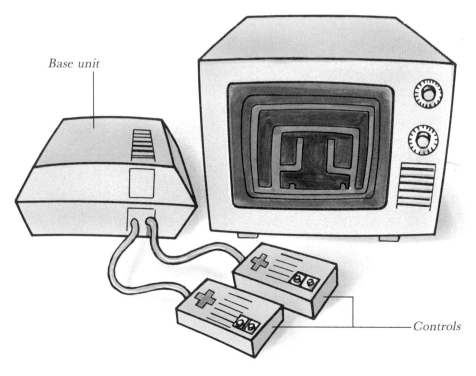

Base unit

Controls

105

VIDEOCASSETTE RECORDER

When you record a program with a VCR, you are not recording it from the TV. A VCR is itself a television receiver. It takes in the signals reaching your home through the air or by cable and records them on magnetic tape (see **Audiotape recording,** page 78). When you play back the tape, these signals are sent to the television set as picture and sound.

In a VCR, the recording head is on a drum that is tilted relative to the tape, so that the recorded material, if you could see it, would appear on the tape as a series of diagonal lines. This arrangement fits more information on the tape than a single line the length of the tape does. The drum also rotates as the tape moves past it. This enables the tape to move across the head faster than it would if the head were stationary, as on an audio tape recorder.

Higher tape speeds produce higher-quality recordings, but they use up more tape. On a VHS recorder, SP (Standard Play) is the highest speed; using this speed, a two-hour event will take up most of a cassette of the T-120 size. The same two-hour event recorded at SLP (Super Long Play) will use only about a third of the cassette.

VCRs of the VHS (Video Home System) type are more common in the U.S. than those of the Beta type. The two systems are not compatible because their cassettes differ in size and the methods of wrapping the tape around the drum are different. Probably the reason that VHS has prevailed in the U.S. is that its tape can run longer—eight hours with material recorded at the slowest speed, compared with five hours for Beta.

When you set the timer to record a program, that information goes to a memory chip. The time you have selected to start recording is continually compared with the real time displayed on the VCR's clock. When the two times correspond, the machine is automatically turned on.

When you put the cassette into the VCR, the machine opens a plastic flap that protects the videotape. Mechanical arms pull out the tape and wrap it around the rotating drum that serves for recording and playback. As the VCR records a television program, the tape moves past the drum, which is slightly tilted.

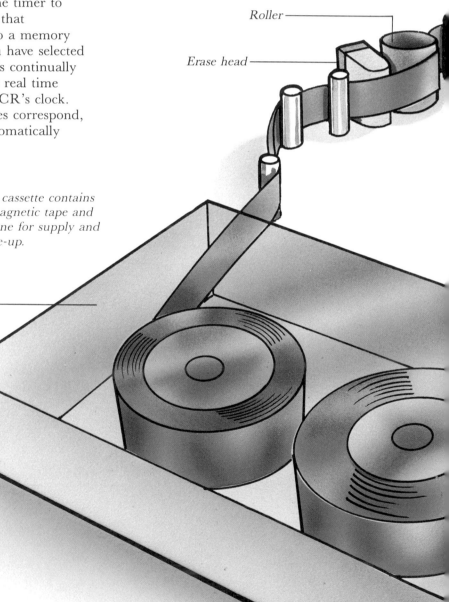

Roller

Erase head

The plastic cassette contains half-inch magnetic tape and two reels, one for supply and one for take-up.

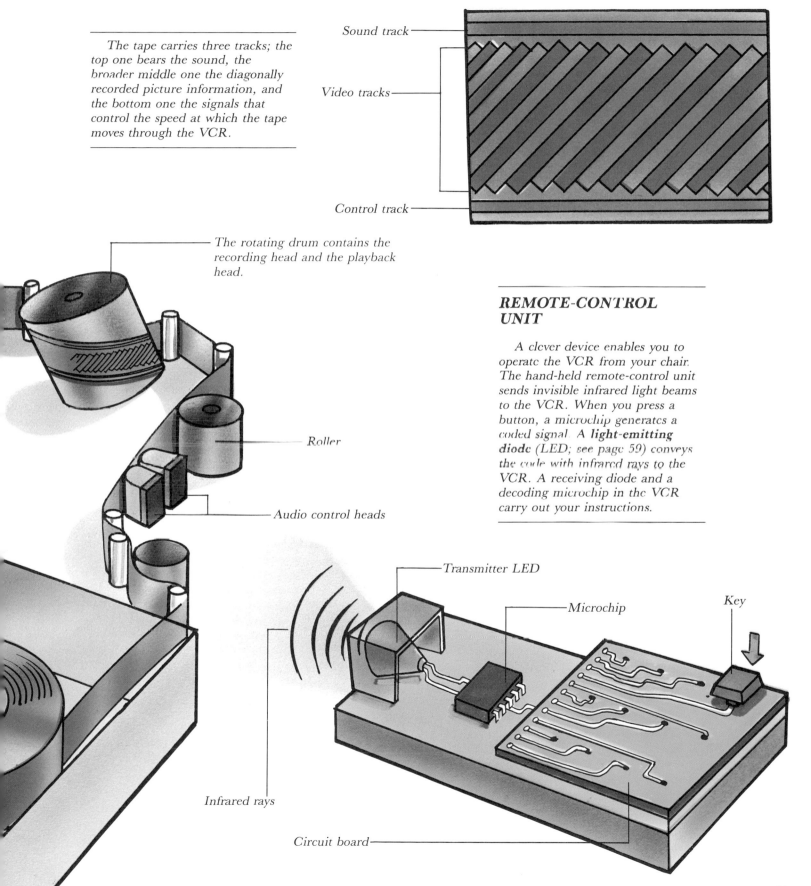

The tape carries three tracks; the top one bears the sound, the broader middle one the diagonally recorded picture information, and the bottom one the signals that control the speed at which the tape moves through the VCR.

Sound track

Video tracks

Control track

The rotating drum contains the recording head and the playback head.

Roller

Audio control heads

REMOTE-CONTROL UNIT

A clever device enables you to operate the VCR from your chair. The hand-held remote-control unit sends invisible infrared light beams to the VCR. When you press a button, a microchip generates a coded signal. A **light-emitting diode** (LED; see page 59) conveys the code with infrared rays to the VCR. A receiving diode and a decoding microchip in the VCR carry out your instructions.

Transmitter LED

Microchip

Key

Infrared rays

Circuit board

COMMUNICATIONS SATELLITES

A communications satellite is a microwave relay station in the sky. It receives microwave signals (see **Radio waves and microwaves,** page 52) from a ground station antenna and retransmits them to other ground stations across a continent or even across an ocean. Since the launch of *Telstar* in 1962, successive generations of communications satellites, with ever more sophisticated technologies and capabilities, have been put into orbit.

Today, hundreds of satellites, in various orbits from a few hundred miles to 22,300 miles above the Earth, relay information to thousands of ground stations spread out among more than 165 countries. They relay telephone calls, TV programs, business and scientific data, facsimile, weather photos, maritime navigation signals, and military intelligence and commands.

Most communications satellites are placed in orbit around the Earth 22,279 miles above the equator. They travel 6,870 miles per hour in the same direction—eastward—that the Earth is rotating.

Despite the pull of gravity, the satellites don't crash into the Earth. Their traveling speeds carry them forward just enough that they "fall" into a circular orbit following the Earth's curvature, and they remain aloft circling the Earth once every 23 hours, 56 minutes, and 4 seconds. Since that is the same time it takes the Earth to turn once on its axis, each satellite moves "in sync" with a point on the equator below it. The satellite is in **geosynchronous orbit** and is sometimes called a **geostationary satellite.** From the ground, it appears to hang motionless in the sky. Ground station antennas pointed at it can send and receive signals without moving.

A geostationary satellite is so high in the sky that it can "see" and transmit signals to a very large area—called a **footprint**—on the Earth's surface.

INTELSAT VI

There are many designs for different communications satellites. They all contain **transponders,** which are radio receivers and transmitters that receive microwave signals from a station on the ground, amplify them, and shift them to another frequency for transmission back to Earth. A satellite may have a dozen or more transponders, each operating on a different carrier frequency. Each transponder can handle one color TV channel or as many as 2600 simultaneous phone conversations.

The Intelsat VI satellite is the largest commercial communications satellite, with a diameter of 12 feet and a length of 38.4 feet. Its 48 operating transponders are interconnectable, making this satellite a "switchboard in the sky," able to relay simultaneously up to 120,000 telephone calls and 3 television programs between specific ground stations. It is expected to function for seven to ten years. Within a few years, five Intelsat VI satellites will be launched.

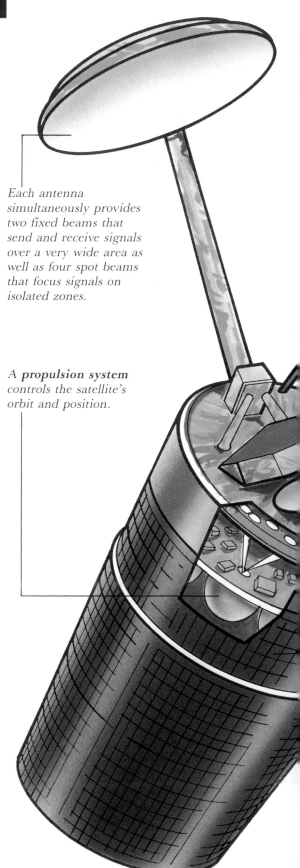

Each antenna simultaneously provides two fixed beams that send and receive signals over a very wide area as well as four spot beams that focus signals on isolated zones.

A **propulsion system** controls the satellite's orbit and position.

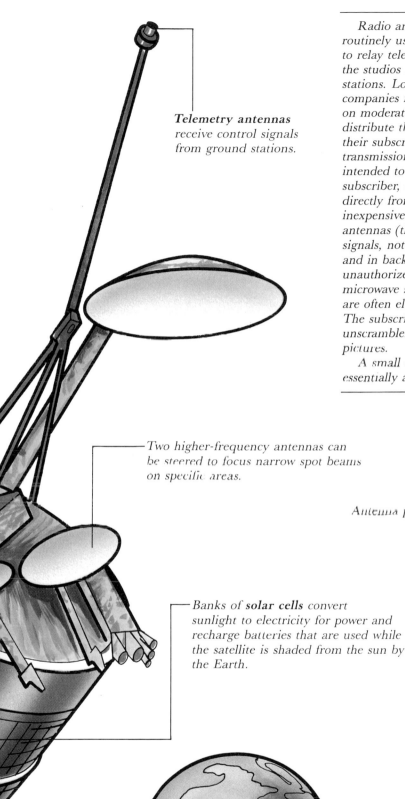

Telemetry antennas receive control signals from ground stations.

Two higher-frequency antennas can be steered to focus narrow spot beams on specific areas.

Banks of **solar cells** convert sunlight to electricity for power and recharge batteries that are used while the satellite is shaded from the sun by the Earth.

Antenna probe

Radio and television networks routinely use geostationary satellites to relay television programs between the studios and the affiliated local stations. Local cable television companies receive these programs on moderate-size dish antennas and distribute them via coaxial cable to their subscribers' TV sets. These transmissions, which are not intended to be seen by anyone not a subscriber, can be intercepted directly from the satellite by small, inexpensive receive-only dish antennas (they can only receive signals, not send them) on rooftops and in backyards. To prevent unauthorized viewing, the microwave signals from the satellites are often electronically scrambled. The subscriber is issued an unscrambler device, to obtain clear pictures.

A small dish antenna is essentially an adjustable reflector that can be pointed at a satellite. It must be positioned so that it has an unobstructed view of the satellite. It collects as much of the satellite's signal as possible and focuses it onto an antenna probe. The extremely weak signals received from the satellite are strengthened about 100,000 times by a low-noise amplifier. Then the high-frequency signals, which are billions of hertz (see **Electromagnetic waves,** page 50), are "downconverted" to an intermediate frequency in the megahertz range (millions of hertz). This is fed to an FM receiver near the TV set, where a tuner selects one of the channels broadcast by the satellite. The carrier signal is removed and the remaining FM video and audio signals are amplified and converted to the frequencies used by the TV set for pictures and sound.

SONAR

Sonar (SOund Navigation And Ranging) originally meant, as its name implies, the technique of transmitting pulses of sound into water and detecting their echoes from such targets as submarines, surface ships, fish, and underwater terrain to locate them. But today sonar refers to all underwater acoustic devices, including those used just for listening to sounds made by ships and marine animals and other undersea noise. Sonar can also be used on land. A pulse is sent into the ground. Echoes come back from different layers of soil and rock, helping geologists locate gas and oil.

When used in sea water, sonar generally uses low ultrasonic frequencies because they get through salt water better than higher ultrasonic frequencies do. (See **Sound waves,** page 48.) The sound waves travel through water by compressing and thinning out the molecules of water. Since water molecules are closer together than air molecules, sound moves almost five times faster in water than in air, and it can travel thousands of miles before dissipating. The velocity of sound in the ocean ranges from about 4,700 to 5,000 feet per second, depending on water temperature, pressure (depth) and salinity (mineral content).

Quirky currents, surface waves, seafloor deposits and seamounts (underwater mountains) can cause sudden changes in these properties at different depths. These changes cause sound transmissions to bend and behave in unexpected ways, thus affecting the sonar reading. To compensate, underwater gauges gather information about temperature and other conditions to help the sonar operator to correct the readings. No longer does the sonar operator merely listen with earphones for the ping of a reflected sound pulse. Sophisticated sonar systems have arrays of extremely sensitive **hydrophones** (underwater microphones) and computers that can detect and analyze the merest whispers of sound amid the cacophony of undersea noises, displaying them on scopes similar to radar screens.

If you take half the time it takes a pulse to go out, be reflected from a submarine, and return to the receiver, and multiply that figure by the known velocity of sound in water, the result is the distance of the submarine from the sonar device. Sonar is also used to locate schools of fish and underwater objects such as sunken ships and reefs.

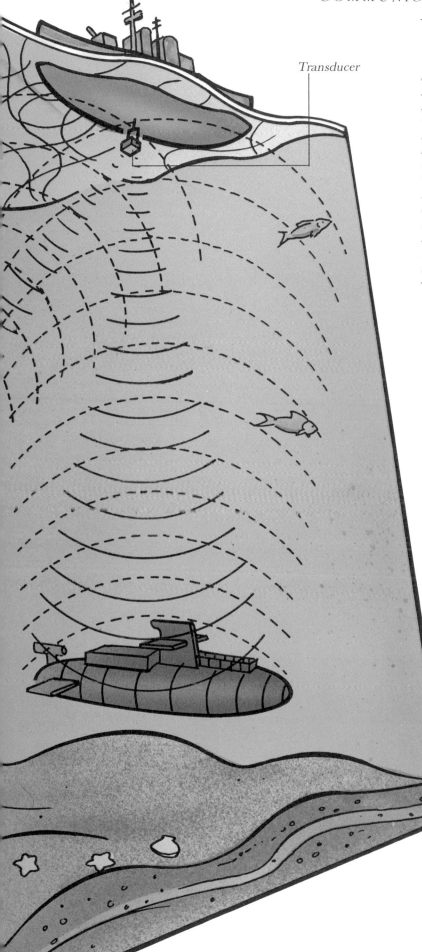

Transducer

TRANSDUCER

A transducer is a device that converts energy from one form to another. Rapidly changing voltages applied across a specially cut crystal will cause it to vibrate. Modern sonar transducers usually consist of an array of these crystals to impart ultrasonic pressure waves into the water. By timing the pulses emitted from each element, the sonar beam can be focused on a target and its bearing can be determined. Echoes and incoming sounds are converted to electrical signals either by the same transducer or a separate one.

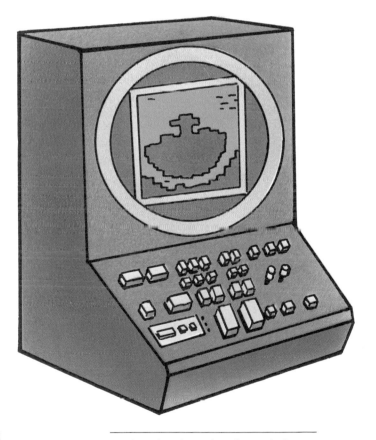

Attack submarines hunt their quarry by listening for sound vibrations generated by propellers and machinery and crew noises originating inside the target. Modern sonar listening systems use extremely sensitive hydrophones to pick up even the weakest sound waves and convert them to electrical signals that are processed by computers and displayed on a scope.

SPEECH SYNTHESIZER

Dolls that cry "Mama," clocks that tell the time, scales that blurt out your weight, ovens that announce "Dinner's ready," cars that nag "Fasten seat belts," and other "talking" gadgets are fun. Some use the simplest form of synthetic speech—playing back a message that was recorded earlier by a human.

Early telephone system announcements ("The number you have reached has been changed to five, five, five, one, three, two, four") used a library of prerecorded message parts—sentences, words, and phrases—and played them back in various combinations. High-quality speech can be produced this way, but the number of different messages that can be created by piecing together prerecorded phrases, no matter how cleverly, is limited by the storage capacity.

The amount of computer memory required can be reduced by having the words digitized—that is, converting the varying sound waves from a human voice into varying electrical currents and then representing them as a series of electrical pulses. To speak, the machine reconstructs the pulses back to varying electrical currents and then to sound waves. The drawback is that the sentences sound unnatural and robotic.

The most successful speech synthesizers use a technique of **phoneme** reconstruction controlled by **algorithms.** Spoken English has 43 distinctive, basic sounds called phonemes. Because the human vocal tract and vocal chords change continuously during speech, a phoneme is affected by the phonemes that precede and follow it. So each phoneme typically has several variations called **allophones.** These are stored in a computer library of elemental sounds that can be smoothly connected to form words and phrases.

Algorithms are rules or formulas that describe the sounds and how people talk. These rules are stored in the speech synthesizer's memory. The algorithms figure out not only word pronunciations, but the intonations and inflections of a sentence, and send the information in the form of digital pulses to electronic circuits that generate speech sounds.

Synthetic speech is used in computers to give instructions and to teach school children some of their lessons. It is used to alert aircraft pilots, astronauts, and automobile drivers to dangerous conditions. And it is widely used in communications to make announcements, answer questions, and transmit voice mail messages.

2. The synthesizer examines the construction, grammar, and punctuation of the sentences and, using a set of rules with a stored dictionary, expands abbreviations and numbers and determines phrasing and stress.

TEXT-TO-SPEECH SYNTHESIS SYSTEM

1. A message can be typed on a keyboard, or it can arrive from a distant source as electrical signals equivalent to text. Some systems can optically scan printed text.

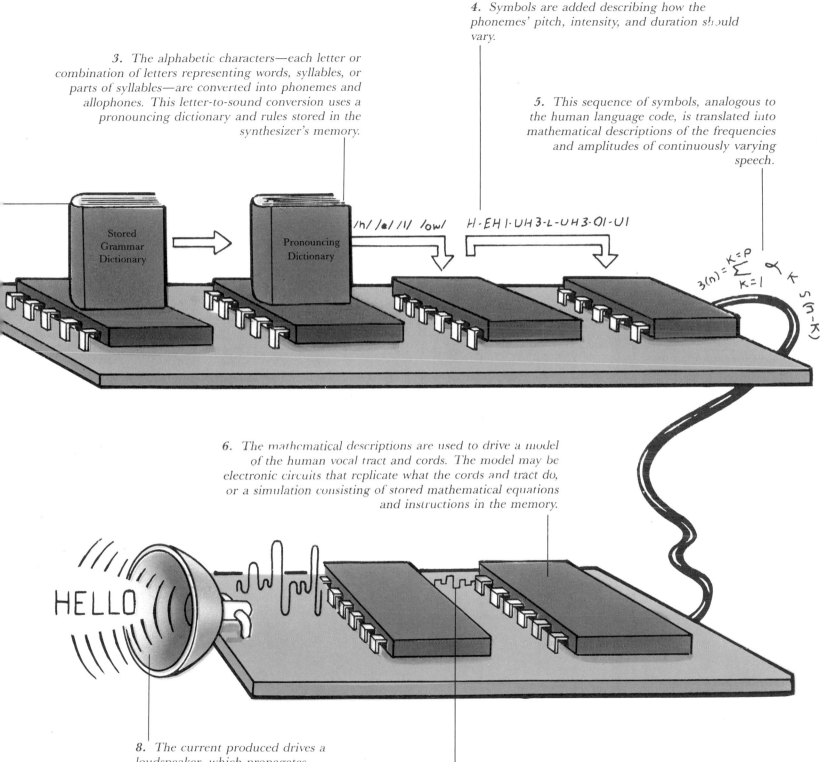

4. Symbols are added describing how the phonemes' pitch, intensity, and duration should vary.

3. The alphabetic characters—each letter or combination of letters representing words, syllables, or parts of syllables—are converted into phonemes and allophones. This letter-to-sound conversion uses a pronouncing dictionary and rules stored in the synthesizer's memory.

5. This sequence of symbols, analogous to the human language code, is translated into mathematical descriptions of the frequencies and amplitudes of continuously varying speech.

Stored Grammar Dictionary

Pronouncing Dictionary

/h/ /e/ /l/ /ow/

H-EHI-UH3-L-UH3-OI-UI

$$3(n) = \sum_{k=1}^{k=p} \alpha_k S(n-k)$$

6. The mathematical descriptions are used to drive a model of the human vocal tract and cords. The model may be electronic circuits that replicate what the cords and tract do, or a simulation consisting of stored mathematical equations and instructions in the memory.

HELLO

8. The current produced drives a loudspeaker, which propagates acoustic waves.

7. The model produces a string of binary numbers that describe the speech sound waves. These are converted from digital to analog form (see page 79).

SPEECH RECOGNITION MACHINES

Factories and the postal system use speech recognition machines on conveyor belts to sort products. These machines are word recognizers that can recognize single words or phrases from a vocabulary of 50 to 100 words. The operator issues one-word commands, such as "accept," "reject," "Boston," or "Texas," which the machine recognizes and uses to control the sorting machinery. Fighter planes are equipped with word recognizers to accept commands such as "fire" or "arm" from pilots. The pilots can issue the commands without taking their hands from controls.

Making a machine that listens and recognizes human speech is more difficult than making one that speaks. The machine must "hear" the sound waves, analyze them, and then compare them to patterns representing words stored in its memory, much as humans do. To complicate the machine's job, the pattern for the same word pronounced by different people, with different accents and different voices, probably won't be the same.

Given the limitations on the memory of a speech recognition machine, it therefore must restrict the number of words it is expected to recognize and the number of people it must listen to.

2. The signals are converted to digital pulses (see page 79) and noise is filtered out.

1. When the person who has "trained" the machine speaks a word into it, sound waves from the speaker's mouth are converted to electrical signals in a microphone (or telephone).

To "train" the machine, the person who is to operate it clearly enunciates the words to be recognized, and they are recorded. Each word is analyzed to determine the distribution of sound energy for the frequencies used in the word. These **spectral features** form a pattern for each word as it is pronounced by that person. The patterns, called **templates,** are labeled and stored in the machine's memory, constituting a vocabulary of voice commands that the machine is expected to recognize.

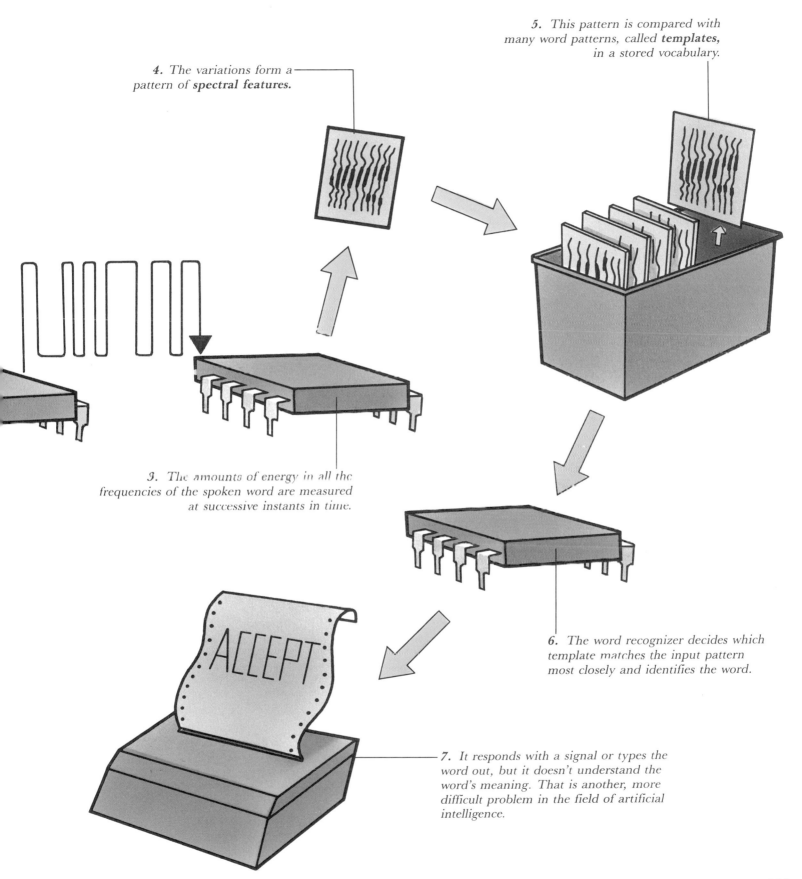

5. This pattern is compared with many word patterns, called **templates,** in a stored vocabulary.

4. The variations form a pattern of **spectral features.**

3. The amounts of energy in all the frequencies of the spoken word are measured at successive instants in time.

6. The word recognizer decides which template matches the input pattern most closely and identifies the word.

ACCEPT

7. It responds with a signal or types the word out, but it doesn't understand the word's meaning. That is another, more difficult problem in the field of artificial intelligence.

VOICE RECOGNITION

No two people speak exactly alike. Therefore **spectrograms** (see **Speech recognition machines,** page 114) of their speech are unique, as distinctive as their fingerprints. Even a professional mimic trying to imitate the voice of another person will produce word spectrograms that are recognizably different. This fact makes possible **speaker**

verification, using the same type of process as speech recognition.

This capability is useful for verifying that the speaker is who he or she claims to be—for example, in banking transactions by phone. The speaker's words are analyzed, not for what is said, but how it is said, providing a check against fraud. One voice verification system that operates over telephone lines, in which a password phrase is spoken twice, has rejected fewer than two out of

a hundred valid users and has accepted no impostors.

Speaker verification systems are being used to prevent impostors from engaging in illegal electronic funds transactions, from accessing computer networks and private information in data banks, and from entering restricted areas. In the future, they will be used to verify credit card and telephone calling card transactions.

1. The person speaks a password—a name or number or an assigned code phrase.

3. They are then compared with a library of word patterns previously recorded.

4. If the differences between the password's pattern and the pattern that person stored previously are significant, the user is requested to speak the password again. The true speaker usually performs better. An impostor is not usually able to mimic the true speaker any better on a second attempt than on the first.

5. If the differences are still significant, the identity claim is rejected; otherwise, the person is accepted.

2. The frequency and amplitude features of the spoken word are analyzed.

LIE DETECTOR

A typical printout from a polygraph shows the measured physiological factors over a period of time. In the indicated area, these factors show distinct changes that champions of the polygraph say are caused by lying.

Body movements
Breathing
Perspiration
Blood pressure and heartbeat

Probable lie

The person being given a polygraph test is connected by tubes and wires to a machine that makes several physical measurements at the same time. These usually include blood pressure, pulse rate, breathing, and changes in the electrical conductivity of the skin. The more moisture—perspiration—on the skin, the greater its electrical conductivity. Pen recorders print out a continuous record of the measurements in graph form.

A lie detector doesn't really detect lies; it detects changes in a person's blood pressure, pulse rate, breathing, and perspiration while that person is being asked questions. For that reason, it is more accurately known as a **polygraph.** The polygraph prints out graphs representing each of the physical changes it is measuring.

Although the most familiar use of the polygraph is in questioning people suspected of committing a crime, it is used with people suspected of not telling the truth in other situations as well. Some businesses use polygraph tests to see if a job-seeker is reliable.

Champions of the instrument argue that when a person is lying, the blood pressure, pulse rate, breathing, and moisture level on the skin change automatically in significant ways. Such changes can be reliably recognized by a trained operator. Opponents say that the stress of being given the test can alone cause such changes. Most tests include questions unrelated to the matter under investigation to determine the normal reaction of the person being questioned. Courts in the United States will not usually permit the use of polygraph evidence in trials unless the lawyers for both sides agree that it is acceptable. Several states have laws forbidding the use of the instrument in screening employees or job-seekers.

TYPESETTING

When Johannes Gutenberg developed movable type in 1440, he was probably not aware that such systems had been used in the Orient hundreds of years earlier. The use of little blocks carved with individual letters that could be used over and over was such a clear advantage over laboriously writing out books by hand that this pivotal invention is sometimes credited with bringing about the Renaissance.

Type continued to be arranged by hand with individual pieces of type for each letter until the nineteenth century, when the Linotype machine was invented. This machine produces an entire line of type on one piece of metal, called a **slug.** The operator types the line at a keyboard. When a key is pressed, a mold for that letter, called a **matrix,** is released from a storage case and added to the other matrices making up the line. When the entire line is complete, it is moved to a casting mechanism, where molten metal is forced into the matrices. The molten metal produces the slug. The matrices go back into their storage case for use in casting the next line.

The slugs for all the lines for a page, together with engravings for photographs and lines of larger type for headlines, are arranged in a heavy steel frame called a **chase.** The words and pictures are mirror images of what will appear on the printed page. Either the chase itself, or a printing plate made from the chase, is used to print by the letterpress process (see **Printing,** page 119).

Although the Linotype is much faster than setting each letter by hand, mistakes in the line must be corrected by typing the entire line over again. Modern typesetting equipment is much faster than the Linotype, and corrections can be made much more easily.

Most typesetting today is done by a photographic process, whereby images of the letters are projected onto paper or film that is sensitive to light. In some systems, the shapes of the letters are stored in a disc. Light shines through the letter on the disc, and lenses focus the image to the right size on the light-sensitive paper. Other systems are **digitized,** which means that the letters are formed from tiny dots or lines. A computer stores the information about each letter's shape as a pattern for that letter. Some typesetting systems generate the letters with a laser, using computer-stored complex mathematical formulas for the curves and straight lines of each letter. The words to be typeset can be typed in at a keyboard, but more commonly computer tape or a floppy magnetic computer disc feeds the information directly.

The words in this book were typed into word processing equipment, which stored them on floppy disks. The disks were inserted into the typesetting equipment, and the operator entered codes to specify the shape of the type, the size, and the width of the columns. The computer in the typesetting equipment converted this information to images produced on special paper.

*This special paper, containing the printed text of the book, was pasted onto sheets of cardboard along with copies of the illustrations. These pasted-up sheets of cardboard are called **mechanicals.** Each mechanical contained everything that appears on a page and its facing page, in exactly the size and position it would appear in the finished book. The mechanicals were then used for making printing plates for **offset lithography** (see page 119).*

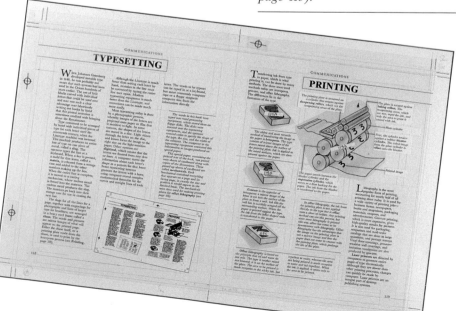

PRINTING

Transferring ink from type to paper, which is what printing is, can be done by many methods. The three most-used methods today are letterpress, gravure, and offset lithography. The differences lie in the formation of the type.

The oldest and most versatile method of transferring the image to the paper is **letterpress.** The type consists of mirror images of the letters raised above the surface of the printing plate. Ink is applied to the raised surfaces, which are then pressed against the paper.

Gravure is the opposite of letterpress; a mirror image of the letter is cut into the surface of the plate to form a well. Ink fills the well but is carefully scraped off the surrounding surface. When the paper is pressed against the type, the ink from the letter-shaped wells is transferred to the paper.

Offset lithography is based on the principle that oil and water do not mix. The type is neither raised nor lowered; it is on the surface of the plate. The area to be printed is made receptive to the sticky ink, but repellent to water, whereas the area not being printed is made receptive to water and ink-repellent. When the ink is applied, it sticks only to the area to be printed.

The printing plate is mounted on a cylinder that rotates against **dampening rollers,** *which wet the nonprinting areas of the plate.*

The plate is rotated against **inking rollers.** *The nonprinting areas, which are wet, repel the ink, and only the area to print is coated with ink.*

Plate cylinder

Next, the cylinder rotates against a rubber **blanket cylinder.** *The inked image from the plate cylinder is* **offset** *onto the blanket.*

Printed image

The paper travels between the blanket cylinder and the **impression cylinder,** *which serves as a firm backing for the paper. The ink from the blanket is transferred to the paper.*

In offset lithography, the ink from the printing plate is not applied directly to the type but is **offset** onto a sheet of rubber. Any printing method can use this process, but since most lithography is printed this way, offset has come to be synonymous with lithography. Offset lithography has the advantages that the image on the printing plate is not a mirror image and that the paper does not come in contact with the printing plate, which prolongs the life of the plate.

Lithography is the most popular form of printing, accounting for nearly half of all printing done today. It is used for a wide variety of printing jobs—business forms, newspapers, books and magazines, packaging materials, coupons, and advertisements. Gravure printing, which is more expensive, gives high-quality results for pictures. It is also used for packaging, magazines and mail-order catalogs that are done in large quantities, and postage stamps. Vinyl floor coverings, pressure-sensitive wall coverings, and imitation woodgrains are also produced by gravure.

Laser printers are directed by computers to generate entire pages of type electronically. Although they are slower than other printing processes, changes can quickly be made by computer. Laser printers are an integral part of desktop publishing systems.

119

PRINTING PICTURES

A black-and-white illustration printed using letterpress or offset lithography (see **Printing,** page 119) produces all the tones of gray between black and white with only black ink. It does this by means of an optical illusion. The illustration is made up of thousands of little dots. Each dot is printed black, but in light areas, the dots are tiny and surrounded by large areas of white paper. In dark areas, the dots are much larger and may touch each other. The white paper acts as a "mixer" to dilute the black into shades of gray. At a distance, the dots can barely be seen; the picture looks as though it has the continuous shades of gray that the original has. This is the **halftone** process, by which the illustration, which has continuous tones of gray, has been converted to an image composed of black-and-white dots.

The original illustration, which includes all shades of gray, is rephotographed through a halftone screen with a pattern of thousands of dots on it. The result is a negative in which the dot pattern on the screen is superimposed on the original picture. Individual dots range in size depending on the amount of light reflected from the original picture. When the picture is printed, dark areas will show as large dots and light areas will show as smaller dots.

The multiple hues of color illustrations are printed using only three colors of ink, plus black. Four slightly different halftone negatives of the same picture are created, one for each of the three colors and one for black.

3. The blue filter produces a negative of only the blue light that comes from the picture.

*4. The positive printing plate made from the negative shows the areas that absorb blue. When blue is taken away from white light, the two remaining primary colors, red and green, become the secondary color **yellow**. The positive from the negative made through the blue filter is the yellow printing plate.*

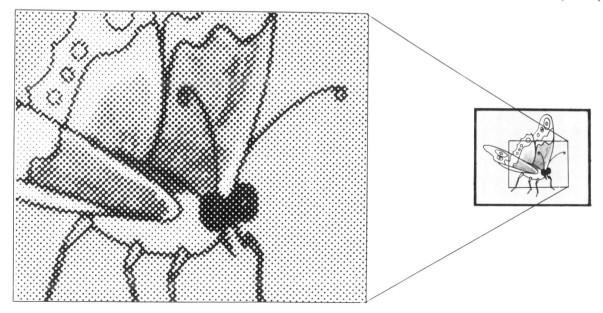

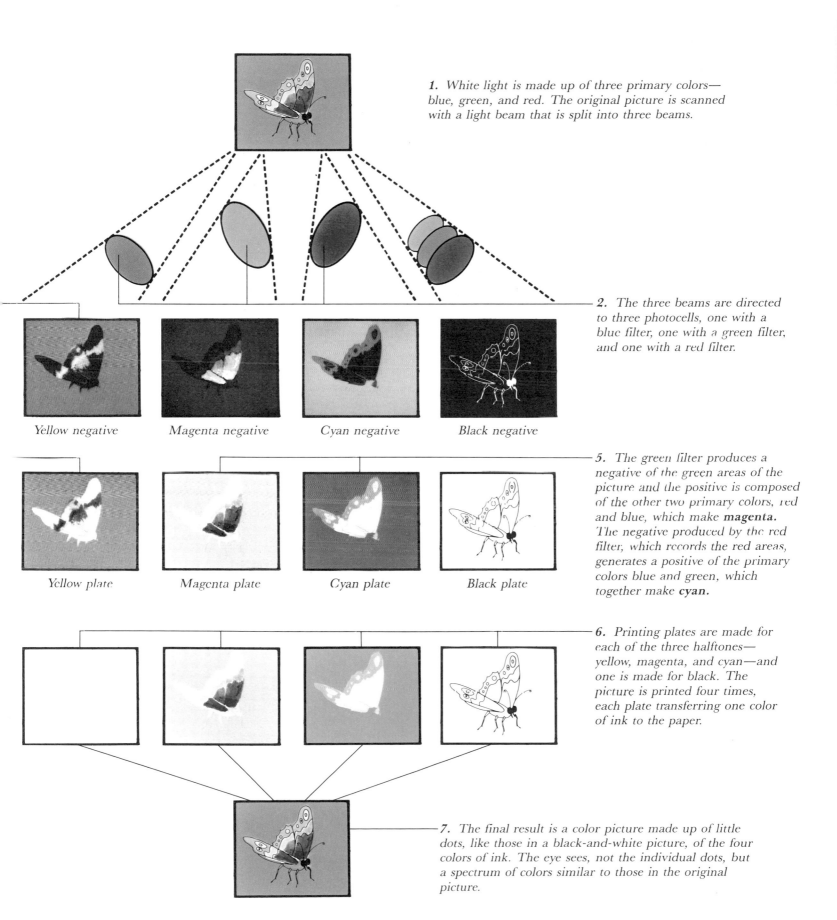

1. White light is made up of three primary colors—blue, green, and red. The original picture is scanned with a light beam that is split into three beams.

2. The three beams are directed to three photocells, one with a blue filter, one with a green filter, and one with a red filter.

Yellow negative

Magenta negative

Cyan negative

Black negative

5. The green filter produces a negative of the green areas of the picture and the positive is composed of the other two primary colors, red and blue, which make **magenta.** The negative produced by the red filter, which records the red areas, generates a positive of the primary colors blue and green, which together make **cyan.**

Yellow plate

Magenta plate

Cyan plate

Black plate

6. Printing plates are made for each of the three halftones—yellow, magenta, and cyan—and one is made for black. The picture is printed four times, each plate transferring one color of ink to the paper.

7. The final result is a color picture made up of little dots, like those in a black-and-white picture, of the four colors of ink. The eye sees, not the individual dots, but a spectrum of colors similar to those in the original picture.

HEARING AIDS

Cupping your hand behind your ear or using an ear trumpet—a kind of reverse megaphone—helps gather in sounds and direct them into your ear, but thanks to modern electronics, more effective types of hearing aids are available. Hearing aids help compensate for hearing impairment, but they cannot restore lost hearing.

Approximately one-tenth of the human population—more than 20 million people in the United States alone—suffer from some degree of hearing loss. No two people have exactly the same hearing impairment. For some people, all sounds are muffled. For some, specific kinds of sounds, such as speech and music, are distorted. Some have difficulty in understanding certain words in speech, particularly in the midst of distracting background noises. Hearing tests determine the exact frequencies and levels of an individual's hearing loss, and a hearing aid must be customized to the needs of the wearer. These adjustments are made in the hearing aid circuits manually by an expert technician or by a computer that is programmed with the results of the hearing test.

All hearing aids have similar components. An **electret microphone** (see page 103) picks up the sounds and converts them to electrical signals; an amplifier increases the strength of the signals; a receiver/speaker reconverts them to sound vibrations; and a custom-made

earmold fitted to the ear or ear canal channels the sound vibrations into the ear. Most hearing aids have an on/off switch and an adjustable volume control, and they are powered by tiny, replaceable batteries. Several forms are available. One type has a separate power supply that is carried in a pocket and is connected to the receiver in the ear with a cord. An extremely small and lightweight type of hearing aid fits directly into the ear with no external parts—all electronics, including batteries, are inside the capsule.

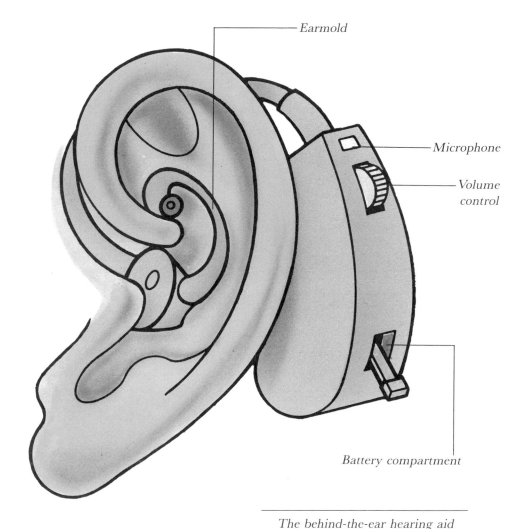

Earmold

Microphone

Volume control

Battery compartment

The behind-the-ear hearing aid has a microminiaturized microphone, amplifier, and receiver/speaker in a small contoured case that fits behind the ear, connected to the earmold by a short plastic tube. Switches for controlling volume and tone and for use with a telephone are on the case. Circuits in the hearing aid automatically suppress extremely loud sounds and boost extremely quiet sounds. The same type of hearing aid can sometimes be modified to fit into an eyeglass frame.

At Home

THERMAL CLOTHING

Insulating your body with thermal clothing works in much the same way as insulating your house. In each case, you help maintain a comfortable temperature by partially blocking the loss of heat to the cold outdoors (see **Conduction and Convection**, page 10). In your house, the heat is provided by the furnace. In your body, the heat is provided by your digestive system, which "burns" the "fuel" you eat.

Your body is continually producing heat and losing it just as continually. Because you lose heat at practically the same rate you produce it, your body temperature remains fairly constant. There's some variation in how rapidly heat is produced —that's why most people eat more in cold weather than in warm weather—but your body's "furnace" can't vary its heat output enough to compensate for the different demands that the weather makes upon it.

Every person has a narrow range of skin temperatures within which he or she feels comfortable. That's usually in the mid-80-degree range. If the skin temperature drops, you feel uncomfortably cold; if it rises, you feel hot. Your body has an arsenal of defenses against changes in skin temperature. Perspiration helps cool the skin in hot weather, and the shivering reflex helps warm you in cold weather. Despite these, humans must rely upon other means, such as wearing proper clothing, to help them function under unfriendly weather conditions.

INSULATING FIBERS

Down has thousands of fibers that trap air, preventing convection from moving heat inside the garment's lining, and at the same time providing a thick but lightweight layer that gives excellent insulation against heat conduction. Feathers work similarly, but not as well as down, because they have more solid material and fewer air-capturing fibers.

Other materials used in cold-weather clothing, such as wool and some synthetics, have fiber structures that can trap air. Certain synthetics have hollow, tube-like fibers that trap air inside them; these work well even if wet. Nature provides this hollow-tube construction in the hair of polar bears and some seals, allowing them to exist in a wet, arctic environment.

Wool fibers have a rough, scale-like surface that traps air between the scales.

Down

Wool

Hollow-structured synthetic

DOWN JACKET

In cold weather, clothing has two major functions. The first is to place a layer of an insulating material between your skin and the outside air. The second is to prevent moving air currents from increasing your convective heat losses. Wearing clothing that traps air inside it fulfills the first function, since still air is a poor conductor of heat. That insulating material needs to be covered with a tightly woven "skin" that blocks the wind.

A down-filled jacket's materials and construction provide excellent insulation against the cold. To keep the down in place, it is usually contained in quilted pockets. Down-filled garments are purposely bulky in order to insulate effectively. A flattened-out down jacket contains less air and doesn't insulate as well. Because down loses bulk and packs down when it gets wet, down-insulated clothing is not effective in wet weather unless it's worn under a water-repellent outer layer.

THERMAL KNITS

Even materials like cotton, which isn't usually considered for cold-weather wear, can be made into insulating fabrics. This waffle-like pattern of knit, if combined with a wind-breaking cover, traps air in its "dimples." A similar effect can be obtained by wearing an open "fishnet" knit undergarment.

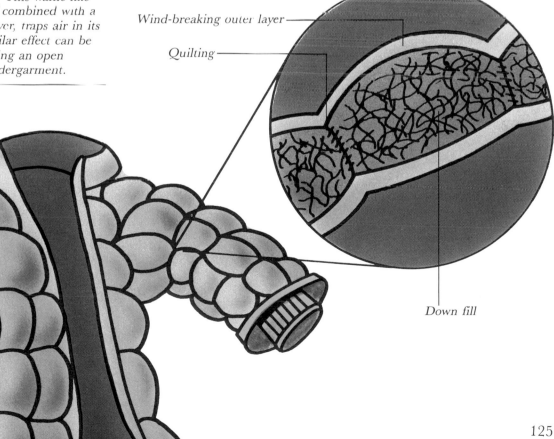

Wind-breaking outer layer

Quilting

Down fill

125

HOOK-AND-LOOP FASTENERS

The hook-and-loop fastener was conceived in the 1940s when George deMaestral, a Swiss inventor, came back from his walk with a collection of cockleburs sticking to his woolen trousers. Struck with how difficult it was to pick them off his clothing, he examined one under a microscope. Each of the burr's barbs is actually a tiny hook, which snagged the fuzzy fibers of the woolen cloth. When a number of hooks engaged the cloth's loops of fiber, the result was a firm attachment. Even though removing a burr required a fair amount of force, neither the burr nor the fiber was damaged by the process.

This combination of characteristics—easy attachment, firm closure, and nondamaging opening—is a description of a good fastener for clothing. After several years of research devoted to recreating this natural effect in synthetic fibers and plastics, the modern hook-and-loop fastener was introduced. These fasteners—commonly called "Velcro," although Velcro is a registered trademark of one leading manufacturer—appear on clothing, wallets, shoes, and watchbands. They're used to attach orthopedic devices and blood pressure cuffs and to hold floor mats and seat covers in cars. A hook-and-loop fastener can be opened and closed three times a day for twenty years without failure. The Space Shuttle program has found literally hundreds of applications for such durable, reliable fasteners, including cable ties that keep complex wiring harnesses in place in zero-gravity conditions.

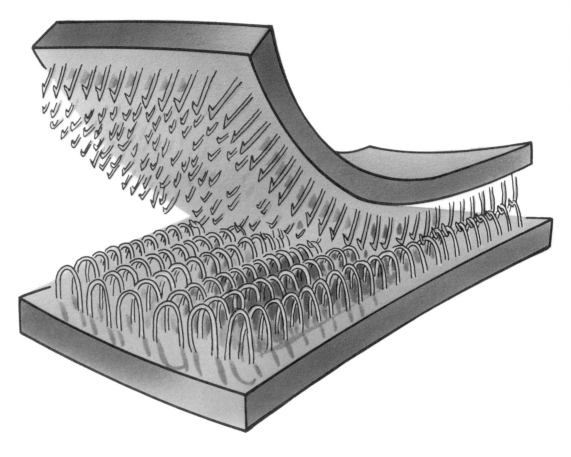

The first commercial hook-and-loop fastener was virtually identical to one type that's in common use today. It consists of two pieces of tape, one with hooks and the other with loops. The hook side is manufactured by weaving an array of tiny loops formed of single-filament nylon fiber into woven nylon tape. The tape is then heat-treated to permanently fuse the woven material into a fixed position. The loops are then automatically cut open to form an array of small, open hooks.

The loop side is made in a similar manner, but instead of regular loops being woven in, a "fuzz" of finer fibers is added to the backing. When heat-set, this fuzzy-faced tape forms the loop side of the fastener. Both loops and hooks can be varied in size and thickness of fiber in order to produce fasteners with the desired strength, ease of opening, and flexibility.

Hook-and-loop fasteners are now made in a variety of plastics, as well as metals and glass fibers to meet such special requirements as heat or cold resistance, electrical conductivity, and ultra-high strength.

ZIPPER

The slide is the key element in making the zipper work. Inside the slide is a Y-shaped channel. When the slide is pulled up, the two rows of teeth are fed together at precisely the right angle so that the teeth lock together.

The zipper is a tight, secure fastener that has the advantage of being flexible and quick to operate. This ingenious device was first patented in the 1890s, but the slide fastener that we know today was not perfected until 1913. Early designs had an unfortunate tendency to pop open. Its first use on garments was in World War I, when the U.S. Navy used slide fasteners on flying suits. Slide fasteners were not christened "zippers" until 1926.

To open the zipper, the wedge shape that forms the Y is forced between the teeth so that they unhook.

The strip of locked teeth go out the bottom of the Y.

The zipper's teeth are metal bars with a protrusion on the top and a matching hollow on the bottom. Some zippers are made of plastic coils instead of metal teeth.

The teeth are staggered along two strips of cloth so that the protrusion of one tooth fits into the hollow of the tooth opposite.

TOASTER

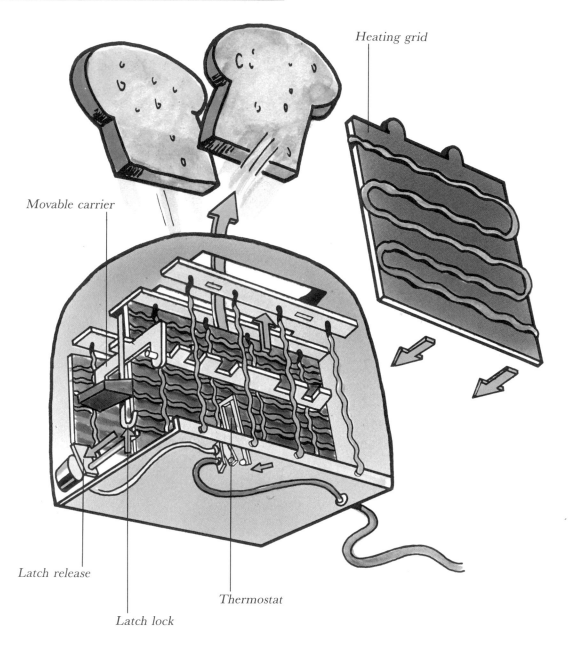

Although toasters have changed considerably in appearance since the first electric ones were introduced, they all function in much the same way. The earliest models used a zigzag grid of resistance wire (see **The Electrical Circuit**, page 18) as a radiating source. When electricity was passed through this wire, it became red-hot, reaching a temperature of 1500°F or higher. Slices of bread were supported in a vertical position an inch or so away from the grid, and the radiant energy from the hot wire toasted it nicely, browning the surface and warming the interior. If you didn't pay attention, that same energy would blacken the bread's surface and fill your kitchen with smoke. Early electric toasters didn't have thermostats or timers.

Because these early models had only one heating surface, you had to turn the bread around to toast both sides. About fifty years ago, toasters with more than one heating surface were developed. This design is still being used today. The toaster has either two grids, one on either side of a slice of bread, or three grids, surrounding two slices. To avoid moving the grids and bending fragile wires when you insert the bread or remove the toast, the bread is placed in a slot in the top, and an up-and-down movable carrier inside lowers the bread into position and pops it up when the toast is done. An adjustable thermostat determines when the carrier ejects the bread.

Heating grid

Movable carrier

Latch release

Latch lock

Thermostat

2-SLICE AUTOMATIC TOASTER

Pushing down the lever that lowers the toast also causes the heating-element contacts to close, thus turning on the toaster. When the toaster's interior reaches a preset temperature, a thermostat establishes an electrical linkage that releases a latch, allowing the carrier to pop up.

HALOGEN COOKTOP

Several tubular halogen lamps are mounted underneath a ceramic cooktop surface. A reflecting surface is applied to the bottom and sides of the housing to direct the radiant energy upward.

A specially designed thermostat rod is mounted over the lamps. This rod has a coating that reflects the radiant energy supplied by the lamps but absorbs heat that builds up in the housing. If the temperature in the housing rises too high, the rod expands and operates a switch that turns off the power to the lamps. It's easy to tell if a halogen element is on—you can see the lamps' glow right through the cooktop surface.

A halogen heating element is more expensive to manufacture than a conventional one; as a result, most commercial halogen cooktops use one or two halogen elements and conventional coils for the others.

Ceramic top

Thermostat rod

Quartz-iodine lamps

Reflecting surface

Halogen cooktops, using radiant energy to heat pots and pans, are one of the newest ideas in cooking. This form of cooktop borrows the technology that has revolutionized automobile headlights over the past decade or so.

Standard electric cooktops are noteworthy for their slow response: They take a long time to heat and a long time to cool off. The newer solid one-piece ceramic cooktops are attractive and easier to clean, but their response is even slower.

This inability to heat or cool quickly is a necessary consequence of the design of most electric cooktops. To avoid the possibility of electric shocks, their heating elements are enclosed in tubular metal coils or inside disc-shaped metal covers. So as to provide a large heating area, these metal housings are massive and therefore take a long time to heat or cool. When a conventional heating element is located beneath a ceramic surface, you need to heat up not only the element, but also the ceramic over it, before heat reaches the pot.

A halogen cooktop sidesteps most of these problems. Quartz-iodine lamps, (see **Halogen lights,** page 59), like the ones used for the ultrabright automobile headlights, radiate a great deal of energy, and they can be designed so that most of their radiant energy is in a frequency range that passes completely through a ceramic cooktop (see **Radiation,** page 14). The name "halogen" cooktop comes from the element iodine, which is a member of a chemical group called halogens. Quartz is used for the bulb because the temperatures inside an iodine-filled bulb are high enough to melt glass.

The radiated energy of a halogen lamp mounted below the surface of a ceramic cooktop passes through the cooktop and is absorbed by the bottom of the pot. Since the lamp's energy heats the pot directly, rather than having to heat a heavy shield or section of ceramic, response time is very short, rivaling that of an open gas burner.

SELF-CLEANING OVEN

There's nothing difficult about heating an oven to a high enough temperature to burn off deposits on the oven walls. However, it takes some care to accomplish this safely, and under enough control to avoid damaging the oven.

Self-cleaning ovens, whether gas or electric, are built with more effective insulation than standard ovens. This increased insulation allows the oven to heat up to cleaning temperatures more readily and keeps the oven's exterior from overheating dangerously during cleaning. It also makes self-cleaning ovens more energy-efficient during normal cooking than their standard counterparts.

Cleaning cycles—typically two to three hours at 800 to 900°F—are controlled by a combination of timers and thermostats. The timer starts and ends the cleaning operation and usually permits you to start the cycle late at night or at any other convenient time. A thermostat controls a door lock that prevents the oven from being opened while it's dangerously hot.

What makes cleaning the oven such a difficult chore is the nature of fats and oils, which spatter the oven's interior during cooking. When fats and oils are heated, they combine with the oxygen in air. This oxidation results in **polymerization,** or the formation of larger molecules from smaller ones. The grease spatters would be easy to clean up if you cleaned the oven right after each use. But if the grease stays there, the next time the oven is heated up, the spatters turn into a kind of varnish much harder than the original grease.

Commercial oven cleaning compounds use strong chemicals to attack the baked-on "varnish." These chemicals can be hard on skin and lungs.

Self-cleaning and continuous-cleaning ovens take a different approach toward dealing with polymerized grease. The **continuous-cleaning** oven has a synthetic nonstick coating applied to the oven's walls. Polymerized grease should not adhere firmly to the oven. The heat from subsequent uses of the oven continues to oxidize the varnish and it will fall off in the form of dust. The nonstick coatings are usually dark-colored, with a dull finish, so that any remaining grease or oil is less visible.

True **self-cleaning ovens** remove the baked-on grease by incinerating it—burning it off. The same chemical process of oxidation that combined the smaller molecules of fats and oils into larger ones can, if carried even further, convert those large

varnish molecules to carbon dioxide gas and water vapor. This complete oxidation occurs most readily at much higher temperatures than are used in cooking: 800 to 900°F is the usual range for cleaning. These ovens heat up rapidly enough, and burn off dirt completely enough, to minimize smoke and odor; the gases formed pass through a platinum-palladium screen that acts like a car's catalytic converter (see page 192) to eliminate virtually all of the smoke.

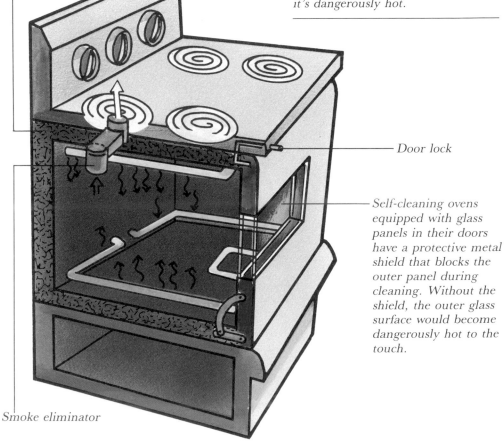

High-density insulation

Smoke eliminator

Door lock

Self-cleaning ovens equipped with glass panels in their doors have a protective metal shield that blocks the outer panel during cleaning. Without the shield, the outer glass surface would become dangerously hot to the touch.

CONVECTION OVEN

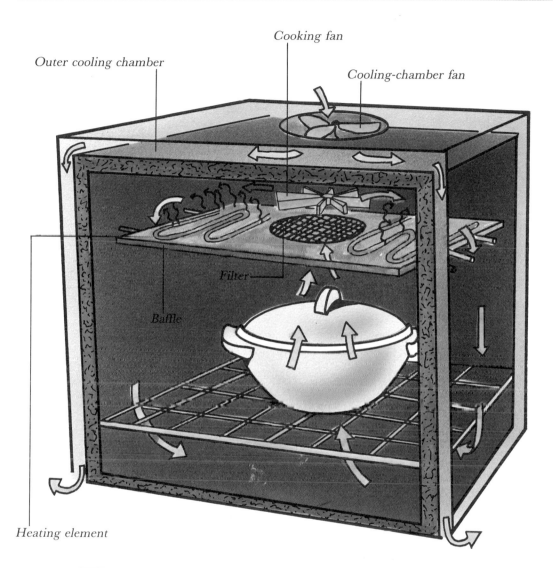

Because conventional gas and electric ovens work by means of convection, the so-called convection oven is really a forced-convection oven.

Any oven is a container in which food is heated and cooked by exposure to hot air. In a conventional oven, air is heated by the burner or heating element and expands, causing it to rise (see **Conduction and Convection,** page 10). When the hot air comes in contact with the food being roasted or baked, it conducts some of its heat to the food, warming the food and cooling the air. The cooled air sinks, and more heated air moves upward to replace it. This process sets up airflow patterns called "convection currents" in the oven.

Air doesn't hold much heat at any one time, and convection currents are gentle. Each time newly heated air passes over the food's surface, it contributes only a small amount of heat to the food. It takes a long time or enough heated air to circulate around the roast or pie to transfer sufficient heat to finish the cooking.

The forced-convection oven speeds up this process by increasing the speed at which the air moves. If you increase the amount of hot air that the food is exposed to during any given time, the food will heat faster, even if the fast-moving air is no hotter than the slow-moving air. In a forced-convection oven, a "wind-heat factor" is created, analogous to the Weather Bureau's wind-chill factor. Depending upon the food being cooked, a forced-convection oven can cook in 30 to 50 percent less time than a conventional oven requires.

Cooking fan

Outer cooling chamber

Cooling-chamber fan

Filter

Baffle

Heating element

The heat source in this forced-convection oven is an **electric heating element.** The heating element is concealed behind a **baffle,** which prevents the heat from being radiated directly to the food. A **cooking fan** blows air across the heating element and circulates it throughout the oven, passing the current of hot air across the food being cooked. The air, cooled by exposure to the food, then passes through a **filter**—to remove grease droplets—and once more is blown across the heater.

The chamber in which the food is cooked is insulated to minimize unwanted heat loss and to keep the outside of the chamber cool to the touch. Some designs have an outer cooling chamber, through which a second fan moves cold air. This feature doesn't affect the cooking itself, but it assures that the oven's exterior will remain cool at all times.

FOOD PROCESSOR

A food processor is the most versatile and least specialized member of the appliance family that includes mixers, grinders, and blenders. It can knead dough, shred cabbage, chop meat, slice carrots, and puree spinach. It accomplishes all this because of its combination of three elements: a powerful motor, a large processing bowl, and an arsenal of extremely sharp blades and cutting discs.

A direct-drive motor is mounted beneath the work bowl, and the food processor's blades and discs attach directly to the motor's shaft.

Two wide, scythe-shaped blades, one mounted higher than the other, form the basic chopping device. The lower blade turns against the bottom of the bowl and cuts the food; the upper blade cuts the food and then throws the pieces of food toward the side of the work bowl, where they fall back down to the center to be cut again.

Each of the slicing and shredding discs is attached to a stem that allows the disc to turn at the top of the bowl. The food to be sliced goes into a feed chute in the food processor's cover. Slices drop down into the work bowl, out of the way of the disc.

The rapidly whirling blades are extremely sharp. A safety switch, which prevents you from turning on the processor unless the cover is completely closed, is actuated by a long rod extending down the side of the bowl. Turning the cover to the closed position presses the rod down to close the switch.

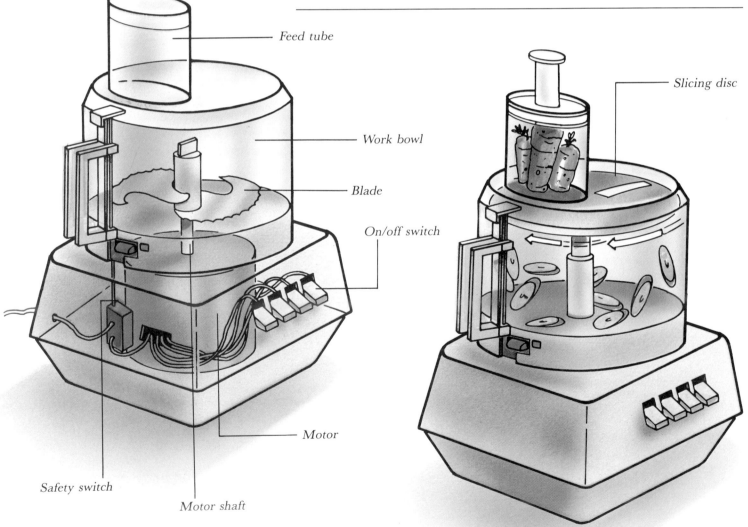

Feed tube

Work bowl

Blade

On/off switch

Motor

Safety switch

Motor shaft

Slicing disc

PASTA MAKER

Basic pasta starts with flour, oil, salt, and, for some recipes, eggs. The ingredients must be mixed and the dough kneaded until it reaches the proper consistency. Then it is rolled out very thin, folded, and cut. Making pasta by hand requires some practice; fortunately, machines are available that make it easy to have fresh pasta.

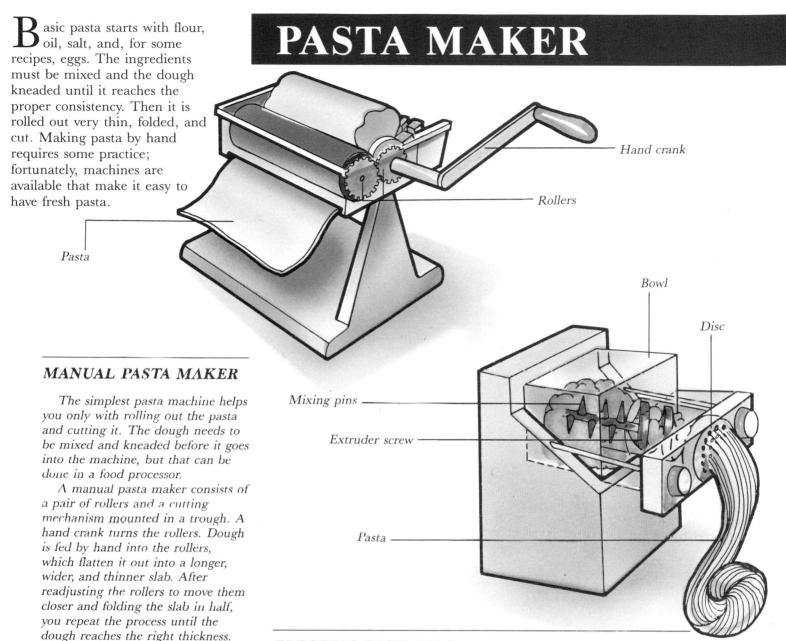

Hand crank

Rollers

Pasta

Bowl

Disc

Mixing pins

Extruder screw

Pasta

MANUAL PASTA MAKER

The simplest pasta machine helps you only with rolling out the pasta and cutting it. The dough needs to be mixed and kneaded before it goes into the machine, but that can be done in a food processor.

A manual pasta maker consists of a pair of rollers and a cutting mechanism mounted in a trough. A hand crank turns the rollers. Dough is fed by hand into the rollers, which flatten it out into a longer, wider, and thinner slab. After readjusting the rollers to move them closer and folding the slab in half, you repeat the process until the dough reaches the right thickness. That usually takes four or five trips through the machine.

The last step consists of moving the hand crank to the cutting slot, threading the ribbon of dough through the slot, and turning the crank. This cuts the pasta into thin strips.

ELECTRIC PASTA MACHINE

More fully automatic and more versatile, as well as more expensive, is an electric pasta maker. The raw ingredients go in and finished pasta comes out without your handling the dough at all.

An electric pasta maker combines a slow-motion mixer and an extruder—a screw that forces the dough under pressure through a disc with holes. Both functions are performed by one motor-driven shaft. Mixing pins are mounted over most of the length of the shaft; an extruder screw is mounted at the end of the shaft.

At the start, the shaft rotates in a direction that directs the extruder screw back into the bowl. When the ingredients are added, the mixing pins blend them and knead the dough. The extruder screw keeps pushing the dough back into the area where the mixing pins are. When the dough is ready, you flip a switch that reverses the direction of rotation. The screw then forces the dough against the end disc, and the pasta is extruded through the holes in the disc, ready to be cut into lengths.

MICROWAVE OVEN

The ability of microwaves to cook food was discovered accidentally when an engineer in a radar laboratory found that an exposure to radar waves caused a chocolate bar in his pocket to melt.

Microwaves are a form of electromagnetic radiation (see **Electromagnetic Waves,** page 50). They will pass through glass, paper, and most plastics, but they are reflected by metals and are absorbed by water and many other components of foods. They can easily be generated by using a compact electronic device called a **magnetron;** this device is the heart of a microwave oven.

In other forms of cooking, heat is applied to the outside surface of the food and then conducted to the interior portions. A microwave oven does the opposite; it cooks the interior of the food directly. When microwaves strike a piece of food, they cause a certain kind of molecule in the food to vibrate. This kind of molecule is found in water and fat. Almost all foods have, dispersed throughout them, either fat or water or both. As a result, microwaves can cook many foods much more quickly than other cooking methods can.

Nevertheless, microwave cooking is far from being a cure-all for cooking problems. Since food isn't always uniform in composition, different parts may heat at wildly different rates. And microwave ovens do not usually brown the surface of a roast or chicken. The outer surface of the chicken is not exposed to high temperatures, as it would be in a conventional oven.

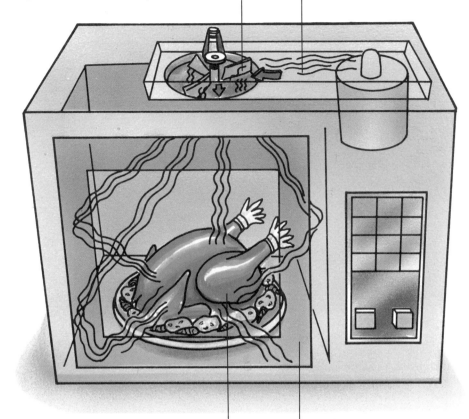

*To keep the microwave radiation well distributed throughout the oven's chamber, a **stirrer,** or metal-bladed fan, reflects the waves.*

*The microwaves generated by the magnetron enter the oven through a **wave guide.***

The waves bounce off in all directions. If they strike the food, they are absorbed; if not, they are reflected by the metal walls of the oven and bounce back and forth until they do strike the food.

A perforated metal grille in the glass-faced oven door keeps the microwaves from escaping.

In a conventional oven, energy heats the air inside the oven. As a result, it takes no longer to bake four potatoes than it does to bake one potato at the same temperature. In a microwave oven, the energy supplied by the magnetron heats the food directly. Those microwaves that miss the food bounce around inside the oven until they strike some portion of the food and are absorbed. Consequently, it takes considerably longer to microwave four potatoes than it does to do one. It doesn't take four times as long, though, because heat escaping from the potatoes while cooking raises the temperature inside the oven, helping to speed the cooking process.

Fancier ovens can have multiple-cycle automatic controls with a vast selection of cycles, timers, power settings, and temperature settings. Although many of these are convenient, the fact is that a magnetron is either "on" or "off." A 50-percent power setting merely cycles the microwave generator off half the time and on the other half.

WHY MICROWAVES HEAT FOOD

Many molecules—water molecules and fat molecules, in particular—have a positive electrical charge at one end and an equal negative charge at the other. Such a molecule is called a **dipole.**

When placed in a strong electromagnetic field such as is caused by microwave radiation, these dipoles try to rotate to line up with the field, oriented so that positive charges swing toward negative ones and vice versa.

Microwaves reverse their directions millions of times per second. As a result, the molecules undergo a very rapid series of back-and-forth rotations; the friction from this movement results in heat.

TOUCH-PAD CONTROL

The touch-pad controls of a microwave oven are a spin-off from computer technology. A microprocessor chip—a small computer—translates the commands made by pressing on the face of the touch-pad into electrical signals.

The face consists of a sheet of flexible plastic with graphics printed on it to resemble push-buttons. On the back of each "button" is a metal foil contact. Behind the sheet of plastic is a rigid board with a second set of contacts on it. When a contact on the plastic sheet touches a contact on the board, a circuit is completed. Between the board and the plastic sheet is a thin, rigid plastic plate with holes at each button. This sheet keeps the contacts separated until a finger presses on the face, which stretches slightly and permits the two contacts to touch through one of the holes.

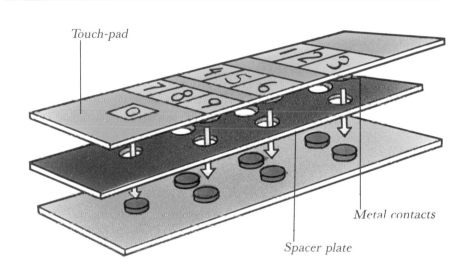

Touch-pad

Metal contacts

Spacer plate

Pressing the 1 connects vertical wire A with horizontal wire D.

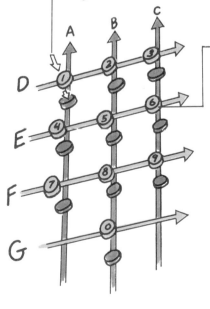

Pressing the 6 connects vertical wire C with horizontal wire E.

Each push-button position corresponds to a particular combination of wires that will be connected. The wires enter a microprocessor chip that converts each signal to a number. If you press 1, 3, and 2 in that order, the microprocessor will set the time for one minute and thirty-two seconds. Other combinations of wire connections are interpreted as commands for different modes of operation, such as "defrost" or "cook."

POPCORN POPPER

Popcorn is not a modern development. Native North Americans started cultivating—and popping—popcorn about 7,000 years ago. Although it's related to sweet corn—the kind we eat off the cob—popcorn is a variety unto itself. Popcorn's ability to explode into large cumulus "blossoms" is unique among cereal grains.

Dried popcorn kernels have a tough, airtight hull surrounding a body composed mostly of starch grains and a little water. When popcorn is heated to about 350°F, the pressure inside the hull reaches as much as 135 pounds per square inch. The pressure finally ruptures the hull and releases the trapped water, which flashes into steam. The rapidly expanding steam inflates some of the starch granules in the kernel and creates air pockets between others. And popcorn is the result.

Any cooking method that heats the kernels sufficiently will make popcorn pop, but since the popping temperature is close to the temperature at which the corn tends to burn, popping takes some care. Too low a temperature leaves unpopped kernels; too high a temperature burns the corn.

Simplest of the corn-popping implements is the traditional one for popping corn over a log fire—a long-handled popping pan with a screen on top. You can also pop corn on a stove using a saucepan—a deep one with a lid. Both implements take time and careful attention to produce good results.

The high-tech approach to popping corn uses either an electric popper or a microwave oven; brands of "microwave"

popcorn come in special packaging. Microwave ovens do a particularly good job popping corn; in fact, popcorn was the first food product to be prepared with microwaves. Electric poppers come in two basic designs: those that cook with hot oil and those that use hot air.

HOT-OIL POPPER

Hot-oil poppers consist of a nonstick metal bowl mounted on a base containing a heating element similar to that used in an electric frypan. A thermostat attached to the heating element switches it off if the oil gets too hot, minimizing the possibility of overcooked (or, conversely, undercooked) popcorn.

Cooking oil is measured into the bowl, and loose kernels of popcorn are added. The purpose of the oil is to conduct the heat evenly around the kernels. The popper usually has a big plastic lid that contains the expanded corn and doubles as a serving bowl once the popcorn is popped.

Cover

Nonstick bowl

Heating element

HOT-AIR POPPER ▶

Hot-air poppers heat the popcorn evenly by blowing hot air around the kernels. A hot-air popper is an upright canister with an upper compartment that holds the unpopped corn. Underneath, inside the base, is a fan that blows air past a heated filament through a screened opening into the popping chamber. Unpopped corn bounces around in the air current until it pops. The popped kernels are not as dense as unpopped kernels; once they have popped, the air current blows them out of the machine through a chute at the top.

Popping chamber

Hot air

Heating element

Fan

Motor

REFRIGERATOR

Home refrigerators, freezers, and air conditioners all work in much the same way. Their operation depends upon three fundamental physical principles: First, heat always flows from high-temperature regions to low-temperature regions. Second, when a liquid is evaporated to a gas, it absorbs heat. Third, if a gas is compressed enough, it turns to a liquid, and when a gas condenses to a liquid, it gives off heat.

The boiling point of a liquid—the temperature at which it evaporates to a gas—depends on how much pressure the liquid is under. By increasing the pressure, the boiling point is raised. A **refrigerant** has a boiling point that is a lower temperature than the temperature we are striving for in the refrigerator. At room temperature it is a gas that can be compressed to the point at which it liquefies (because, under compression, its boiling point is raised). Fluorocarbons are the most common refrigerants today.

The essential components of a refrigerator or freezer are a compressor to compress the gaseous refrigerant; a condenser, where the refrigerant becomes a liquid and gives off heat; an evaporator, where the refrigerant boils into a gas and absorbs heat; and the refrigerant itself.

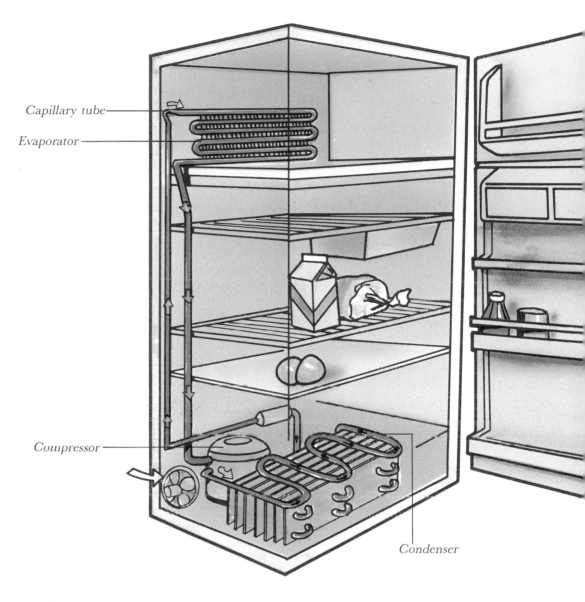

Capillary tube

Evaporator

Compressor

Condenser

THE REFRIGERATION CYCLE

In the **evaporator,** the refrigerant is under relatively low pressure and is allowed to evaporate. In doing so, it absorbs heat from the air inside the refrigerator, and the air inside the refrigerator becomes colder.

The gas then travels to the **compressor,** which is located underneath the refrigerator. There it is compressed until it is quite warm. The warm gas is then pumped to the **condenser,** where it gives off heat to the kitchen. Because of the increase in pressure and because of the loss of heat, the refrigerant liquefies. It is forced up through a thin tube called a **capillary,** which maintains the pressure on the liquid. Through this tube, it goes back to the evaporator, which has larger tubes that relax the pressure on the refrigerant and it evaporates into a gas once more, starting the cycle over again.

DISHWASHER

A dishwasher is one of those rare devices that saves both work and money. It uses considerably less hot water than hand-washing does—enough less so that the savings in hot water more than make up for the dishwasher's purchase price over the life of an average dishwasher. You can save even more if your dishwasher has an "economy dry" setting, or if you interrupt the dishwashing cycle after the final rinse and let the dishes dry without using the dishwasher's heating element.

Despite the bewildering array of controls, cycles, dials, and touch-pads available on many models, dishwashers are simple devices. They all work in much the same way. A dishwasher uses high-pressure sprays of hot water, combined with strong detergents, to remove soil and grease from dishes and cooking utensils. Rather than rely upon the water pressure in the house's water lines, a pump in the dishwasher pumps the water to the spray arms, which rotate, like a lawn sprinkler, owing to the thrust of the water spray. The hotter the water, the better they work. Some dishwashers further heat the hot tap water during the wash cycle.

A dishwasher has four basic functions: It fills, washes and rinses, drains, and dries. All four are controlled by a timer, which is set on the dishwasher's control panel and activated once the door is locked.

Fill: *When the door is locked, the timer opens a valve that lets hot water flow into the dishwasher. After a predetermined time, the timer shuts off the water, and the unit is ready for a wash or rinse. An overflow protection switch shuts off the water valve in case the dishwasher should overfill.*

Wash or rinse: *The timer starts the pump, which forces water into the spray arms. The water leaves the spray arms as a large number of high-speed jets of water; these create the thrust that turns the spray arms and helps "scrub" the dishes when the jets strike them. Wash cycles and rinse cycles work the same way, except that during a wash cycle the timer also opens a compartment that holds the detergent.*

Drain: *With the pump still running, the timer opens a drain valve that allows water to enter the drain. Since the drain offers less resistance to the flow of water than do the spray arms, the greater part of the pump's output goes to pumping the used water down the drain.*

Dry: *After the tub is empty, the timer either starts another fill or starts to dry, depending upon how the controls have been set. In the dry part of the entire cycle, the timer turns on the heating element, which heats the air in the dishwasher to dry the dishes.*

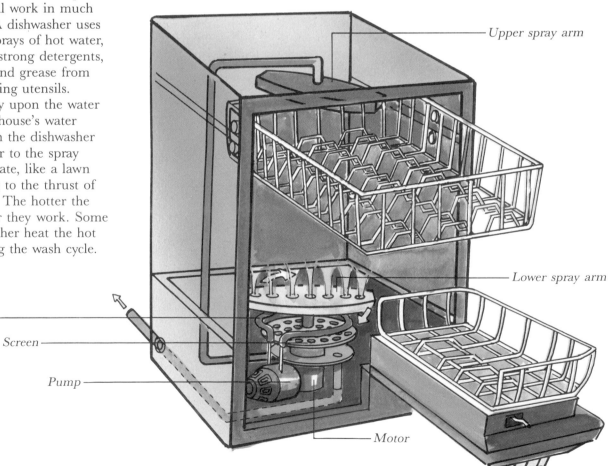

Upper spray arm

Lower spray arm

Heating element

Screen

Pump

Motor

GARBAGE DISPOSER

A sink-mounted garbage disposer makes most food wastes disappear down the drain of the kitchen sink. Garbage disposers can be used in virtually every house where the sewage or septic system is able to handle the extra load of solid materials.

The disposer is mounted below the sink and supported from the sink's bottom by heavy screws. All the water that drains from the sink passes through the disposer; if the kitchen also has a dishwasher, the outflow line from that appliance is also routed through the disposer.

When you use the disposer, garbage falls through the sink drain opening into the disposer's hopper, where it is ground up into small particles. At the same time, cold water from the sink passes through the unit and flushes the particles down the drain.

Food scraps drop down from the sink to the disposer's hopper. There they strike a spinning turntable driven by a powerful electric motor. On the turntable are mounted impeller blades, which pass close to the cutting edges of a stationary shredder ring. Centrifugal force slings the scraps to the outer edge of the turntable, where they are caught between the impeller blades and the shredder ring and ground or shredded into small particles. These particles are small enough to fall through holes in the turntable; they are carried down the drain by a stream of cold water from the tap. The disposer works with cold water because hot water melts grease, which can then clog the drain.

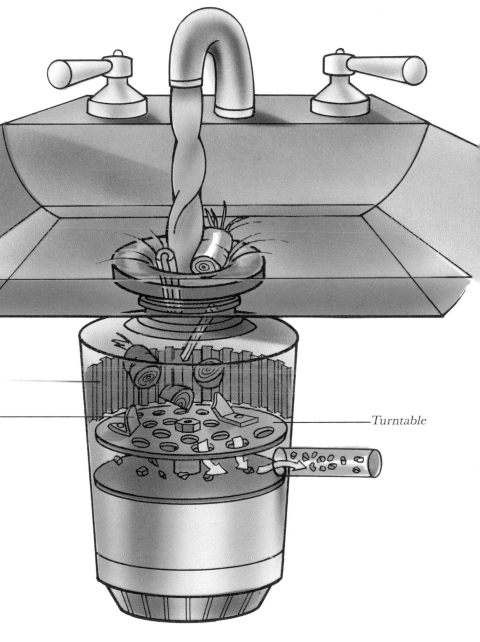

Shredder ring

Impeller blade

Turntable

WATER PURIFIER

S everal types of purifiers for drinking water exist; each of them works on a different type of contaminant. At this time, more than 70,000 different contaminants have been identified that can pollute drinking water. All these contaminants can be divided into four classes: biological impurities (largely bacteria), dissolved organic chemicals (including PCBs and other halogenated compounds), heavy metal salts (such as lead and mercury salts), and suspended solids (including asbestos fibers).

It is possible to buy a water purification system that can handle all four classes of pollutant, but it would be needlessly expensive if only one or two were present. Determining if contaminants are present in the water supply and what the contaminants are requires the services of a water testing laboratory.

Many people consider the reverse osmosis, or RO, units the most versatile type of water purifier available. Using technology originally developed to make sea water drinkable, these devices use a cellophanelike membrane as a filter. Water is forced under pressure against the membrane. A portion of it passes through, leaving the impurities behind to be carried out in a waste stream. RO models can be effective in dealing with all four classes of pollutant, although they are often combined with an activated charcoal filter to help remove dissolved organic compounds.

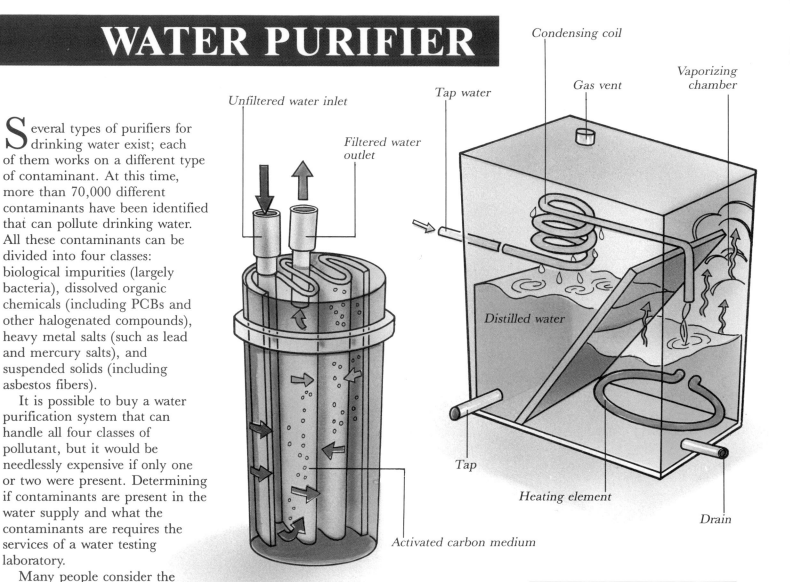

Unfiltered water inlet

Filtered water outlet

Tap water

Condensing coil

Gas vent

Vaporizing chamber

Distilled water

Tap

Heating element

Drain

Activated carbon medium

ACTIVATED CARBON FILTER

The most common device for home water purifying is the activated carbon filter. This system makes use of the fact that many chemical compounds will stick to the surface of a particle of treated carbon. Passing water through a bed of such carbon particles will do a good job of removing dissolved organic chemicals. If the device includes a good filter, it will also remove suspended solids. Activated carbon filters are not effective against most biological contaminants or dissolved heavy metal salts.

DISTILLATION SYSTEM

Home distillation systems, or stills, are a time-honored approach to water purification. A still boils the water, cools the resulting steam until it condenses, and collects the condensed water vapor. Since heavy metal salts and suspended solids do not vaporize, they are effectively removed, as are bacteria. Some organic compounds with high boiling points are removed, but many others are boiled off and condensed with the water.

AIR CLEANERS

As pollutants become more readily identifiable and as our homes and offices become more tightly sealed, we become more aware of the quality of the air we breathe. Mold, mildew spores, and dust not only cause odors and increase household cleaning, but also aggravate allergies and contribute to respiratory illnesses.

The simplest way to get rid of these problems is to filter them out. Heating and cooling systems using air ducts have always contained filters, usually of glass fiber, to filter out dust. Nowadays, high-efficiency air filters remove more than dust and lint—they can even trap smoke particles and spores.

The two basic types of high-efficiency filters are the entrapment type and the ionization or "electronic" type. The entrapment filters are more efficient materials for catching the particles; the electronic type filters them out using electric charges. The electronic models are more efficient at the start, but as they begin to fill up, their efficiency drops off sharply. The nonelectronic types are not quite so efficient in the beginning, but their efficiency actually increases as the filters load.

ENTRAPMENT FILTER

The entrapment type of high-efficiency air cleaner includes the ordinary panel filters, but it also incorporates mesh filters and pleated-paper filters. These present a large surface area to the air flowing through them. They catch even tiny spores that attempt to pass through. Particles in the air can be caught as they attempt to pass through the material or they can be trapped as they brush against the surface of the filter.

Pleated-paper filter

Fan

Filter

Charcoal filter

Negative grid

Positive grid

Power source

ELECTRONIC FILTERS

Electronic filters also contain a conventional filter to trap large particles. Smaller particles are dealt with by electrically charged high-voltage grids. As the remaining particles pass across the first grid, they are given a positive electric charge. Next they come to a negatively charged grid, and here they meet their doom. The difference in charge causes the particles to be attracted to the grid, where they are trapped. In some versions, the filtered air is passed through a charcoal filter for odor removal before being returned to the room.

THERMOSTAT

A furnace, like most appliances that heat or cool—ovens, air conditioners, irons—has a thermostat, which is a device that holds a constant temperature by turning the furnace on and off. When the furnace has first turned off, the stored heat in the heating system continues to heat up the room. The furnace will also be slow to start heating up when the furnace is switched on again. This would cause the room's air temperature to fluctuate widely between uncomfortably great temperature differences. To compensate for that, thermostats incorporate an **anticipator,** which is a small

electric heater that fools the thermostat into behaving as though the room were warmer than it really is. The thermostat will shut off the furnace before the room reaches the desired temperature.

The heart of most thermostats that control household appliances is a **bimetallic element,** a coil made of two different metals layered together. The two metals expand and contract at different temperatures, causing the element to be more loosely coiled when it's cool than it is when it's heated (see **Thermometer,** page 262). The end of the coil activates a switch that turns on the furnace.

The bimetallic element in most house thermostats is fastened securely at one end. The switch contacts are sealed behind glass to protect them from dirt. As the temperature drops, the bimetallic element starts to uncoil. The force exerted by this uncoiling separates a stationary steel bar from a magnet at the end of the coil. The magnet drops down close to the glass-enclosed contact, pulls up the contact arm inside the tube, and causes the contacts to close. This completes the electrical circuit and turns on the furnace.

At the same time, the anticipator turns on. The anticipator heats up the bimetallic element, and it begins to coil more tightly. Eventually the hold of the magnet on the contact arm is broken; the arm drops, breaking the circuit and turning off the furnace. The magnet attaches itself to the stationary bar, which holds it away from the contact arm until the room cools.

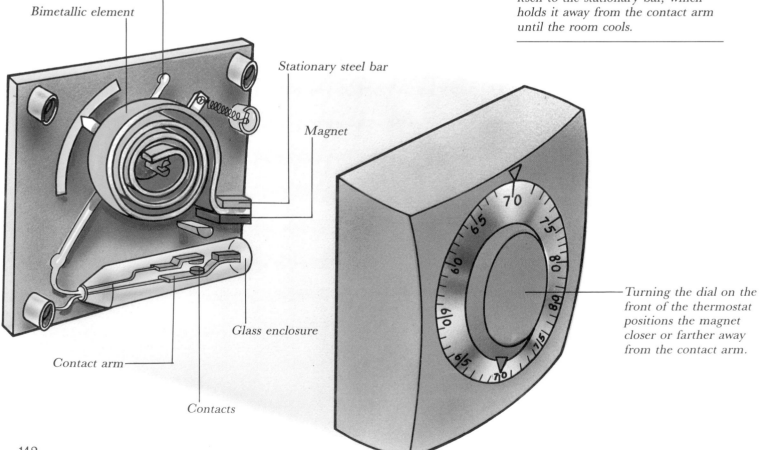

Anticipator

Bimetallic element

Stationary steel bar

Magnet

Glass enclosure

Contact arm

Contacts

Turning the dial on the front of the thermostat positions the magnet closer or farther away from the contact arm.

SMOKE DETECTOR

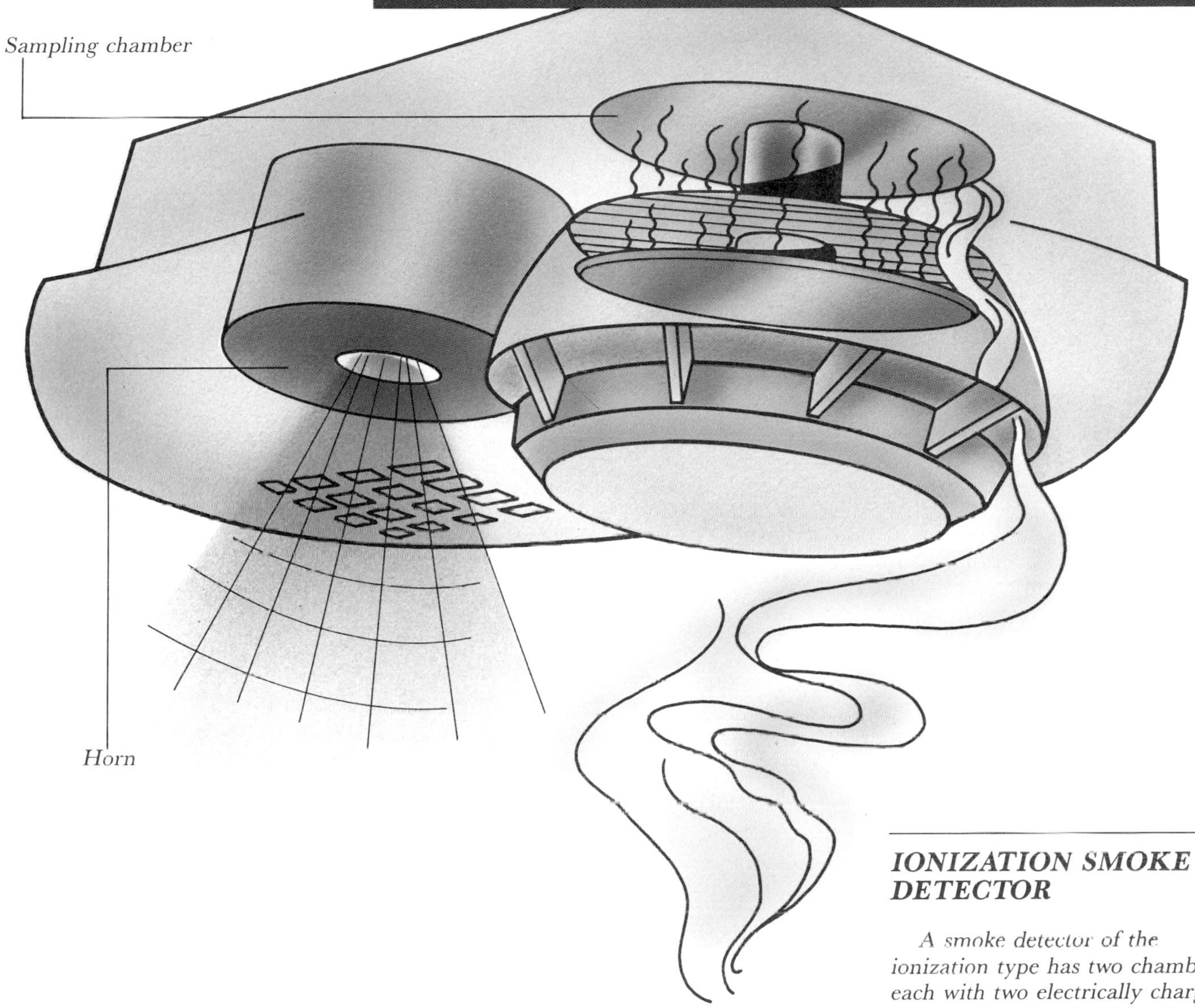

Sampling chamber

Horn

IONIZATION SMOKE DETECTOR

A smoke detector of the ionization type has two chambers, each with two electrically charged plates. A source of radioactive particles—harmless alpha particles—ionizes air molecules. Ions are atoms or groups of atoms that have an electrical charge because they have lost or gained free electrons. The presence of the ions causes a constant flow of electric current between the two plates. When combustion products enter the outer chamber, the flow of electrons is reduced. The drop in conductivity triggers the alarm. The sealed inner chamber acts as a reference source. It stabilizes the detector at all levels of atmospheric pressure, temperature, and humidity.

M any home fires smolder for long periods before they generate enough heat to trigger an alarm of the thermal type. A sleeping resident could be overcome by smoke long before such an alarm sounded. But smoke detectors detect the presence of the products of combustion before a fire actually starts.

Two types of smoke detectors are available—photoelectric detectors and ionization detectors.

The photoelectric type contains a small light focused on a photoelectric cell. If smoke particles enter the chamber, they reduce the amount of light reaching the photoelectric cell. This in turn reduces the voltage being generated by the cell. The detector circuitry senses this and sounds the alarm. The ionization type uses ionized air molecules to create an electric current. Combustion products—not necessarily smoke, but other, invisible products of combustion as well—interrupt the electric current.

HUMIDIFIERS

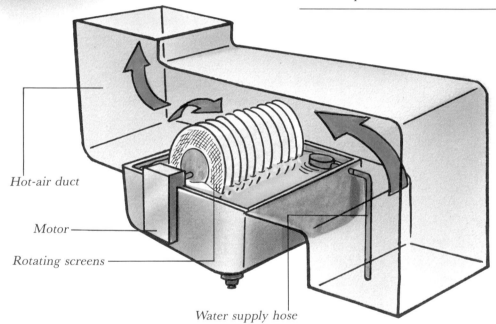

Fan

Belt motor

Roller belt

Hot-air duct

Motor

Rotating screens

Water supply hose

ROOM HUMIDIFIER

Room humidifiers usually need to be larger than furnace models. Because the air circulating through it is air from the room, which is much cooler than the air emerging from the furnace, the water doesn't evaporate as readily.

*A common type of room humidifier is the **roller-belt** humidifier. A wide belt of flexible porous foam rotates over a pair of rollers. The bottom of the belt passes through a pan of water. A fan blows room air through the porous belt, and the water evaporates. The pad provides a large surface so that a reasonable amount of water can be evaporated even at relatively low room temperatures.*

I n the wintertime in cold climates, indoor air becomes dry because the cold outside air cannot hold a great deal of water, even at 100 percent relative humidity. When that air is heated for indoor warmth, its relative humidity drops and the air becomes drier. The colder the outside air and the more it is heated, the drier it becomes.

If anything containing water is exposed to dry air, some of that water will evaporate. Your skin, your potted plants, even the wood in your furniture lose moisture to the air, becoming dry in turn.

Several different appliances work to put some water back into dry air. Even though a relative humidity of around 50 percent would alleviate almost all the problems associated with dry air, 50 percent is not a practical level to maintain in the winter. Air at that humidity coming in contact with colder window or walls will condense and cause water drips and indoor "rain." At an outdoor temperature of 10°F, a double-glazed window starts to sweat at an indoor relative humidity of 35 to 40 percent; a single-glazed window will sweat at 20 percent.

FURNACE HUMIDIFIER

*With forced-hot-air heat, the whole house can be humidified with a furnace humidifier. A **rotating-screen** humidifier is usually mounted underneath a horizontal section of hot-air duct. A stack of round screens mounted on its side projects up into the air stream; the bottom portion of these screens are immersed in a pan of water. The screens rotate, picking up water from the pan and carrying it up*

into the duct, where the warm air evaporates it. A float valve, much like the one in a toilet tank, allows water to flow into the pan to keep it at a constant level. A small electric motor rotates the screens.

Other designs of furnace humidifiers spray a mist of water directly into the air stream; others incorporate a separate small fan that blows over a saturated pad and directs the moist air into the duct.

ULTRASONIC HUMIDIFIER

The latest development in devices that add moisture to dry air is the ultrasonic humidifier. Unlike a furnace humidifier, an ultrasonic humidifier doesn't depend upon a stream of moving air to evaporate water—it adds water directly to the surrounding air in the form of a mist of tiny droplets. And, it doesn't boil the water to form a cloud of steam—its mist is cool.

The key to an ultrasonic humidifier's operation is a **piezoelectric transducer,** a device that changes the alternating electric current into mechanical vibration. The vibration agitates the water in a cup-shaped cavity. The agitation of the water bounces droplets of water from the surface of the water to the air.

Piezoelectric devices make use of crystals that generate an electric current when flexed, and flex when exposed to an electric current. (See **Quartz watch,** page 31). If such a crystal is placed at the bottom of a container of water and vibrated slowly, the water level would rise and fall as the crystal flexed and unflexed. In an ultrasonic humidifier, the rate of vibration is increased to the point at which the inertia of the water prevents it from following the movement of the crystal. Some of the water at the surface is still traveling upward while the remainder is traveling downward. The result is that tiny droplets of water move upward and leave the container as a fine mist.

This action begins to occur at ultrasonic frequencies, which are too high for the human ear to detect (see **Sound,** page 48). To obtain electric currents that alternate at such high frequencies, an **oscillator** is used. This is a device that produces current at a specific frequency.

If the water level becomes too low, the valve admits water from the **water reservoir.**

The mist of air and water droplets leaves the humidifier through a nozzle in the mist outlet.

Water droplets are propelled up into the **mist chamber,** where they mix with air.

When the power is switched on, an oscillator forces the **nebulizer** to vibrate at a very high frequency.

The water level in the cup over the nebulizer is controlled by a **float valve.**

Air is sucked into the **air intake** by a small fan.

A **humidistat** lets you select the level of humidity you want.

In the ultrasonic humidifier, the piezoelectric transducer is the **nebulizer,** a device that reduces the water to a fine spray. If you use the humidifier with hard water, you can encounter a problem that other types of humidifiers don't generate—dust. Conventional humidifiers evaporate the water inside the unit, so that the minerals that make the water hard are left behind in the unit. The ultrasonic humidifier evaporates the water into the air, which can leave a fine white dust of calcium and magnesium compounds on room surfaces near the humidifier. A demineralizer can be used to treat the water fed to the nebulizer. Demineralizers are built into some ultrasonic humidifiers; they require the purchase of replacement cartridges.

AIR CONDITIONER

An air conditioner has two different, though related, jobs to do: cooling indoor air and removing unwanted moisture from the air. Both tasks require taking energy from indoor air and dumping it outdoors. An air conditioner pulls energy, in the form of heat, out of the air, and the air's temperature drops. In the same process, it removes the air's moisture.

If air—at any temperature—is holding all the water vapor it can, it's said to be "saturated," or to have a relative humidity (RH) of 100 percent. If it contains half that amount, its RH will be 50 percent. The colder air gets, the less moisture it can hold. If you were to lower the temperature of the air without adding or removing any moisture, its capacity for holding water vapor

would drop, and its RH would rise. If the temperature drops low enough, the air can no longer hold all its moisture, and the water condenses. That's why air conditioners sometimes drip water. And finally, if you reheat the air, it will have a lower RH because water was removed when the air was cold.

An air conditioner lowers the indoor relative humidity by cooling some of the room's air until some of its moisture condenses out and allowing the colder air to mix with the warmer room air to reheat it, thus providing a lower RH. The cold air leaving the air conditioner is at about 100 percent RH until it warms up a little.

ROOM AIR CONDITIONER

An air conditioner works in almost exactly the same way as a refrigerator (see **Refrigerator,** page 137), and it has the same basic parts: condenser, compressor, and evaporator. In a room air conditioner, all the parts are mounted on a single chassis. Usually a room air conditioner is mounted in a window, with the cold evaporator coils inside the room and the hot condenser coils outside. In most room air conditioners, a single motor drives both the blower that circulates room air past the evaporator and the fan that sucks in outdoor air and blows it across the hot condenser coils to cool them. The compressor has its own, larger motor.

When the warm, moist room air is chilled by the evaporator, water is condensed. This water drips down the evaporator, is collected in a drip pan, and passes (by gravity) through a tube that leads to another pan at the bottom of the condenser. The water then helps to cool the condenser; any excess water drips outside.

Most room air conditioners allow you to circulate the air in the room, cooling it somewhat each time it passes through the evaporator coils. By opening an exhaust vent, you can let some of the room's stale air be drawn out by the condenser fan.

Condenser fan

Compressor

Condenser coils

Blower

Evaporator coils

VACUUM CLEANER

The electric vacuum cleaner has changed remarkably little since its introduction more than 80 years ago. The 1908 version would be recognized as a vacuum cleaner today, though its rugged, virtually all-metal construction caused it to weigh a hefty 60 pounds. In comparison, today's cleaners, with extensive use of plastics and lightweight alloys, typically weigh between 11 and 30 pounds.

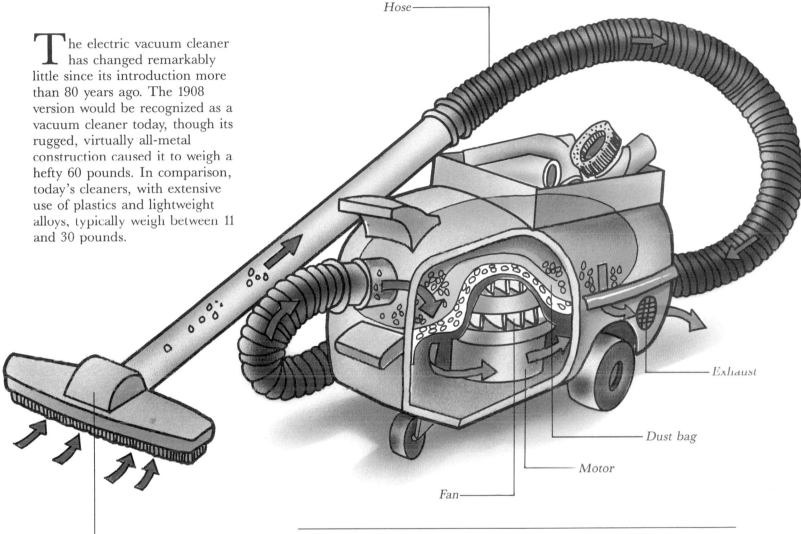

Hose

Exhaust

Dust bag

Motor

Fan

Floor nozzle

All vacuum cleaners work by the same principle. An electric motor drives a fan that blows air through the unit. This creates a vacuum at the cleaning end of the appliance. Air rushes in to fill the vacuum, bringing with it the dirt and dust that are being cleaned. The dust-filled air is forced by the fan into the bag, where the dust remains and the air is blown out of the cleaner.

Both upright and canister vacuum cleaners come with attachments to make them more versatile. Long hoses and wands extend the reach of the vacuum cleaner and enable you to clean in awkward spaces that the cleaner is too big to fit into. Various shapes of nozzles can concentrate the air force for special jobs such as crevices.

An upright vacuum cleaner is good for cleaning carpets because it has a rotating brush that loosens the dirt, so that it is more easily swept up into the machine. Some canister vacuum cleaners have a carpet attachment that includes a beater brush.

SHOWER TEMPERATURE REGULATOR

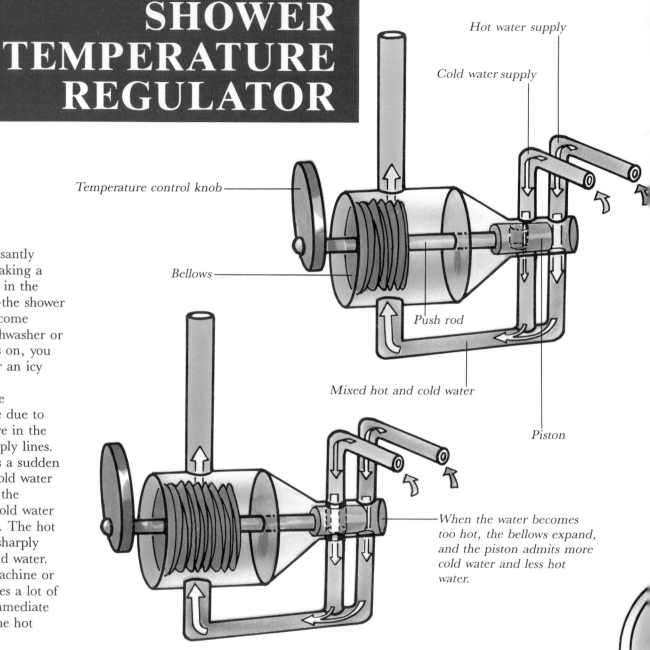

Hot water supply

Cold water supply

Temperature control knob

Bellows

Push rod

Mixed hot and cold water

Piston

When the water becomes too hot, the bellows expand, and the piston admits more cold water and less hot water.

You can be unpleasantly surprised while taking a shower if someone else in the house flushes a toilet—the shower water will suddenly become scalding hot. If the dishwasher or washing machine turns on, you can find yourself under an icy spray.

These uncomfortable temperature swings are due to changing water pressure in the hot and cold water supply lines. Flushing a toilet makes a sudden large demand on the cold water supply, and as a result the pressure drops in the cold water line feeding the shower. The hot water is mixed with a sharply reduced quantity of cold water. Similarly, a washing machine or other appliance that uses a lot of hot water creates an immediate fall-off in pressure in the hot water line.

To solve this problem, two types of water temperature regulators can be installed in the shower. One type directly controls the water temperature; the other senses pressure variations in either the hot or cold water line and adjusts the flow in the other line to compensate.

THERMOSTATIC MIXING VALVE

A regulator that directly controls temperature uses a liquid-filled bellows as a temperature sensing element. The bellows is located directly behind the shower temperature control handle. The mixed hot and cold water flows around the bellows. When the water temperature rises, the liquid in the bellows expands, opening the bellows, concertina-fashion. The bellows pushes on a piston that controls the flow of hot and cold water through the valve. The farther the piston is pushed, the more cold water—and less hot water—is supplied to the shower. The temperature control handle can be set to select the piston's initial position; after that, temperature control is automatic.

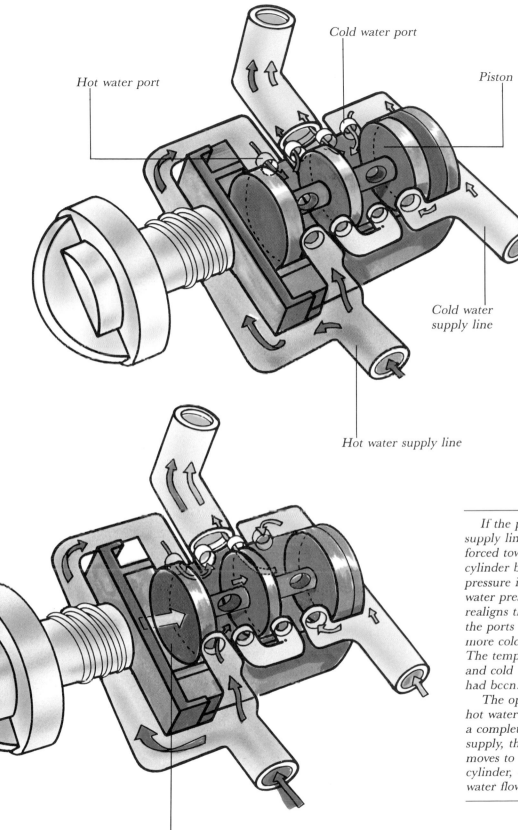

Hot water port

Cold water port

Piston

Cold water
supply line

Hot water supply line

The piston moves toward the cold water end and reduces the
opening at the hot water end, reducing the hot water pressure
to equal the cold water pressure.

PRESSURE-BALANCING REGULATOR

The type of regulator that
compensates for pressure variations
does not use a thermostat. This type
of regulator consists of a piston that
can slide freely back and forth in a
horizontal cylinder. Pressure in the
cold water supply line pushes it in
one direction, while pressure in the
hot water line pushes it in the other.
Both hot and cold water to the
shower pass through holes in the
piston that line up with ports in the
cylinder. When the hot and cold
water pressures are equal, the piston
remains centered in the cylinder.

If the pressure in the cold water
supply line drops, the piston is
forced toward that end of the
cylinder because the hot water
pressure is higher than the cold
water pressure. That movement
realigns the holes in the piston with
the ports in the cylinder, admitting
more cold water and less hot water.
The temperature of the mixed hot
and cold water stays the same as it
had been.

The opposite action occurs if the
hot water pressure drops. In case of
a complete failure in the cold water
supply, the piston immediately
moves to the extreme end of the
cylinder, completely blocking hot
water flow.

HOT TUB

I t's been known for thousands of years that soaking in hot water relaxes the muscles and calms the spirit. Around the world, spas with natural hot springs have been famous gathering spots for both the infirm and the wealthy. The social life centering around the baths was an important adjunct to the supposed therapeutic effects of the baths themselves, which were thought to cure virtually any malady that afflicts mankind.

The medical value of hot baths became more clearly understood as medical science grew more sophisticated. Although the effectiveness of spas in curing most internal disorders is negligible, their value in orthopedic medicine, for sore muscles and joint problems, is better recognized. For injuries to professional athletes, in particular, it became obvious that an artificial spa would be a helpful innovation. The whirlpool bath, combining hot water and a massage from water movement, was the result.

Ski resorts and other vacation centers quickly recognized that the relief obtained from immersion in hot, swirling water could benefit people other than those with athletic injuries. Larger, heated tubs, with pumps to circulate water, were installed for the pleasure of their guests. Now, of course, hot tubs can be found in many homes.

A typical hot tub consists of an acrylic tub contoured with seating areas and set in a redwood base. Water is heated and circulated throughout the tub, providing a hot, swirling, bubbly bath to soak away tensions and aches. Many include air blowers and bubblers to provide a different effect from that of the circulating water jets. The water is completely recirculated by the system.

While the tub is being used, the water temperature is normally 100°F; the tub is usually maintained with the water held at a lower temperature to reduce the rate of heat loss. A switch or an automatic timer turns on the heater to bring the water up to temperature by the time you want to use it. A unit combining heater and circulating pump circulates and heats the water. The water temperature is controlled by an adjustable thermostat, and the air-on/off and water-jet speed controls adjust manually.

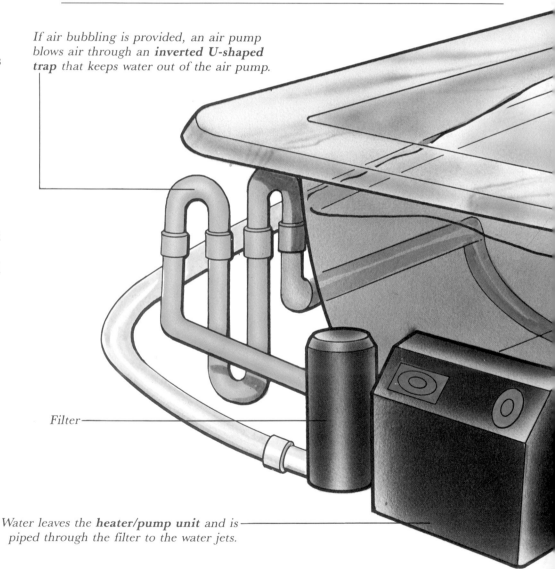

*If air bubbling is provided, an air pump blows air through an **inverted U-shaped trap** that keeps water out of the air pump.*

Filter

*Water leaves the **heater/pump unit** and is piped through the filter to the water jets.*

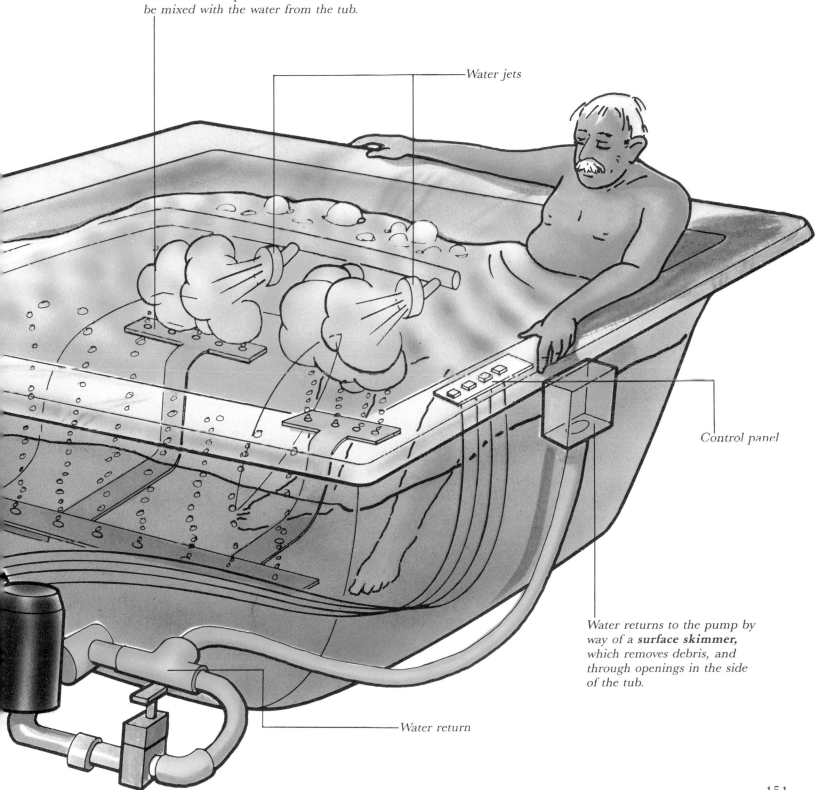

The air goes through a series of channels with holes that bubble the air up from the bottom of the tub. Air can also be mixed with the water from the tub.

Water jets

Control panel

Water returns to the pump by way of a **surface skimmer,** which removes debris, and through openings in the side of the tub.

Water return

LOCKS AND KEYS

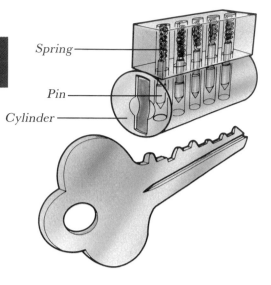

Spring
Pin
Cylinder

Your front door lock presents the same appearance to all keys, yet it opens to just one. The ingenious design for the pin-tumbler lock was developed by Linus Yale in 1848 and is called the Yale lock. Used almost universally on outside doors of buildings, it is probably the most familiar lock and key system in the world. The series of peaks and valleys on the upper edge of the key can be varied in almost unlimited ways, establishing the uniqueness of your own key.

YALE PIN-TUMBLER LOCK

A row of pins connects the cylinder to the lock and prevents the cylinder from being turned. The pins are held in the holes of the cylinder by springs. Each pin is actually in two pieces, and where the break occurs is different for each pin. When the right key is inserted, the serrations raise all the pins to just the right level so that the lower part of each pin is flush with the cylinder. The key can turn the cylinder, thus unlocking the door.

If the wrong key is inserted, the pins will be raised, but not the right distance. One or more of the pins will still prevent the cylinder from turning.

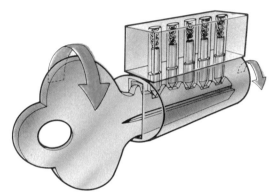

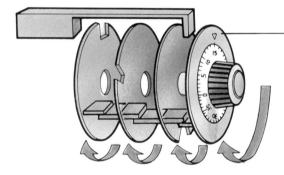

If the combination is 15–25–35, by rotating the dial three times and stopping at 15, the third ring is in position with its notch correctly aligned.

COMBINATION LOCK

The simple combination lock has the advantage of not needing a key to open. The type seen on gym lockers works on the same principle as the large combination locks used on safes.

Three rings turn freely and independently on the same axis. The knob also rotates on that axis. Each ring has a notch cut into its rim. When the notches are aligned, the lock is opened.

The dial has a pin extending from it toward the first ring. As the dial is rotated clockwise, the pin will encounter an arm within the first ring, so that the first ring will also rotate. The first ring has a similar pin extending toward the second ring, which also has an arm. Thus with another revolution of the dial, the first ring starts turning the second ring. A similar hookup causes the second ring to turn the third ring. After three rotations of the dial, it is turning all three rings at once.

You then rotate the dial in the reverse direction, clockwise. The third ring is now free of the pin from the second ring. The first and second ring rotate together for two turns and you stop the dial at 25. This aligns the second ring.

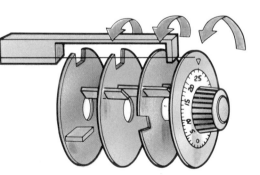

By reversing the dial again, turning counterclockwise, only the first ring moves. By stopping the dial at 35, its slot is correctly aligned, and the lock can open.

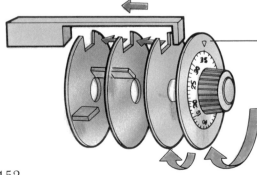

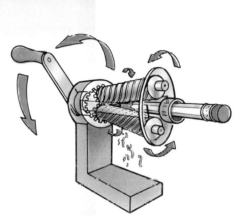

At Work

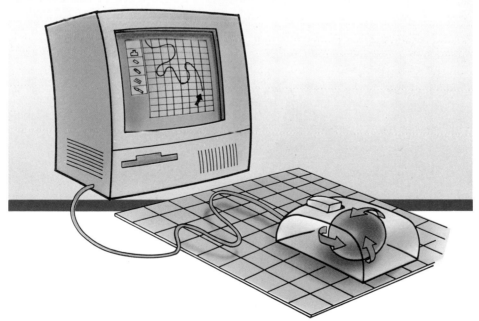

PENCIL SHARPENER

A pencil sharpener has two rollers with raised and sharpened ridges on them that shave thin slivers off the pencil point. The rollers can freely spin from a yoke, which is connected to a set of gears and a crank (see **Gears, chains, and belts,** page 40, and **Cams and cranks,** page 38).

The pencil is inserted at one end through a hole in the yoke, between the two rollers. The rollers are slanted so that they come together at the opposite end, next to the crank. At that end, each roller has a gear affixed to it. The two gears mesh with a larger gear inside the housing of the pencil sharpener. That larger gear is an **internal gear**—its teeth are not on the outside of the disc, but face inward toward the center.

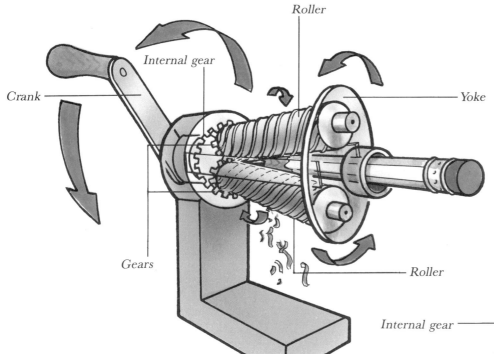

Roller

Internal gear

Crank

Yoke

Gears

Roller

The crank handle turns the yoke, which causes the two rollers to rotate around the pencil. The gears at the opposite ends of the rollers turn inside the internal gear, which is stationary. This set of gears makes the two rollers rotate on their axes. As the rollers are rotating around the pencil, they are also turning against the pencil's surface. The sharp ridges of the rollers shave the pencil to a sharp point.

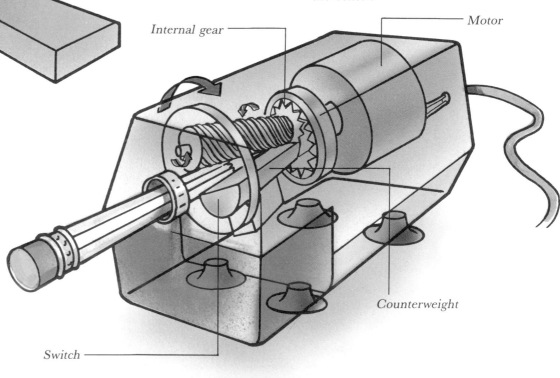

Internal gear

Motor

Counterweight

Switch

ELECTRIC PENCIL SHARPENER

An electric pencil sharpener has only one roller; opposite it is a rotating counterweight with a switch at the end where you insert the pencil. The pencil presses against the switch, turning on an electric motor that is attached to a drive shaft. The motor continues running as long as the pencil holds the switch closed.

CALCULATOR

A hand-held electronic calculator does arithmetic almost instantaneously. Like a computer, it uses microchips to perform automated operations and it has a memory. The memory in a calculator contains the instructions for performing mathematical operations. The simplest calculators usually perform addition, subtraction, multiplication, division, and square roots; some will do trigonometry for engineering calculations. Most hand-held calculators contain a limited memory that can hold a frequently used number for further calculations.

Numbers are entered on a keyboard. Pressing a key closes a set of contacts that sends electrical signals to the **microprocessor,** a set of circuits where the calculations are performed. The numbers are also routed to the display screen.

Calculators are sometimes powered by solar batteries, which can also be charged by being exposed to the artificial light in a room. No matter what the power source, very little of the energy is needed for the actual calculating. Most of the power used by a calculator goes to the numerical display screen, which is usually a liquid crystal display (LCD). Some calculators use LEDs, or **light-emitting diodes** (see page 59), which use more energy. The remaining power goes to the microprocessor.

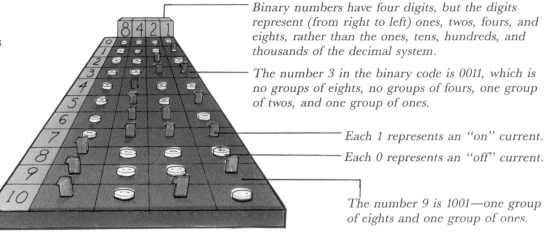

Binary numbers have four digits, but the digits represent (from right to left) ones, twos, fours, and eights, rather than the ones, tens, hundreds, and thousands of the decimal system.

The number 3 in the binary code is 0011, which is no groups of eights, no groups of fours, one group of twos, and one group of ones.

Each 1 represents an "on" current.

Each 0 represents an "off" current.

The number 9 is 1001—one group of eights and one group of ones.

BINARY CODE

Normally we count using the decimal system, in multiples of ten. Since computers and calculators operate with electrical impulses that are either "on" or "off," they convert letters and numbers to a binary code that stands for the presence or absence of an impulse. The binary code uses only two digits, 0 and 1. Each digit in the code is called a **bit** (short for binary digit). Calculators, which deal only with numbers, use four bits for each number. Computers, which process letters and symbols as well as numbers, usually use an eight-bit code.

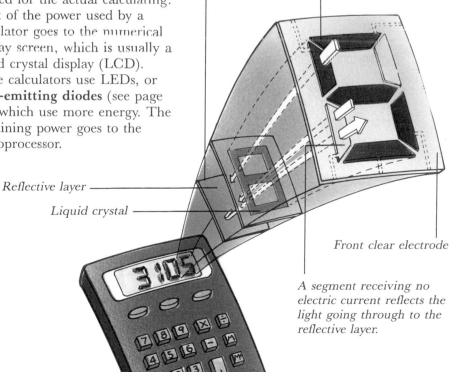

Rear electrode

Reflective layer

Liquid crystal

When an electric current is directed to one of the segments, the molecular structure of the liquid crystal is changed. Light is not reflected, and that segment appears dark.

Front clear electrode

A segment receiving no electric current reflects the light going through to the reflective layer.

LIQUID CRYSTAL DISPLAY

LCDs are sandwiches of transparent electrodes with a layer of liquid crystal between them. Behind the sandwich is a reflective layer. When no current is present, light is reflected through the electrodes and liquid crystal to the reflective layer and back. But when an electric current is passed through the liquid crystal, its molecular structure changes in such a way that light is prevented from reflecting back from the reflective layer. A pattern of seven segments can form all the numbers from 0 to 9, depending on which of the segments receive the electric current. Those segments will reflect no light and appear dark.

155

TYPEWRITER

The first typewriters, which were constructed in the 19th century, were actually slower than handwriting. A practical design was finally achieved in 1867 by an American inventor, Christopher Latham Sholes; with the improvements that were made over the next few years, the basic form of the typewriter remained unchanged for a century. The principle of touch-typing—typing without looking at the keys—enabled the typewriter to produce letters and documents much faster than could be written by hand, and more legible ones, too. Today, of course, the basic typewriter is rapidly being supplanted, both in offices and in homes, by electronic typewriters and word processors (see **Computer Peripherals** page 160), which are even faster and more versatile.

MANUAL TYPEWRITER

Paper is fed around a roller called the **platen.** *The individual letters are on type bars, which are slim rods attached by a series of levers and springs to the keys. Pressing a key causes the type bar for that letter to fling upward and strike an inked ribbon, pressing the ribbon against the paper and leaving behind an inked impression of the letter.*

After the key hits the ribbon and paper, it returns to its home and the carriage holding the platen moves to the left one letter space, so that the next letter struck will fall directly to the right of the first letter. At the end of a line, the typist pulls a lever that rotates the platen to the next line and moves the carriage to the right so that a new line of paper is in position.

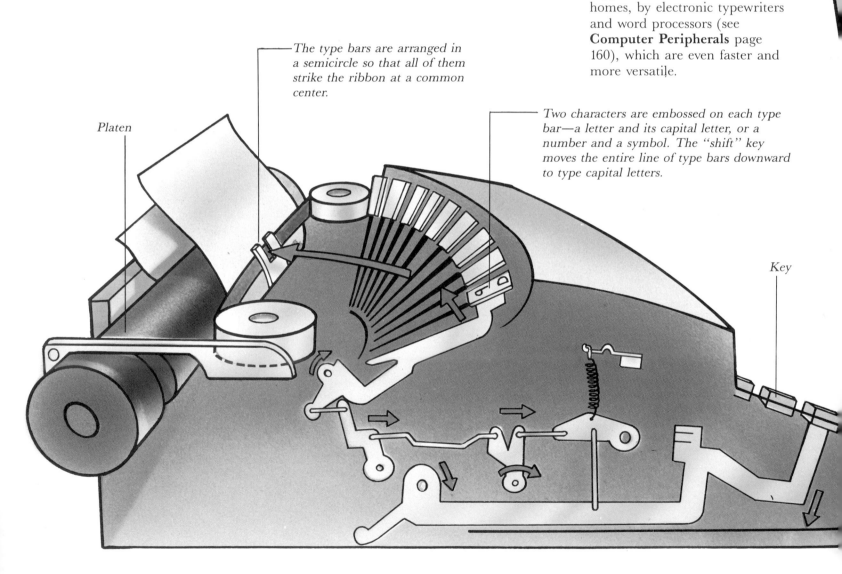

The type bars are arranged in a semicircle so that all of them strike the ribbon at a common center.

Two characters are embossed on each type bar—a letter and its capital letter, or a number and a symbol. The "shift" key moves the entire line of type bars downward to type capital letters.

Platen

Key

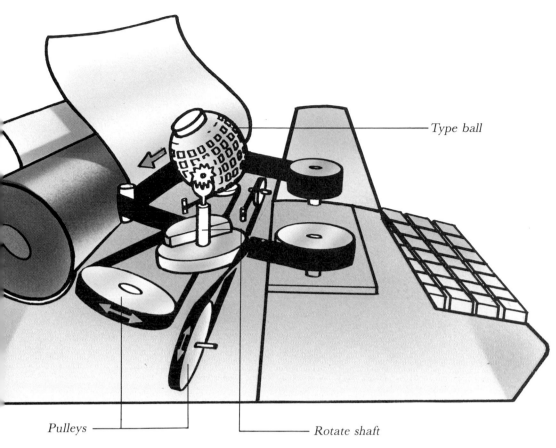

Type ball

Pulleys

Rotate shaft

ELECTRIC TYPEWRITER

Typing on an electric typewriter is less tiring, because the typist presses the key only a short distance and an electric motor activates the lever system that raises the type bar. In the 1960s, a typewriter design was introduced that had the characters embossed on a type ball rather than on bars. The ball tilts and turns as the keys are pressed, presenting the desired character to the ribbon. It travels along the length of the platen; the carriage does not move. Each key has a selector lever under it with projections, which are arranged differently for each key. The combination of projections activates a complicated system that tilts and rotates the ball so that the right letter is in striking position. One advantage to this design is that the keys cannot jam if the typist goes too fast—theoretically, a person could type 930 strokes per minute. Another advantage is that by simply removing the type ball and replacing it with another one, you can change the style of the type.

ELECTRONIC TYPEWRITER

Electronic typewriters look like their older cousins in that they have a keyboard and a platen. But the mechanical parts have been replaced with microchips, which makes an electronic typewriter not only more powerful, but quieter.

An electronic typewriter shares many of the features of a computer. Pressing a letter key "codes" the letter in the typewriter's computer chips. A memory in the typewriter stores enough information for at least a page of text. Many electronic typewriters have a built-in screen that shows a line or two of what is being typed. Mistakes can be corrected before the letter is printed.

*The printer is often a **daisy-wheel** printer that has the characters embossed on the ends of lightweight plastic spokes radiating out from a hub. The wheel rotates as the keys are struck, stopping at the letters that have been encoded. A hammer moves to strike the spoke against the ribbon and paper.*

PHOTOCOPIER

1. The document to be copied is placed face down on a glass surface.

2. A bank of lights illuminates the image on the paper.

Within seconds, a photocopier can deliver a copy of a document that is often indistinguishable from the original. This seems miraculous to those who remember the drudgery of using carbon paper, but the principle behind a photocopier is quite simple.

Certain substances called **photoconductors** conduct electricity when exposed to light. A large drum is coated with a photoconductor and then treated so that it has an electric charge. The document to be copied is illuminated and the light from it is reflected onto the drum. The dark areas retain the electric charge, but the photoconductor allows the charge to leak away in the light areas. The result is a picture of the original on the drum, not in ink, but in invisible static electricity.

Black toner powder that has been given an opposite electric charge is passed over the drum. The toner clings to the electrically charged area of the drum, corresponding to the image. The drum next comes in contact with the paper, which is also given an electric charge. The toner is attracted to the paper. The paper is heated to retain the image, and the copy is made.

*5. The drum rotates against a **corona,** which covers the entire surface with a positive electrical charge.*

10. After the image has been transferred, the drum rotates to a cleaning area. Untransferred toner and the remaining electric charge is taken off, preparing the cylinder for the next copy.

9. The thermoplastic resin in the toner melts when heated. The paper passes between heated rollers, fusing the image and making it permanent.

8. To transfer the toner from the drum to the paper, the paper is given a higher electrical charge than the image on the drum. Another corona between the paper and the drum delivers this charge, pulling the toner particles onto the paper.

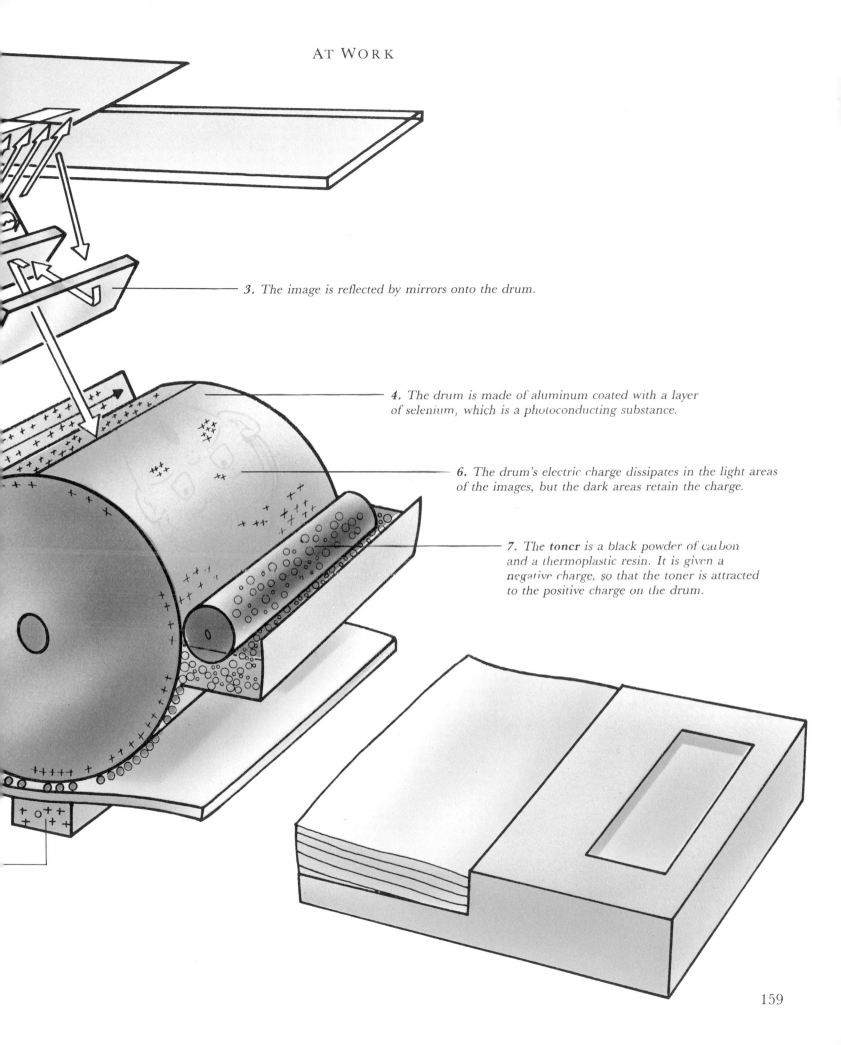

3. *The image is reflected by mirrors onto the drum.*

4. *The drum is made of aluminum coated with a layer of selenium, which is a photoconducting substance.*

6. *The drum's electric charge dissipates in the light areas of the images, but the dark areas retain the charge.*

7. *The* **toner** *is a black powder of carbon and a thermoplastic resin. It is given a negative charge, so that the toner is attracted to the positive charge on the drum.*

COMPUTER SYSTEM

A computer is a device that performs logical operations; therefore an abacus, a slide rule, and a calculator are computers. Modern electronic computers can operate without direct human involvement, although a human must tell the computer what to do.

Large, expensive **mainframes** and smaller **minicomputers** can manage vast amounts of data for businesses or government. A **personal computer (PC)** fits on a desktop and costs from $100 to more than $10,000.

A computer system consists of hardware, software, and a human operator. **Hardware** refers to items that you can touch—the computer, printer, video monitor, and so on. **Software (program)** refers to instructions written in a computer language that tell the hardware how to perform its tasks. The human operator determines which tasks should be performed.

All computers, large and small, have three basic types of hardware—an **input unit,** such as a keyboard; a **central processing unit;** and an **output unit,** such as a video monitor or printer.

KEYBOARD

*The PC receives information chiefly from a keyboard, which resembles a typewriter keyboard except that it has additional, specialized keys. Each key sends a unique electrical signal that enters the CPU as a series of on/off electrical impulses (see **Binary code,** page 155). The CPU then takes the appropriate action, which may be to display a letter on the video monitor or perform a specific task.*

Without specific instructions, the computer can do nothing. Software programs give the computer its instructions. For personal computers, they are usually on a disk or cartridge that are put in use when needed.

The most familiar software is an application program, which the computer uses to accomplish specific tasks. The three most popular applications are word processing, database management, and spreadsheets.

COMPUTER SOFTWARE

__Word processing software__ allows you to write, revise, format, and print text for letters, reports, and other documents. __Database management software__ lets you easily store, update, organize, locate, and report on specific information concerning such subjects as inventories and prices of products. __Spreadsheet software__ is used to test hypothetical situations that involve numbers—for example, "How does a price increase affect profits?" Numbers and formulas to relate the numbers to each other are entered in a grid of rows and columns. When you change any information, the spreadsheet recalculates any related data.

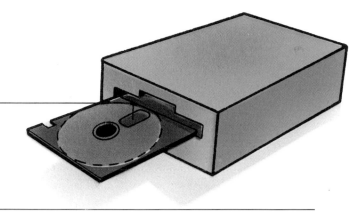

The chief attribute that distinguishes a computer from other data processors is its memory. One type of memory resides in a microchip, and the other is recorded on a tape or disk.

Read-only memory (ROM) is in a microchip in the CPU and normally cannot be erased or changed. It may contain instructions allowing the computer to "understand" or manipulate information entered into the computer. While the computer is in use, part of a program and the associated data that the CPU is using is temporarily stored in the **random-access memory (RAM)** chips. Unlike ROM, RAM constantly changes and is usually erased when the computer is turned off. To save the data contained in the RAM, you can record it on a tape or disk.

Open slits in the cover allow the head of the disk drive to contact the surface of the floppy disk.

CPU

*Most of the computer's "thinking" occurs in the **central processing unit (CPU)**, which is also called the **microprocessor**. The CPU is a microchip (see page 30) that may have over one million microscopic electronic parts. In addition to sorting out all the information (data) it receives and directing the flow of data from one part of the system to another, the CPU performs mathematical and logical operations.*

After the data is received from the input device, the CPU performs any required calculations on the data and sends it to an output device.

*The CPU is in the main circuit board (motherboard) in the PC's main housing—the **system unit.** The motherboard also holds the RAM and ROM and other parts. Normally, the system unit also houses a power supply, a cooling fan, one or more disk drives, and **ports** or **interfaces** that let you attach devices such as a printer to the computer.*

DISKS

*A **floppy disk** or **diskette** looks like a small, flexible phonograph record that is made of plastic that is coated with magnetic particles. This circular disk is encased in a square cover.*

A new floppy disk has to be formatted, which means that the computer encodes data onto the disk that tells it how to arrange the stored data. Because the information on the disk is stored in magnetic patterns, it can be damaged by exposure to static electricity or a magnetic field.

*A **hard disk,** which is sealed to protect it from dust, can store much more information than a floppy disk; hard disks that hold 20 to 100 megabytes of data are common, whereas floppy disks normally store between 360 kilobytes and 1.44 megabytes. A **byte** is a group of bits (see **Binary code,** page 155) that make up a character. A kilobyte is made up of roughly one thousand bytes, and approximately one million bytes make up a megabyte.*

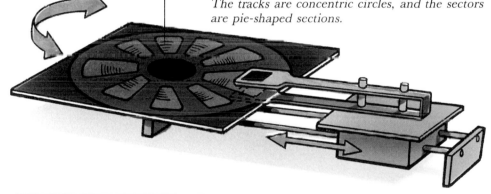

Floppy disks are divided into tracks and sectors. The tracks are concentric circles, and the sectors are pie-shaped sections.

DISK DRIVES

In order to be accessed, a disk must reside inside a disk drive, which usually lies inside the system unit and works like a phonograph. When the CPU needs to access a disk, the drive rapidly spins the disk while the drive's magnetic read/write head moves to the part of the disk that contains the desired data or program. When it writes data on the disk, the read/write head converts electrical signals from the computer into magnetic patterns, which are "written" onto the disk by arranging the magnetic particles on the surface of the disk in a specific pattern. The same head "reads" the information on the disk into the computer by determining the arrangement of the magnetic particles on the disk's surface and creating corresponding electrical signals. Note that a disk drive is both an input and output device.

A hard disk drive spins a stack of hard disks and uses multiple heads to provide greater amounts of storage and faster speed than a floppy disk drive.

COMPUTER PERIPHERALS

MOUSE

*A **mouse** is an input device that is named for its appearance. It is a fat, oblong gadget whose "tail" is a wire that attaches to the computer. Most mice have a ball mounted on the bottom. As you move the mouse across your desktop or on a special pad, it moves an arrow around your computer screen in the same direction that the mouse moves.*

*The mouse can free you from typing complex commands in order to interact with the computer. A "user-friendly" computer uses an **operating shell** that simplifies the use of the computer. Some operating shells use on-screen graphics or **icons,** which are tiny images that represent computing concepts. For instance, an icon of a typewriter or a pen writing on a piece of paper might represent a word processing program. Move the mouse until the arrow points at the desired icon, press a button on top of the mouse, and a series of preprogrammed signals will start up the word processing software. With a simple motion of your hand and a click of a button, you have input several commands.*

A **peripheral** is any device that attaches to a computer for the purpose of outputting or inputting data or programs. One example is a **monitor** or **video display terminal (VDT),** which receives data from the CPU and displays it on a screen. Desktop computers usually employ a televisionlike display called a monitor, whereas most portable computers utilize a liquid crystal display (LCD; see page 155). Other popular peripherals include a **modem,** a **printer,** and a **disk drive**.

PRINTERS

*Printers use various methods to form text characters and/or graphic images on paper. The most common type is an **impact dot-matrix printer,** which uses metal pins to strike a ribbon against the paper. This creates dots that are arranged in a matrix (grid) to form the letter or image.*

*The operation of a **laser printer** is similar to that of a photocopier (see page 158), except that the image is formed by a laser beam. The laser printer offers faster, quieter, and better-looking printing than most other types of printers, but it is a great deal more expensive.*

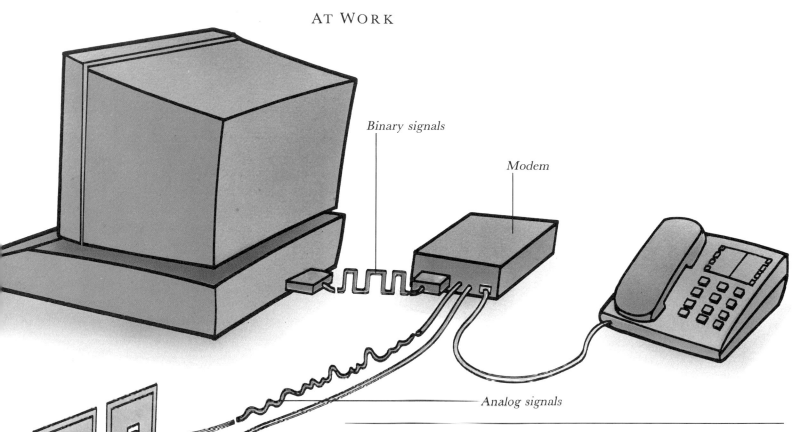

Binary signals

Modem

Analog signals

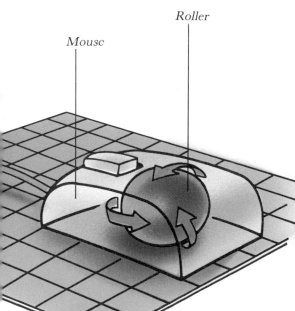

Icon

Roller

Mouse

MODEM

One reason that a PC is much more powerful than a plain word processing machine is that the PC can communicate with other computers. This ability allows your system to call up a solitary PC or an electronic information service that uses a mainframe computer, which allows you to access vast amounts of knowledge; send and receive electronic mail; obtain a current stock quote; check airline schedules and make reservations; shop in electronic "malls"; and even send and receive software.

The only problem is that PCs use **binary code** (see page 155), which produces signals that are either "on" or "off," and telephone lines employ analog signals, which vary in strength. To overcome this difference, a **modem** is needed.

Modem is an acronym for MOdulator-DEModulator. A modem modulates (alters) the sending computer's digital signals to an analog state so the signals can travel over a telephone line. When the signals are received, the remote

computer's modem demodulates the analog signal back to a digital form so the computer can understand the message.

For PCs, a modem can be either internal or external. An internal modem fits directly into an expansion slot in the system unit. The acoustic coupler is an external modem that uses two rubber cups into which a telephone handset's mouthpiece and earpiece are inserted. Because room noise interferes with the transmission, acoustic couplers are now rarely used. Newer external modems (and all internal modems) connect directly to the telephone line, bypassing the handset.

The speed at which data is sent is often called the **baud rate;** bits per second (bps) is more accurate. Typical speeds are 300, 1,200, and 2,400 bps, which are roughly equal to 30, 120, and 240 characters per second. A fast modem can save money if you connect to an information service that charges for each minute it is being accessed.

ROBOT

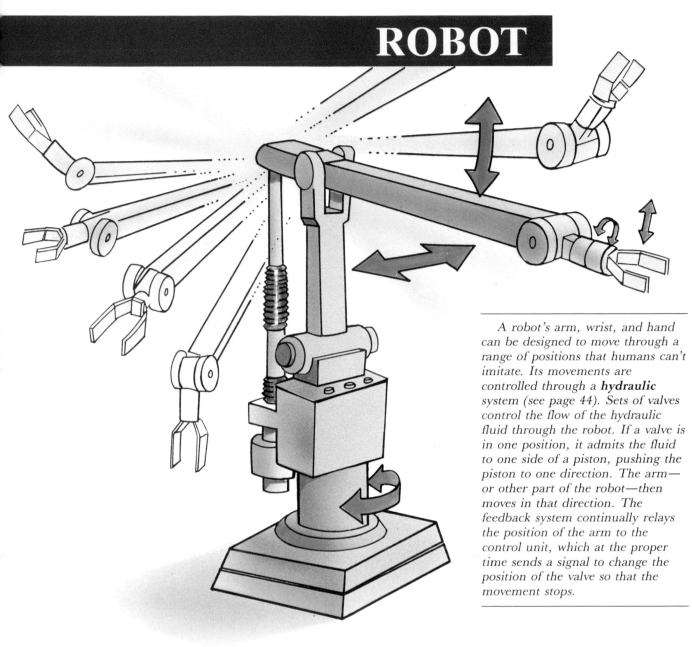

A robot's arm, wrist, and hand can be designed to move through a range of positions that humans can't imitate. Its movements are controlled through a **hydraulic** system (see page 44). Sets of valves control the flow of the hydraulic fluid through the robot. If a valve is in one position, it admits the fluid to one side of a piston, pushing the piston to one direction. The arm—or other part of the robot—then moves in that direction. The feedback system continually relays the position of the arm to the control unit, which at the proper time sends a signal to change the position of the valve so that the movement stops.

As workers, humans have their limitations—they need time off for food, sleep, and recreation; they complain about working in excessive noise or heat; and they are easily bored by routine, repetitive tasks. Robots, which have none of these disadvantages, have become stalwart workers in many industrial situations. Most of the major automobile manufacturers use robots to spot-weld, to spray-paint, and for other tasks.

An industrial robot is a device that is programmed to go through various motions in order to perform a task. The typical design consists of one arm stationed on a platform. Its proper sequence of movements has been programmed into the memory of a computer, which controls the sequencing of its movements. The robot has a feedback system—when the robot's arm has moved to the correct location as specified in the memory, that information is relayed to the control unit. The

control unit then sends signals to go on to the next step, such as tightening a fastener.

Some robots are programmed by leading them through the motions required. The movements are then fixed in the robot's memory as a series of very small individual actions. Jobs that need great accuracy of speed and placement often use this type of robot.

BAR CODE SCANNER

Those rectangular patterns of black and white stripes on the boxes and cans you buy at the supermarket are the Universal Product Codes (UPC), in which the dark bars and white spaces stand for individual numbers. Each product has its own code number.

When the checkout clerk passes the code across the glass in the counter, a laser beam (see page 55) reflects the light and dark areas to a photodetector. The intensity of the reflected light varies with the width of the bars and whether they are black or white. These varying signals are converted to a varying electrical current that represents the item's UPC number.

The store's computer system translates the electrical current into the correct number and finds the description and price of the item, which are rung up on the cash register. The customer gets a receipt showing a complete list of all the purchases. Not only is the bar code scanner fast and accurate, but also the computer can be programmed to keep track of the store's inventory and tell what products are selling well.

In the UPC symbol, each number is represented by two dark bars and two light spaces that vary in width. The first five numbers are the code number for the manufacturer of the product, and in a 10-number bar code the second five numbers stand for the individual product. In addition, the UPC symbol includes control numbers that ensure that the code is scanned accurately.

The busy supermarket clerk might not take the time to place the UPC symbol flat against the glass so that the laser can scan the code at the best possible angle. A holographic scanner prevents errors in scanning that might result. It uses a rotating disc of holograms (see page 72). The holograms show three-dimensional images of bar codes; their purpose is to focus the laser beam. The mirrors direct the beam through the glass plate onto the symbol. The reflected beam varies in intensity according to the light and dark areas in the code.

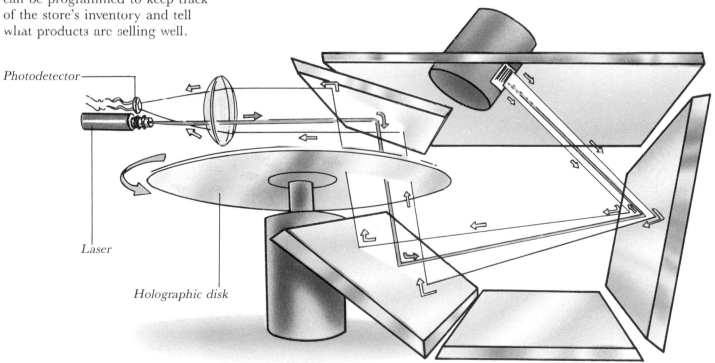

Photodetector

Laser

Holographic disk

VENDING MACHINE

I n the first century A.D., an inventive genius known as Hero of Alexandria described the first vending machine, a coin-operated automatic temple holy-water dispenser. It used many of the same principles as today's coin-in-the-slot marketers. In Hero's invention, a coin that was dropped through a slot near the top of a water urn fell on a rocker arm that pivoted like a teeter-totter. The weight of the coin pushed the rocker down at one end and up at the other. That lifted a plunger, which let water out of a spigot. When the coin slid off the rocker, the plunger fell back and stopped the flow of water.

Today, vending machines still depend upon levers (see page 34) to activate a vending mechanism. Before those levers operate, however, the coins are tested for authenticity.

ELECTRONIC COIN TESTER

Mechanical coin testers, which reject bogus coins—slugs—on the basis of weight, are being replaced with solid-state electronic devices with few moving parts. An electronic coin tester consists of a sloped track with a series of sensors along it to test the coin in certain ways.

Surviving coins then roll down the track through the poles of a strong magnet. The magnet creates currents in conductive metals, which slow down the rolling coin. The alloy the coin is made of determines how much its rolling speed is slowed down.

First an electric current passes through the coin to determine its metal content and size.

Wrong-size or nonmetallic coins activate a reject latch and leave the track.

An array of optical sensors note the coin's diameter, measure its rolling speed, and compare that information with information on good coins stored on computer microchips.

If the coin purports to be a quarter, but its speed and diameter don't match those of a quarter, a chute opens and rejects the coin. Genuine coins in the proper amounts trigger electronically controlled vending release mechanisms and the change-making device.

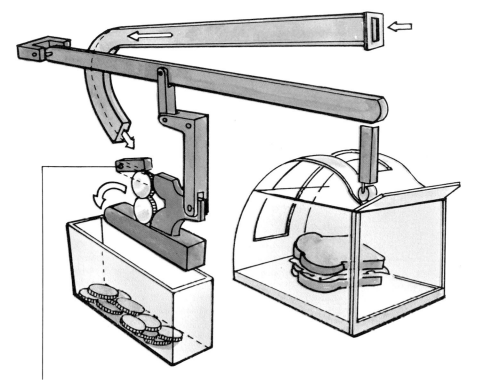

VENDING MACHINE

All vending machines, regardless of design, perform the same basic functions. When coins are inserted in a coin slot, they travel through a coin tester. If the coins are accepted, they are directed to a coin hopper. If the proper number of coins have been inserted, the stack of coins raises a stop pin. This permits a toggle to push the coins back toward the coin drop. A push-rod over the compartment door is connected by a series of links to that toggle. Opening the door pushes up the push-rod, which moves the toggle, which pushes the coins into the coin drop. If no coins have been inserted, or not enough coins, the stop pin is still in place, preventing the toggle from moving. If the toggle cannot move, the push-rod remains down, and the door will not open. Opening any door automatically locks all the others. Once the coins have dropped into the collecting box, the entire mechanism is reset.

The stop pin rests on top of the coins, allowing the toggle to move if the door is opened.

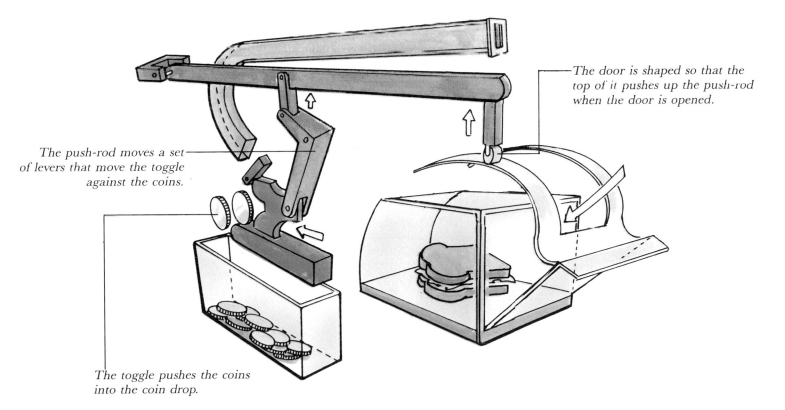

The door is shaped so that the top of it pushes up the push-rod when the door is opened.

The push-rod moves a set of levers that move the toggle against the coins.

The toggle pushes the coins into the coin drop.

ELECTRONIC EAVESDROPPING

Elaborate devices for eavesdropping on other people's conversations can be a help to law enforcement agencies in detecting crime, but they can also be used in industrial espionage, to the dismay of companies trying to protect secret formulas or processes. Using microphones to listen in on oral conversations is called "bugging"; intercepting telephone conversations is called "wiretapping."

Since World War II, great numbers of new surveillance devices have been created. Microminiaturized microphones can now be made small enough to fit into buttons, watches, or jewelry. Wiretapping a telephone can be done very simply with a "bug box," available in most electronics stores, which relays part of the telephone wiring to the phone company's connecting box and part to a hidden tape recorder. Standard video cameras can be outfitted with infrared lenses to detect the presence of people in the dark of night. Such cameras detect heat, instead of light. Small television cameras have been developed that can function in a room that is nearly dark.

Wiretapping without a judicial warrant is a violation of Federal law punishable by prison sentences and fines, and many states have laws prohibiting recording of conversations without the knowledge of the other person.

Conversations can be overheard without using microphones. Ultrasound or radar beams can be directed to the window of the room being monitored, and a portion of the beams will be reflected back. Any speech in the room causes the window to vibrate, and those vibrations cause changes in the reflected beams. A special receiver can read the vibrations in the reflected beams and change them to electrical signals that represent the sound of the conversation.

RADIO BUG

Transistors and microchips make possible a combination microphone and radio transmitter so small that it can be built into an ordinary item such as a pen or calculator. This radio bug does not need to be wired into the listening device; the conversation being monitored is sent by radio waves to a receiver (see page 52). The smallest radio bugs, however, are not very powerful, so the receiver needs to be quite close to pick up the signals.

Radio bugs need a source of electric power. Sometimes batteries are used, although very small batteries need to replaced after a few hours. If battery replacement is not feasible, hidden wiring from the building's electrical supply can be used. Placing the radio bug in a telephone is very efficient. The bug not only listens to the phone conversations, but uses power from the telephone circuit to pick up conversations in the room.

AUTOMATIC TELLER MACHINE

Anyone who has waited half an hour in line at the bank can appreciate the advantages of automatic teller machines. They're fast and convenient, and they are usually open even if the bank is closed.

Automatic teller machines, or ATMs, have a lot in common with personal computers (see page 160). They both base their operations on a microprocessor, they have similar amounts of computing power, and they both use keyboards, disk drives, printers, and modems.

To use the ATM, you need a magnetic card that is much like a credit card. The ATM's card reader reads the card's magnetic strip, which contains your account number and personal identification number (PIN). After you enter your PIN on the keypad, the ATM's computer compares the number you've entered with the number on the card. If the numbers match, the ATM begins your transaction by displaying instructions on the screen.

When you make a withdrawal, the computer instructs the money dispenser to pull a certain number of bills off the stacks of money stored in the machine. Most ATMs stock only one or two denominations—usually tens and twenties—to make counting simpler. A set of suction cups comes down onto the stack. Then a vacuum activates, lifting only the top bill off the stack. The mechanism carries the bill over to a separate stack, and when that

stack reaches the total to be dispensed, friction wheels push the money out through the dispenser slot.

For a deposit, the machine uses another set of friction wheels to pull your envelope in through the deposit slot, and then dumps the envelope into a locked metal bin. As it enters the bin, the envelope triggers a mechanical switch or an infrared detector, which tells the ATM that you've made your deposit. Some depositories have a printer for time and date stamping or numbering the envelopes.

Often you can use an ATM away from your own bank, because the machine can call the mainframe computer in your bank that controls all ATM transactions involving accounts with your bank. Through a modem (see page 163) and a dedicated phone line, the mainframe tells the ATM how much money it can give you.

At the end of the day, bank employees open the machine to replenish the money supply and the paper for the receipt printer. Most ATMs allow the bank employees to reprogram the machine, using a floppy disk drive.

Depository

The ATM identifies legitimate bank customers with a magnetic card reader. Using friction rollers, the reader pulls the card in through the slot, passing the card's magnetic strip over a head similar to the playback head on a tape recorder (see page 78). The information—your account number and your PIN—travels from the head to the ATM's computer.

If the number you enter on the keypad fails to match the number provided by the card reader, the computer gives you another try. If you enter the wrong number too many times, the card reader's rollers eject the card into a bin, where it stays until a bank employee can remove it.

When your transaction is finished, a printer like the one on a cash register types out a receipt.

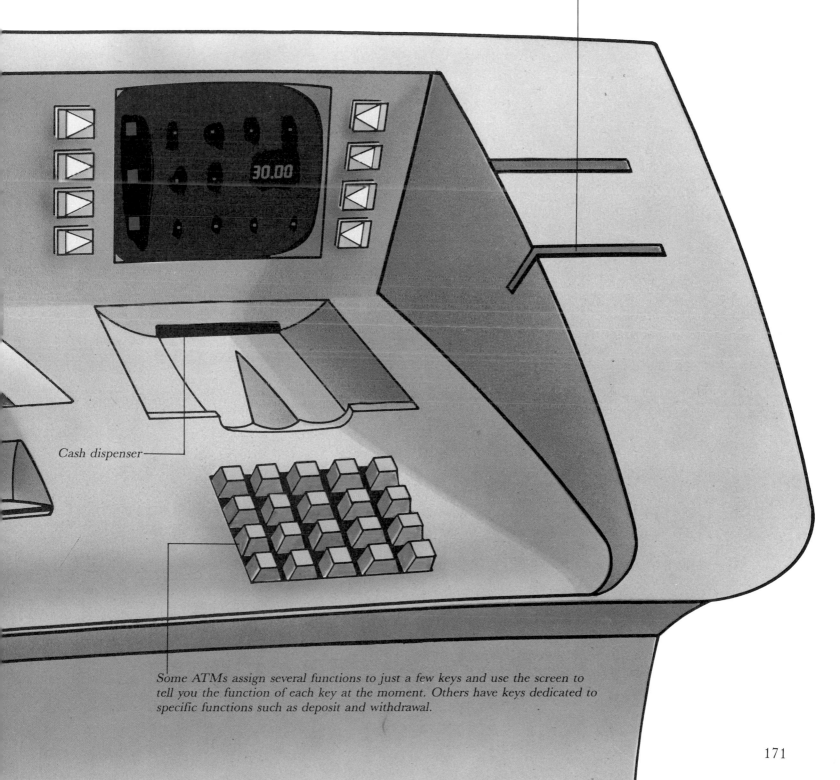

Cash dispenser

Some ATMs assign several functions to just a few keys and use the screen to tell you the function of each key at the moment. Others have keys dedicated to specific functions such as deposit and withdrawal.

ELECTRONIC CASH REGISTER

An electronic cash register can be thought of as an electronic calculator with a few extra components. These parts usually include a receipt printer; a journal printer, which produces a printed record for the store's bookkeeper or manager; and a cash drawer.

The calculator section works much like any other electronic calculator (see page 155), using a microprocessor to arrive at sales totals. Many modern cash registers have perhaps 50 separate keys, each one programmed with the price of a certain item, so that the cashier seldom has to enter numbers. Other registers use a standard numeric keypad; some of these use codes for each product, so that the cashier simply enters the code and the machine provides the price. Price information is stored in a memory and can be changed if actual prices change. Many registers connect to a bar code scanner (see page 165).

The most advanced registers can interface with a computer, either through a cable or a modem (see page 163) and phone line. The computer then stores the information and uses it to calculate inventory and sales statistics.

Once the prices are entered, the cashier pushes the "total" key, which tells the microprocessor to add the prices, calculate the sales tax, and combine the price total with the tax. The cashier enters the amount the customer has given and presses the "amount tendered" key. The microprocessor subtracts the total from the amount the customer has given, tells how much change to return to the customer, and then triggers an electronic release on the cash drawer.

At the end of the day, the manager uses a key switch to change the machine's mode. One mode totals the day's receipts. The microprocessor adds up all the transactions, calculating the total cash taken in during the day. Then the manager usually turns the key to a second mode that "zeros out" the register, resetting all totals to zero to prepare for the next day. On most registers, a journal printer, which usually prints the same information as the receipt printer, provides the manager with the details on each transaction.

Locomotion

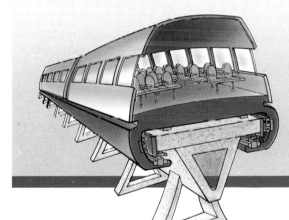

BOBSLED

Modern, slickly designed bobsleds are the late-20th-century version of one of the oldest vehicles known to man. The operation of a bobsled is still as simple as that of a soap-box derby car. The two front steering runners are connected and pivot on a vertical axle. The driver steers by pulling on ropes attached to the front runners. The brakeman, in the rear, controls the speed of the sled with a simple lever brake. The riders start their run by grasping handles on the sled and pushing the sled forward, jumping in after the sled is moving fast enough. Bobsledding competition features separate events for two- and four-person bobsleds; on the four-person sled, the two riders in the middle serve only as extra muscle at the start and extra weight during the run.

A bobsled runs in a U-shaped track. One or more runners can lose contact with the ice in the turns, reducing the driver's control over the bobsled. On newer sleds, an articulation ring joins the front half with the back half, allowing the two sections to turn independently. The runners retain full contact with ice.

Newer modifications reduce weight in the sled. Bobsled designers recently abandoned

Beneath the ice, a layer of concrete covers several one-inch refrigeration pipes, which carry pressurized ammonia gas. This refrigeration system can keep ice on the track at outside air temperatures up to 70°F. Steel wires reinforce the concrete.

Between the pipes and the track's fiber-reinforced plastic shell, a layer of urethane insulation helps keep the track cool.

fiberglass and steel construction in favor of a molded shell made from a lightweight sandwich of carbon fiber and Kevlar (the same material used in bulletproof vests). Substituting titanium for steel in the frame cuts weight even further. Bobsled teams always run their sleds at the maximum weight allowed by the rules (827 pounds for two-person sleds, 1,389 pounds for four-person sleds), but by cutting weight in the shell and the frame, they can add weight wherever they want it in order to enhance control and speed.

The driver's feet rest on the bar connecting the runners, steadying them and giving the driver firmer control over the sled.

The rear rider controls the brake, slowing the sled when going into turns and at the end of the run.

To engage the brake, the rear rider pulls on a lever that forces a toothed metal piece into the ice.

To turn right, the driver pulls on the right steering rope, and the tension travels through the rope and a series of pulleys. This tension pulls the left-hand runner to the right. Because the front runners are connected and pivot on a vertical axle, the tension also pulls the right runner. The bobsled turns right.

LUGE

A luge is nothing more than a sled designed to carry the luge rider down an icy bobsled run at 70 miles per hour. The rider lies on a platform on top of two wooden runners with steel strips called steels along their bottoms. Each runner curves upward at the front of the sled. To steer the sled, the luger presses his or her feet against these curved sections, bending the runners.

Luges are actually simpler than an ordinary sled. The International Olympic Committee feels the sport won't be as challenging if participants can use computer-designed steering systems and space-age materials in their luges.

Races are often won by only hundredths of a second. To get to the top of their sport, lugers must become experts at filing their steels. Rounded steels let the sled run faster, but sharper steels give the luger more control. Lugers must file their steels for the perfect balance, for their size and weight, between control and speed. They finish their steels by polishing them with diamond paste.

To start the run, the luger is seated on the sled at the top of the run. The luger plants spiked gloves firmly in the ice, then pushes the sled back and forth rapidly to build up momentum. The sled hurls down the run; the luger lies back and rips through a dangerous array of tightly banked turns and smooth straightaways. At the end of the run, the same spiked gloves used to get that fast start dig in and brake the luge.

The biggest enemy lugers face isn't the sled or the ice, it's the wind. Wind resistance is one area in which the sport's governing bodies permit lugers to use modern technology to their advantage. Lugers wear skin-tight plastic suits to turn their bodies into smooth projectiles that cut cleanly through the crisp winter air. Gloves and boots are made of the same slick material.

Despite the frighteningly high speeds that lugers attain, they almost never look where they're going. By keeping their heads down, lugers cut their drag even further. Lugers memorize the course and picture it in their minds as they ride, digging in with their steels when they know a turn's coming up. To fine-tune their riding positions, Olympic-class lugers have even climbed into wind tunnels with their sleds.

Skin-tight plastic suit

Spiked gloves

Steels

Lugers must rely mainly on their own skills rather than the advantage of high technology. East German lugers once tried cone-shaped helmets, which calmed the swirl of air behind the moving sled. The helmets worked perfectly, but the Olympic Committee banned them soon afterward.

SNOWMOBILE

With steerable skis in front and a motor-driven track taking up the rear, a snowmobile works like a combination of a Sherman tank and a sled. The track's cleats give the machine a tanklike grip on the snow, and the skis let it glide in the direction you choose.

Driving a snowmobile—snowmobilers call them sleds—is much like driving a car that has automatic transmission. You turn the throttle to go faster, steer with the handlebars, and stop with a hand brake.

The two front skis hang from leaf springs or shock absorbers, which smooth out the ride. Each ski pivots on its own axle, but they work together because they're mechanically tied in with the handlebar. By turning the handlebar to one side, you turn the tips of both skis to that side.

SNOWMOBILE DRIVE TRAIN

The power from the sled's two-cycle engine goes through a system of pulleys and sprockets to drive the track. The snowmobile sports its own version of automatic transmission, called a **torque converter.** The two pulleys automatically change the drive ratio to meet the running conditions of the snowmobile.

The driven pulley is adjustable. An increased workload on the track, such as when the sled is just starting to move, puts more stress on the belt. This stress pulls the two sides of the pulley apart. The pulley's diameter is decreased, making each crankshaft revolution do less work.

The drive pulley turns the driven pulley through a toothed belt. As the engine speeds up, the two sides of the pulley are forced together, and the belt rides higher on the pulley. The pulley's diameter is increased, and the snowmobile is driving in a higher gear.

The chain-driven sprocket is on the same shaft with the sprocket that drives the track.

Ski

Handlebar

A sprocket on an adjustable shaft to the rear of the track keeps the track tight and running straight.

The track-drive sprocket meshes with a series of holes in the track, so that the track works like a chain.

Track

Bogie wheels between the track sprockets support the sled's weight.

SKATEBOARD

The nut can be adjusted to regulate the ease of turning.

The wheels, which are made of polyurethane, spin on sealed roller bearings, one on each side of the wheel.

The **trucks** are usually made of aluminum.

Most boards, or **decks,** are made of laminated maple. The tail slopes upward for kick turns.

Rails along the underside of the deck give the skater a convenient handhold.

Skaters cover the tops of the decks with nonskid tape and the bottoms with stickers or brightly painted designs.

Despite the complex maneuvers performed by the most daring skateboarders, there is nothing mysterious about their equipment. A skateboard is a board with two axles mounted to it, with wheels attached at both ends of each axle.

The **trucks** are the assemblies that attach the axles to the board and allow the rider to steer. When the skateboarder presses on one side of the board with his or her feet, the T-shaped part of the truck, which holds the axles, turns. Since the trucks are mounted opposite each other,

they pivot in opposite directions, and the skateboard turns.

Since today's top skaters perform wild spins and aerial maneuvers on wooden ramps, they've substituted plastic for the older rubber dampers on the trucks, so that the trucks barely move. Instead of pushing down on the side of the board to turn, skaters now push down with their back feet, raising the front of the board off the ground so that they can spin the board around on its back wheels.

A heavy pin supports one end of the T-shaped piece of the truck. The main bolt, which supports the other end, is isolated from the T-shaped piece by rubber dampers. The truck flexes on these points.

If the truck were straight, it would bend only toward the board when the skater applies pressure. But the bolt end of the truck hinges at a point farther from the board than the pin end. For this reason, the side of the truck under pressure bends not only toward the board, but also toward the bolt end of the truck. The wheels under pressure move toward each other, and the wheels on the other side move away from each other. The board will turn until the skater lets up on the pressure.

COG RAILWAY

The smooth rails and wheels of a train do not work well on hills. The wheels slip on the tracks, and the weight of the cars behind the engine makes it slip backward. A railroad that must go up a mountain—rather than around it or through it in a tunnel—needs a different kind of design.

To solve this problem, European engineers developed the cog railway, which uses a rack-and-pinion system to give the train a better grip on the track. A cog railway runs on standard rails, but between them is a third rail that has teeth in it (the rack). A cog wheel—the pinion—is fitted to the locomotive and sometimes on the cars as well. The turning cog moves the train along the track. Since the cog and rack form a solid mechanical connection, the train is prevented from slipping on an inclined track.

Many passenger trains in mountainous areas use a combination of cog and standard drive. The train's normal drive wheels propel it up to 40 miles per hour over flatter terrain. When a steep grade approaches, the train slows to a walking pace and engages its cogs until it's over the hill.

Cog railways are most practical for limited use such as going up and down one specific mountain. Such a cog train is usually a one-car unit. The best of these trains can climb a 22.5-degree angle.

This railway's rack is like a ladder set between the standard rails. Some cog railway tracks use two rails, set so that the teeth are staggered, and two cogs run on the same axle. This system minimizes the possibility of a loose connection between gears, which would result in excess play and premature wear.

Rack

Cog

ELECTRIC TRAINS

Commuter trains that run through tunnels in the ground have several clear advantages: The noise of the train is contained in the underground tunnel, the trains are out of the way of city traffic, and there are no elevated tracks to doom entire city blocks to eternal shade. Electric trains are universally used in subway systems, because they generate no smoke.

The electric power gets to the motor in the train through a third rail in the track. One line of the power supply connects to the normal rails and the other line connects to the third rail. The third rail, which is either between the two normal rails or off to the side, can have 500 to 700 volts of electricity running through it. A connecting shoe on the train car runs along the third rail, picking up the electricity and transferring it to the motor, the lights, and the other electrically powered features of the car.

While riding in the train, you'll notice that every few minutes, the lights go out momentarily. This happens when the train changes circuits. The system is composed of many separate sections, each connected to an independent power supply. If one power supply malfunctions, the whole system is not affected. A break in the third rail forms the boundary between the circuits.

Trolleys and streetcars, which are still used in some cities, use an overhead power line so that the high voltage is out of the way of pedestrians and passengers.

SUBWAY TRAIN

The **trucks** of the subway train house the axles and are attached to the chassis on a pivot so that the axles can turn. Motor configurations differ depending on the train, but most have at least one motor in each truck. Gears transfer the power from the motor's shaft to the wheels.

The motorman, who operates the train, controls the motor speed with a simple deadman control. By pushing the control forward, the motorman sends the train forward, but when the control is released, the power shuts off and the brakes are applied automatically. A switch lets the motorman send the train into reverse. Usually, each car contains its own set of controls, but the motors, doors, and brakes are connected together so that they all can be run from the front car.

Conducting shoe

Truck

Third rail

HIGH-SPEED TRAIN

Cross-country trains in Europe and Japan zip along at up to 160 miles per hour, using electric power or gas turbine engines. Although their design is sleek, the real secret of their breakneck speed lies in the power train and the rails.

The engines are considerably more powerful than conventional diesels or electrics, and the power is not limited to an engine in the front. Usually, every axle on the train has its own motor.

At such breakneck speeds, misalignment of the rails by as little as a sixth of an inch can cause derailment. Many high-speed railroads need rail adjustment nightly. The slightest bump in the track is magnified by the high speed; sophisticated air suspension systems cushion the ride.

This French high-speed train, the Atlantic Train à Grande Vitesse (TGV), is the world's fastest train. It makes its run from Paris to France's west coast at speeds up to 186 miles per hour. The front and rear electric locomotives run on 25,000 volts. With nearly 12,000 horsepower at the engineer's command, this train can reach its high speeds as easily as a car accelerates onto the highway.

MAGLEV

Tomorrow's railroads probably won't run on rails—they'll ride on a magnetic field. In Germany and Japan, you can find working versions of these trains. They're called maglevs, short for "magnetically levitated train."

Powerful **electromagnets** (see page 24) inside the maglev lift the train off its track, so that it actually rides on the magnetic field (see **Magnetism,** page 17). There's no mechanical friction and no rails or wheels to wear out. The train has only the wind to fight against.

The power of magnetism also propels the train. Maglevs rely on a **linear motor** for propulsion. The guideway that the train rides along houses electromagnetic coils. In the sides of the train are propulsion magnets. A computer senses the train's position and sends current only to those coils in the guideway over which the train is running. As the electromagnetic force travels along the guideway, it pulls the train along with it. And with only the friction against the air, the trains can reach jetlike speeds. Prototypes with 200 seats have run at more than 250 miles per hour.

A portion of the magnetic energy from the guideway passes through an extra set of coils under the train, generating electricity that recharges the train's batteries.

In the experimental stage are maglevs that use **superconductivity** to lower electrical resistance to negligible levels (see page 32).

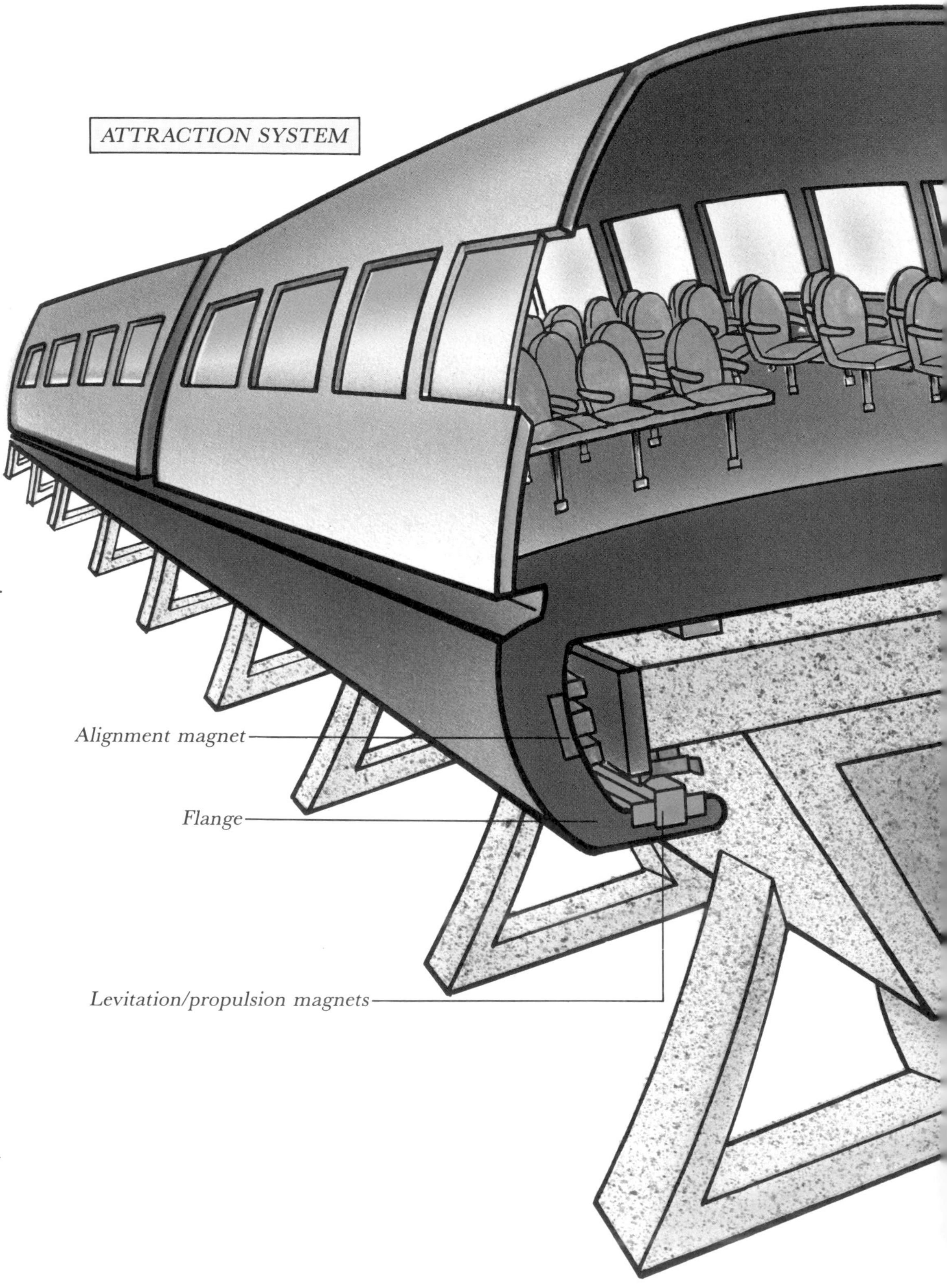

ATTRACTION SYSTEM

Alignment magnet

Flange

Levitation/propulsion magnets

Most of the maglevs being tested today use an **attraction system.** On each side of the train, a flange wraps around the guideway, which is the track the maglev runs on. Electromagnets line the inside of the flange. These magnets, powered by batteries on board the train, pull the lower lip of the flange toward the guideway, levitating the train. A second set of electromagnets, mounted on the sides of the flanges, keeps the train aligned with the guideway.

Because the attraction force of the magnets increases as the train gets closer to the guideway, it's difficult to keep a consistent gap between the two. Attraction maglevs need computer sensors to measure the distance between the train and the guideway and to adjust the magnet's power so that the proper gap is maintained. And the guideway itself must be constructed within close tolerances.

Repulsion maglevs use their electromagnets, mounted underneath the train, to push the train away from the track rather than pulling the train toward it. In this system the magnetic force also increases as the train nears the guideway, but the increased force works in the train's favor by keeping it away from the track.

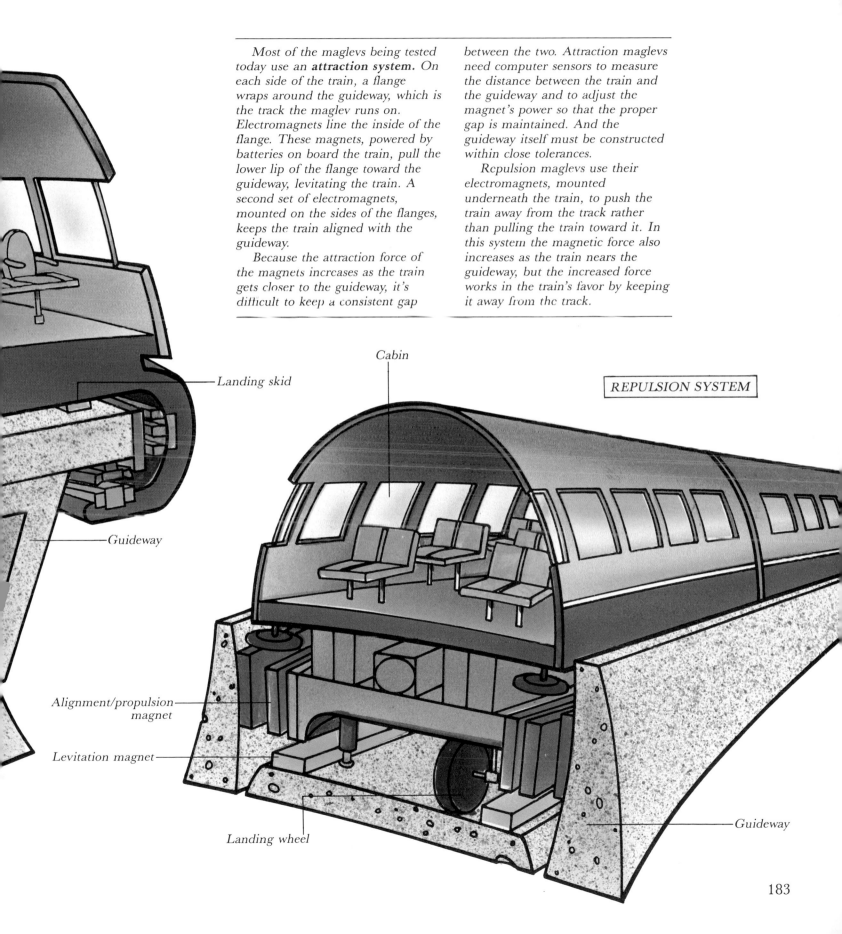

Landing skid

Guideway

Cabin

REPULSION SYSTEM

Alignment/propulsion magnet

Levitation magnet

Landing wheel

Guideway

THE COMPLETE CAR

Muffler

Coil spring

Shock absorber

Differential housing

Most cars have front disc brakes and rear drum brakes.

Fuel tank

When Siegfried Marcus drove the first-ever horseless carriage through the streets of Vienna, Austria, in 1875, his vehicle resembled the horse-drawn carriages along his path in every respect except one. It was equipped with a snorting, banging, smoke-pouring engine that drove chains wrapped around the rear wheels to make the wheels, and thus the vehicle, move without benefit of the power of a horse.

In two respects today's cars haven't changed since Marcus's: All power to move a car comes from the engine, and that power is applied to the wheels to get a car rolling.

In a modern car, the **transmission** transmits that power from the engine to a **differential** to get two of the car's four wheels to turn. The differential is so called because it allows these two wheels to turn at different speeds so that the car can go around corners and negotiate curves without skidding.

The two wheels that are driven by the differential drive the other two wheels, the same way that pedaling a bicycle gets its rear wheel going, which in turn gets the front wheel going.

REAR-WHEEL DRIVE

Although transaxle-equipped front-wheel-drive cars now prevail, cars that have the engine, transmission, and differential separated and laid out from one end of the car to the other are still being made. The object is to have the rear wheels as the prime movers.

From the vantage point of the front seat of a rear-wheel-drive car, the engine is ahead of you. The transmission lies just behind the engine. The engine and the transmission are connected by large gears. The engine gear is called the **flywheel.** If the car is equipped with an automatic transmission, the connection to the transmission is called the **torque converter.** If the car is equipped with a manual transmission, the connection to the transmission is called a **clutch.**

A long metal tube called a **drive shaft** connects the transmission to the differential, which is in the rear of the car midway between the two rear wheels. Rotating shafts called **axle shafts** project from the differential to the rear wheels.

Engine power flows through the transmission, which turns the drive shaft and causes the differential to turn the axle shafts that turn the rear wheels. The front wheels follow along.

FRONT-WHEEL DRIVE

A front-wheel-drive car has the engine, transmission, and differential coupled together into a single unit called a **transaxle.** The transaxle lies transversely across the engine compartment. It's positioned like this so that it can get the front wheels to move first. The expression "front-wheel drive" means that engine power is transmitted by the transmission through the differential to the front wheels to get them rolling. As the front wheels move, the rear wheels follow along.

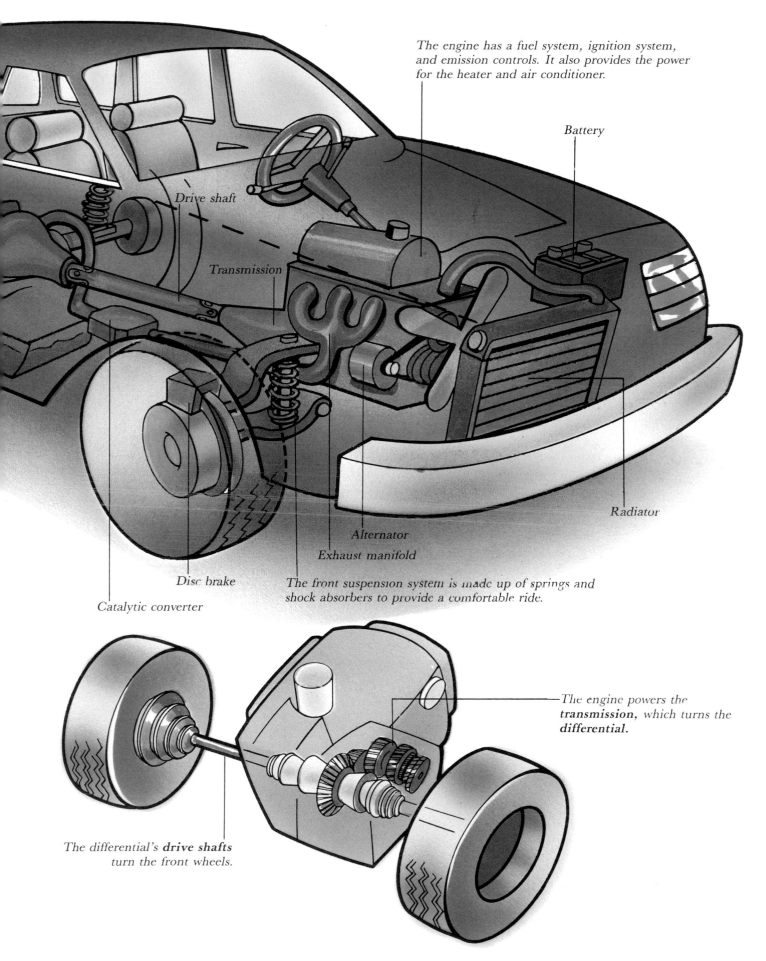

The engine has a fuel system, ignition system, and emission controls. It also provides the power for the heater and air conditioner.

Battery

Drive shaft

Transmission

Radiator

Alternator

Exhaust manifold

Disc brake

The front suspension system is made up of springs and shock absorbers to provide a comfortable ride.

Catalytic converter

The engine powers the **transmission,** which turns the **differential.**

The differential's **drive shafts** turn the front wheels.

185

GASOLINE ENGINE

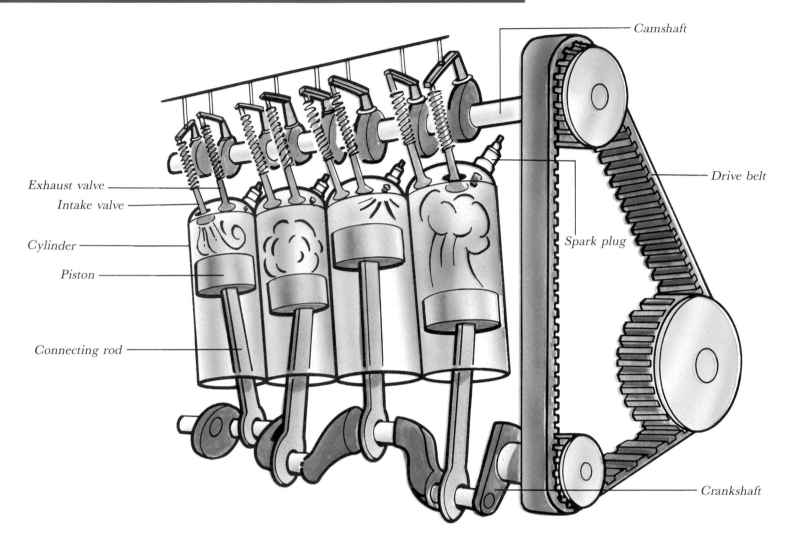

Camshaft

Drive belt

Spark plug

Exhaust valve

Intake valve

Cylinder

Piston

Connecting rod

Crankshaft

The single most important function of your car's engine is to produce a rotational or twisting force, called **torque.** The car's engine is a torquing machine.

Your car gets power from a gasoline engine or a diesel engine (see page 188). A gasoline engine converts a combustible fuel mixture of air and gasoline into mechanical energy. The mixture gets packed into cylinders (also called combustion chambers) and is ignited by an electric spark.

The gasoline engine is an internal combustion engine (combustion takes place inside the engine). It is also called a "reciprocating internal combustion engine" because of the pistons, which reciprocate, or move back and forth, both to compress the fuel mixture and to transfer power produced by fuel as it burns.

Without a **compressed fuel mixture** that is **ignited** by a spark, an engine remains a piece of dead metal. It takes all three—fuel, compression, and spark—for an engine to spring to life and start to torque.

Once the fuel mixture begins to burn, the energy it imparts drives pistons, which are attached to rods. The other ends of the rods are secured to a shaft; because they connect the pistons to this shaft, they are called **connecting rods.** As the pistons are driven, the connecting rods move to turn, or crank, the shaft, which is called the **crankshaft.** The turning force is the torque. The rotating crankshaft transfers power to a **transmission,** which, in turn, passes it on to a **differential,** which then passes it on to the wheels inside their tires, which propel you down the highway.

A gasoline engine is a "four-stroke cycle engine." Four piston strokes cause a single charge of fuel to produce one power impulse. This life cycle is a fraction of a second long.

We can divide the car's gasoline engine into three major units: cylinder head, cylinder block, and crankcase. The **cylinder head** contains two sets of valve mechanisms, intake and exhaust. When the intake valves are opened, the fuel mixture is drawn into the cylinders. When the exhaust valves open, they allow waste products generated during combustion to leave the cylinders.

The valves are opened and closed by a system of cams (see **Cams and cranks,** page 38) on a rotating shaft (the camshaft). The camshaft is connected to the crankshaft by a drive belt or chain. Thus, as the crankshaft turns by the action of the pistons and connecting rods, it turns the camshaft, which opens and closes the valves. More fuel is admitted to provide energy to move the pistons. The entire system is interconnected.

The cylinder block houses the cylinders, which are surrounded by passages cut into the block. A cooling agent flows through these passages to carry away the heat of combustion. Otherwise, the cylinder block would melt (temperature inside the cylinders gets as high as 4,500°F).

The crankshaft rotates in the crankcase, which is also called the oil pan, because it holds oil and a pump that gets the oil to engine parts. Without the protection of this lubricant, very hot parts rubbing against very hot parts would weld themselves together. As the crankshaft turns, it imparts torque to a big gear at the back end of the engine called a **flywheel,** which, in turn, transfers power out of the engine so that it can end up at the wheels.

1. *Starting with a downstroke by the pistons called the* **intake stroke,** *the camshaft opens the intake valve and closes the exhaust valve to draw fuel into the cylinder.*

4. *As the piston is pushed down and starts to rise again, the final stroke of the cycle begins. This is the* **exhaust stroke,** *during which the camshaft opens the exhaust valve and the piston pushes the waste products of combustion out of the cylinder.*

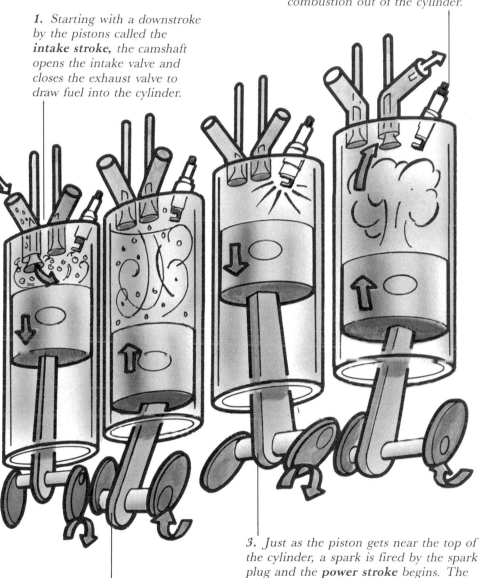

3. *Just as the piston gets near the top of the cylinder, a spark is fired by the spark plug and the* **power stroke** *begins. The fuel mixture ignites and begins to burn, releasing energy that drives the piston down.*

2. *As the piston reaches the end of the downstroke and starts its upswing, the intake valve closes to seal the cylinder so that fuel can't be pushed out of the cylinder. This is the beginning of the* **compression stroke.** *As the piston drives up, it squeezes the fuel mixture into a space that's about 10 percent of the overall volume of the cylinder.*

DIESEL ENGINE

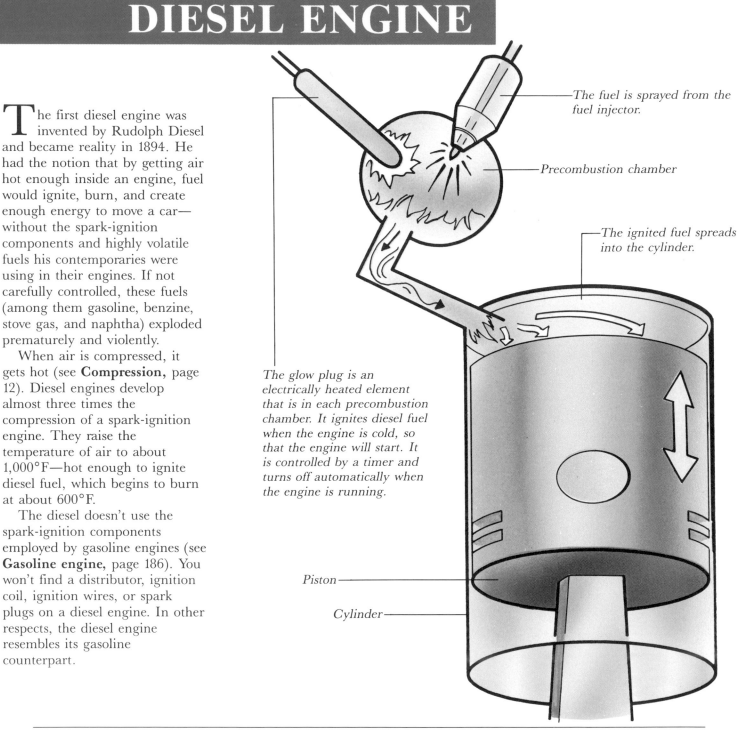

The first diesel engine was invented by Rudolph Diesel and became reality in 1894. He had the notion that by getting air hot enough inside an engine, fuel would ignite, burn, and create enough energy to move a car—without the spark-ignition components and highly volatile fuels his contemporaries were using in their engines. If not carefully controlled, these fuels (among them gasoline, benzine, stove gas, and naphtha) exploded prematurely and violently.

When air is compressed, it gets hot (see **Compression,** page 12). Diesel engines develop almost three times the compression of a spark-ignition engine. They raise the temperature of air to about 1,000°F—hot enough to ignite diesel fuel, which begins to burn at about 600°F.

The diesel doesn't use the spark-ignition components employed by gasoline engines (see **Gasoline engine,** page 186). You won't find a distributor, ignition coil, ignition wires, or spark plugs on a diesel engine. In other respects, the diesel engine resembles its gasoline counterpart.

The fuel is sprayed from the fuel injector.

Precombustion chamber

The ignited fuel spreads into the cylinder.

The glow plug is an electrically heated element that is in each precombustion chamber. It ignites diesel fuel when the engine is cold, so that the engine will start. It is controlled by a timer and turns off automatically when the engine is running.

Piston

Cylinder

The notable differences between a diesel engine and a spark-ignition engine is that a diesel engine has **precombustion chambers** and **glow plugs,** and it does not have spark plugs projecting into the cylinders to ignite the fuel mixture.

A diesel engine, like a gasoline engine, is a four-stroke cycle engine. What happens inside a diesel engine during three of the four piston strokes—intake of fuel, compression of air, and exhaust of waste gases—duplicates what happens in a gasoline engine during the same three strokes. But the power stroke differs.

The power stroke begins at the completion of the compression stroke, when the piston is at its zenith in the cylinder and is squeezing air into the smallest possible space. At about 1,000°F, that air is now hot enough to ignite the diesel fuel being sprayed into the precombustion chamber by the fuel injector. The flaming fuel spreads to the main body of the cylinder, and the energy it creates pushes the piston down to turn the crankshaft.

CRUISE CONTROL

Although cruise control systems installed throughout the years have varied in makeup and components, they all do the same job. They keep the speed of the vehicle at a level set by the driver without the driver having to maintain pressure on the gas pedal.

The driver brings the car up to a desired speed and turns on the cruise control. Next, the driver programs that speed into the system by pressing a button. An electronic sensor measures the speed at which the vehicle is traveling. A computing device differentiates between the information it receives about the actual speed and the desired speed that the driver programmed into it. The computer then tells a **servo** to set the throttle so that the speed of the engine is at the level needed to have the car travel at the desired speed. A servo is an automatic control device.

When the driver turns off the engine or shuts off the cruise control system, the electronic regulator is disconnected and the speed setting in its memory is erased. But stepping on the brake pedal does not erase the speed setting. A switch wired between the brake system and the electronic regulator temporarily interrupts the signal between the electronic regulator and the servo. The car can be slowed to accommodate the traffic condition. Pressing a button reestablishes the signal, automatically bringing car speed up to the level it was before the brakes were applied.

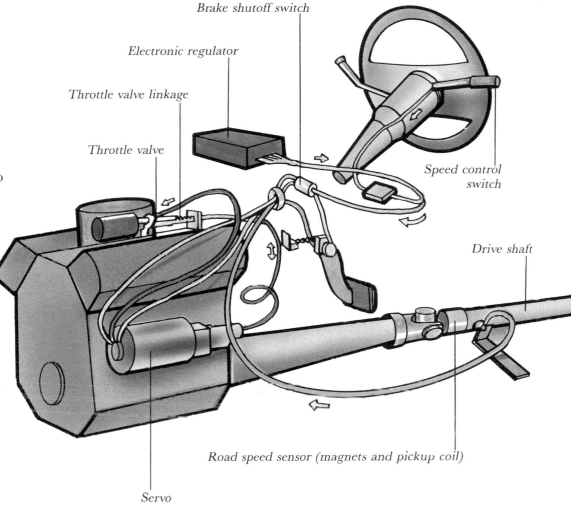

Brake shutoff switch

Electronic regulator

Throttle valve linkage

Throttle valve

Speed control switch

Drive shaft

Road speed sensor (magnets and pickup coil)

Servo

The **road speed sensor** consists of magnets and a pickup coil. (See **Electromagnet,** page 24.) The magnets—two of them in most installations—are positioned 180 degrees apart on the vehicle's drive shaft. As the magnets turn with the shaft, they pass across the face of the pickup coil, permitting the pickup coil to sense the speed at which the drive shaft is revolving.

This data is sent to the **electronic regulator,** which compares drive shaft speed (which is directly related to the speed of the vehicle) with the desired speed programmed into it by the driver.

When the two are not the same, the electronic regulator sends a signal to the **servo** to alter the speed.

Without cruise control, you alter the speed of the car by pressing down or letting up on the gas pedal. This activates the **throttle valve,** which allows the engine to take in more or less fuel. With cruise control, the servo causes the throttle valve to do the same thing without the driver having to use the gas pedal. The servo is connected by a cable to the throttle valve linkage of the carburetor or fuel injection system; the cable moves the throttle valve.

FUEL INJECTION

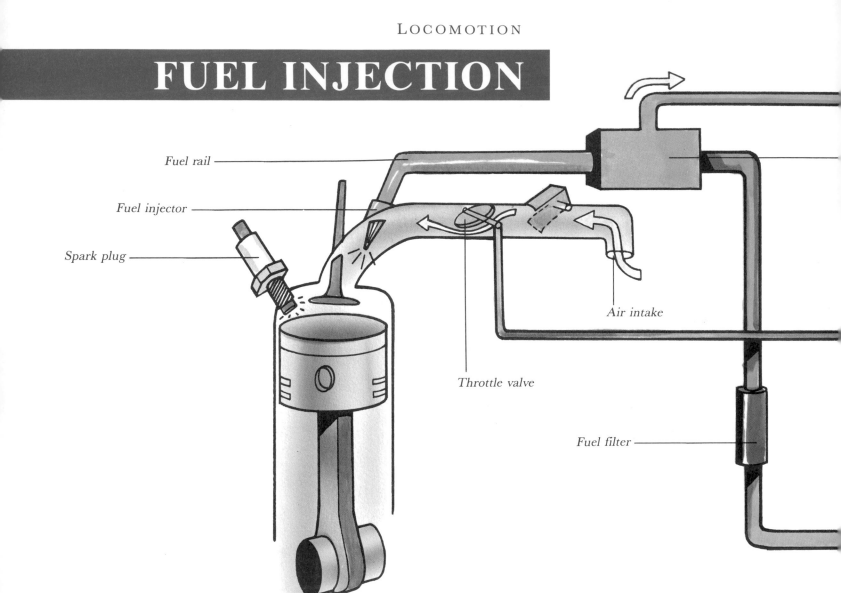

Fuel rail

Fuel injector

Spark plug

Air intake

Throttle valve

Fuel filter

Gasoline won't ignite and burn inside the engine without air. The amount of gasoline and the amount of air must be kept in proper balance as driving conditions vary. On cold mornings, for example, an engine needs a mixture richer in gasoline (about 1 part of gasoline to 7 parts of air) in order to start. When the engine is warmed up and the car is cruising, it requires a much leaner mixture—perhaps 1 part of gasoline to 14 or 16 parts of air—otherwise, the engine would flood.

A perfect balance of 14.7 parts of air to 1 part of gasoline allows the fuel to burn almost completely (see **Combustion,** page 8). The mixing of air and fuel used to be the job of the carburetor, but automobile manufacturers have retired the carburetor in favor of electronic fuel injection (EFI). EFI surpasses the carburetor in reliability and efficiency; it possesses far fewer parts that go bad.

But the real reason EFI has replaced the carburetor is to keep polluting elements produced by gasoline engines at a minimum. The ideal fuel mixture of 14.7 parts air and 1 part gasoline is almost entirely consumed in the cylinders of a gasoline engine. An engine that consumes fuel almost entirely is an engine that produces the least amount of exhaust to pollute the atmosphere. Furthermore, such an engine provides its owner with maximum fuel economy. Carburetors are not as precise at maintaining a good balance of air and gasoline. Electronic fuel injection uses a number of sensors to determine the proper fuel mixture needed for the car's driving conditions.

Although there are two kinds of EFI systems, the most efficient and widely used system is called **multiport fuel injection.** It has one fuel injector for each cylinder in the engine. Fuel injectors are the parts that spray gasoline into the engine.

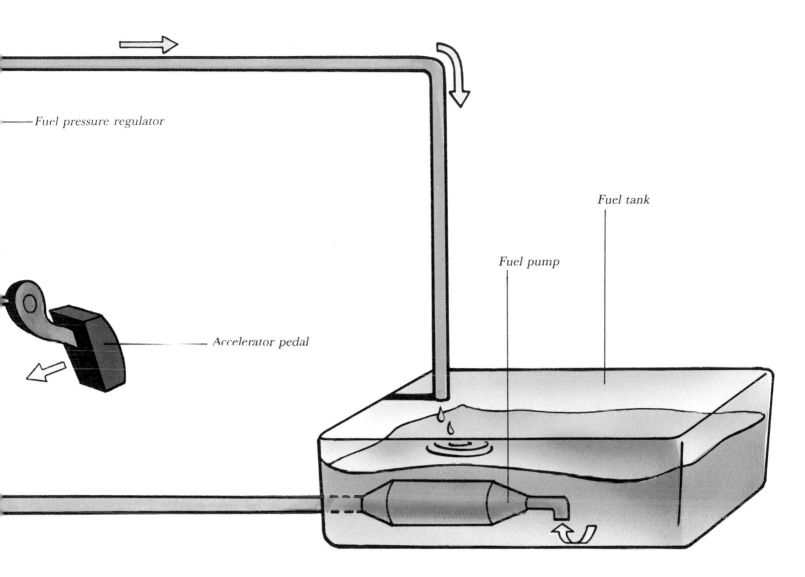

Fuel pressure regulator

Fuel tank

Fuel pump

Accelerator pedal

Air is scooped into an air intake and flows through a duct into an air flow meter—a hinged plate that opens and closes as necessary to maintain the correct quantity of air flowing into the engine. The key to maintaining the air flow at precisely 14.7 parts air to 1 part gasoline is the throttle valve, which you control with the accelerator pedal.

As you vary pressure on the accelerator pedal with your foot, a sensor records the position of the throttle valve. This data is sent to the engine computer, which "tells" the air flow meter to open or close as much or as little as necessary to let in more or less air.

With electronic fuel injection, the fuel system is delivering a constant supply of gasoline, so that in a situation when you need more power, you must feed the engine more air. Pressing the accelerator pedal opens the throttle valve and air flow meter plate, allowing more air to come through. More fuel is then injected into the engine to create the right fuel mixture.

An electrically operated pump inside the gasoline tank sends gasoline through the fuel delivery system at a constant pressure—for example, 36 pounds per square inch. This pressure doesn't vary one iota when you accelerate rapidly, drive up a hill, or in any other situation in which you would expect

the machine to consume more gasoline. More than enough gasoline is always being delivered. What gasoline the fuel injectors don't spray into the engine is diverted back to the gasoline tank by a **fuel pressure regulator.**

On its way to the fuel injectors, gasoline flows through filters to trap dirt that could clog fuel injectors. Gasoline then flows into a pipe called a fuel rail. The fuel rail holds the fuel injectors, whose nozzles are aimed at a point above the engine intake valves where the gasoline mixes with air. The mixture of air and gasoline then passes through the intake valves into the cylinders.

CATALYTIC CONVERTER

Primary chamber

oxic agents that escaped from automobiles not equipped with control systems, as was the case in cars that were manufactured before the early 1960s, harmed the environment and caused respiratory and other illnesses in humans. The passage by Congress of laws to protect us from these agents led to research programs by the automobile manufacturers.

They came up with several ways to reduce pollution, including the positive crankcase ventilation (PCV) system, exhaust gas recirculation (EGR) system, and fuel evaporation emission control system. These are effective in blocking the escape of emissions from such areas of the car as the carburetor, fuel tank, and crankcase.

Another system was needed to neutralize hydrocarbon (HC) and carbon monoxide (CO). These compounds are created as fuel is burned in an engine's cylinders and are thrown into the atmosphere with the exhaust. The catalytic converter was developed to neutralize these compounds; it has been installed on practically every vehicle having a gasoline engine that's been manufactured since 1975.

A catalyst is a substance that affects a chemical reaction without changing its own chemical composition. The catalytic converter contains catalysts that cause HC and CO to react more rapidly with oxygen. In neutralizing these two,

a third air-polluting agent is created—oxides of nitrogen (NO_x). The catalytic converter also deals with NO_x. It changes all these into water vapor, nitrogen, and carbon dioxide (CO_2), which were believed to be harmless agents when the converter was developed. Now there is a question as to whether carbon dioxide is as innocent as previously thought. Excessive CO_2 in the atmosphere is suspected of trapping heat and causing a gradual rise of the earth's temperature. This theory is called the greenhouse effect.

Converting exhaust gases is necessary even with engines that are outfitted with computers. Even though a function of the computer is to maintain a 14.7:1 air:fuel ratio (see **Fuel injection,** page 190), no computer is able to maintain the perfect ratio each moment that an engine runs. Therefore, some harmful gases are always being produced that would escape into the atmosphere.

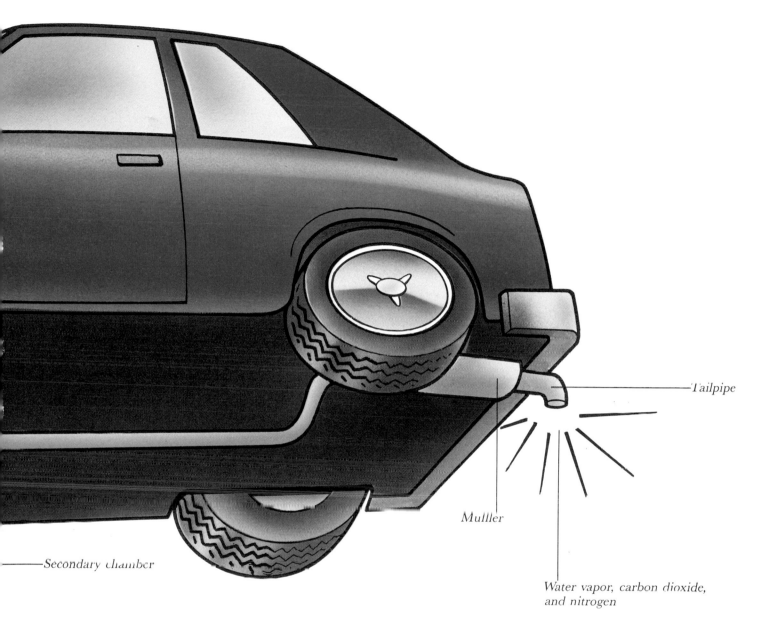

Tailpipe

Mulller

Secondary chamber

Water vapor, carbon dioxide,
and nitrogen

The catalytic converter is part of a vehicle's exhaust system. The illustration shows a system consisting of two chambers, a primary chamber and a secondary chamber. Some cars are equipped this way. Others have two separate converters that perform the same tasks as the two chambers.

As the engine burns the fuel mixture, exhaust is created and drawn from the engine into the **exhaust manifold;** from there it goes into the primary chamber (or the primary catalytic converter in

vehicles that have two catalytic converters). Here, catalysts—usually platinum and palladium—convert HC and CO into water vapor and carbon dioxide, respectively. Extreme heat is needed to make this conversion, but extreme heat reacts adversely with nitrogen in the exhaust. The result is the creation of the third toxic element, NO_x, which is the primary contributor to atmospheric smog. This is why the secondary chamber (or second catalytic converter of a two-catalytic-converter system) is needed.

As the partially treated gases exit the primary chamber (or primary catalytic converter), they are cooled by air from an air pump connected to and operated by the engine. The gases then enter the secondary chamber (or secondary catalytic converter), where conversion is completed. A third catalytic agent—usually rhodium—transforms NO_x back into harmless nitrogen. The three agents—water vapor, carbon dioxide, and nitrogen—are then expelled through the rear exhaust pipe, muffler, and tailpipe.

STEERING SYSTEMS

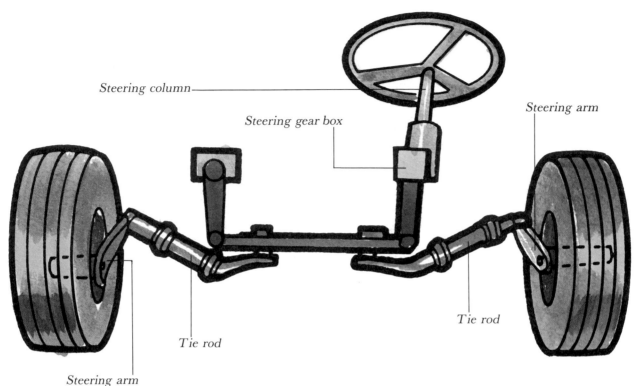

Steering column

Steering gear box

Steering arm

Tie rod

Tie rod

Steering arm

The driver of a car needs a **mechanical advantage** (see page 35) to overcome the inertia of the heavy car and the friction between the road and the tires. Automobile steering systems are designed so that the relatively small steering wheel turns a number of revolutions to move the wheels from side to side. Thus it takes less force to turn the steering wheel than it would if the wheels were being turned directly.

Rack-and-pinion steering first appeared on a Cadillac in 1905. It and the parallelogram steering

system, which was developed by a French carriage maker in 1878, are the only kinds of automobile steering systems in use. Parallelogram steering, which is used only on rear-wheel-drive vehicles, is named for the geometric shape its three main components form. Rack-and-pinion steering possesses fewer parts and is less complicated than a parallelogram steering system. It also weighs less and takes up less space, making it ideal for today's small, lightweight cars.

PARALLELOGRAM STEERING

Turning the steering wheel causes the steering shaft, which extends into the steering gear box, to rotate. The mechanism inside the steering gear box is a complex arrangement of gears and ball bearings that translate the rotary movements of the steering column to a side-to-side movement that can operate the front wheels. The **Pitman arm,** which is a metal arm connected to a shaft inside the gear box, moves like a clock hand as the steering wheel is turned. The other end of the Pitman arm is connected to the relay rod, which transfers the movement of the Pitman arm to the tie rods, steering arms, and wheels.

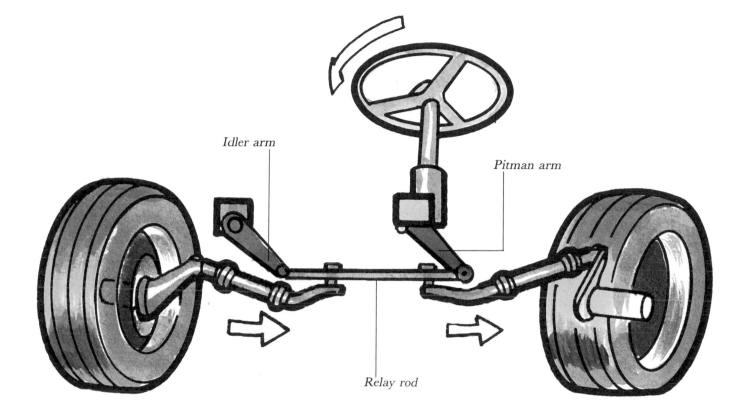

Idler arm

Pitman arm

Relay rod

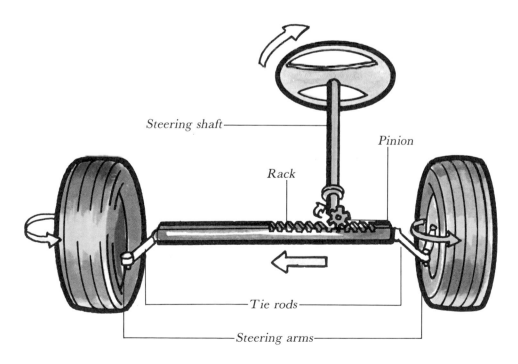

Steering shaft

Pinion

Rack

Tie rods

Steering arms

RACK-AND-PINION STEERING

Rack-and-pinion steering is by far simpler. It has no steering gear box. A pinion gear on the end of the steering shaft moves a rack as you turn the steering wheel. The rack is a horizontal bar with gear teeth that mesh with the pinion gear. The rack is connected to the front wheels by means of tie rods and steering arms. As you turn the steering wheel, the steering shaft and pinion gear rotate together. The rotating pinion gear activates the rack, which moves the tie rods and the steering arms holding the wheels.

Since steering-wheel maneuvers don't pass through a relay rod and parallel arms before reaching the wheels, steering is more responsive.

195

AUTOMATIC SEAT BELTS

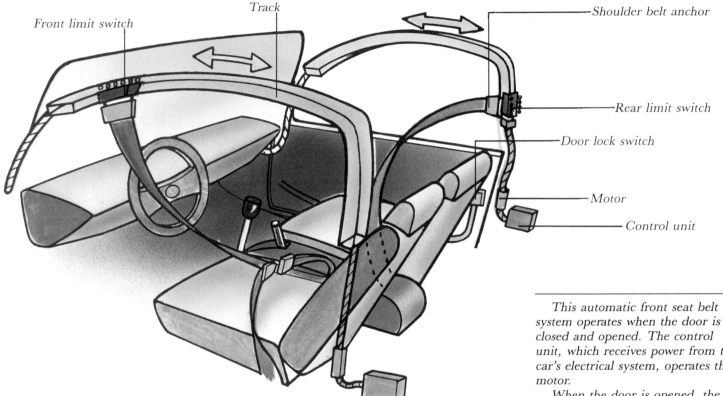

Front limit switch

Track

Shoulder belt anchor

Rear limit switch

Door lock switch

Motor

Control unit

The next new car you buy will have a front-seat automatic restraint system. Federal safety standards require that, beginning in 1990, all new vehicles must have this feature. That system can be air bags or seat belts that fasten automatically around occupants of front seats. Because of cost, most manufacturers have chosen to install automatic seat belts.

Some automatic seat belts are not motorized. The lap and shoulder belts connect to the doors. People entering the car slide in under the belts, which have a spring system that automatically tightens the belts around them when the doors are closed. In some arrangements the lap belt is not automatic.

In a motorized system, each of the two front shoulder belts are operated by an electric motor. Closing the door and/or turning on the ignition activates motors to move front seat belts along a track, tightening the belts around the driver and front-seat passenger. Some cars have an arrangement that allows the belt to be detached from the door frame and reattached as a manual shoulder belt.

This automatic front seat belt system operates when the door is closed and opened. The control unit, which receives power from the car's electrical system, operates the motor.

When the door is opened, the door lock switch transmits an electric signal to a relay in the control unit. The control unit activates the motor, which moves the shoulder belt anchor forward. When the belt anchor hits the front limit switch, the switch turns off the motor and the anchor stops moving. The person entering the car can now sit down.

Closing the door activates another relay inside the control unit. This causes the motor to move in a direction opposite to the way it moved before, bringing the shoulder belt anchor back along its track until it hits the rear limit switch. This turns off the motor just as the belt comes to rest in a position that will restrain the occupants in case of accident. The shoulder belt anchor moves along its track from one end to the other in about two seconds. Separate lap belts are buckled and unbuckled manually.

AIR BAGS

Air bag systems that cushion automobile drivers in accidents have seen extensive use in cars driven by government and insurance industry personnel to see just how effective they are in preventing death and injury. The results seem to reinforce the position that air bags are very effective life-saving devices.

For example, 5,300 cars owned by the Federal government were equipped with air bags. Sixty-six of them were involved in highway accidents severe enough to have caused death or injury. Yet not one person driving these cars was injured badly enough to require hospitalization.

A major insurance company installed air bags in 2,900 of its vehicles. Twenty-two of them were in accidents in which bags deployed. Seven of these cars were totaled. Only two of the drivers had to go to the hospital.

Results such as these make it appear likely that air bags will become a dominant automotive safety device in the years ahead. It's estimated that by 1995 approximately 20 million cars on America's roads will be equipped with them.

1. Air bags are designed to deploy only in a direct frontal or front-angle crash that is equivalent to hitting a wall at 12 miles per hour or more. Since the violence in a crash that causes death or injury is over in one-eighth of a second, the air bag must deploy to provide full protection in much less time. It does this in just one-twenty-fifth of a second, which is faster than you can blink an eye.

2. The force of the impact is detected by an electronic sensor and causes the electrical contacts to close. The electric charge that results triggers a small explosive— sodium azide—which "detonates" a canister of harmless nitrogen (78 percent of the air we breathe is nitrogen). The nitrogen is expelled into a big fabric bag. The bag, which is contained in a module in the center of the steering wheel, inflates and billows out in front of you.

3. In that split second as the bag fills with nitrogen, the car comes to an abrupt halt from its contact with the other car or object, and you are propelled forward toward the steering column, the area above the windshield, or the windshield itself. If you hit any of these areas, you could be critically injured or killed. But that billowy, soft bag stops you and keeps you from impact.

4. When the air bag reaches maximum inflation—and all this is taking place in one-twenty-fifth of a second—gas escapes through the fabric and the balloon deflates in less than one-fiftieth of a second. Instant deflation is designed into the system to make sure that your face will not be forced into the pillow, possibly causing suffocation.

ANTILOCK BRAKING SYSTEM

The number of cars equipped with an antilock braking system (ABS) has been increasing each year. This extraordinary safety system prevents the wheels from locking, or going into a skid, before the vehicle stops. If the driver slams on the car's brakes, it's possible for the wheels to stop spinning before the car stops moving, which can result in a skid. The driver loses both braking effectiveness and steering control of the car, which continues to move in the same direction regardless of how the wheels are turned or where the front of the car is pointing. The only way to come out of a skid is to release pressure on the brake pedal enough to allow the wheels to start spinning again so that the driver regains steering control and braking effectiveness.

If the wheels continue to turn, as they will with ABS, the driver can continue to steer and stop the car while applying maximum braking force. The driver can slam on the brakes and keep them fully applied, knowing that the system will allow steering control to be maintained.

Antilock braking uses sensors to detect when the speed of one of the wheels is starting to differ from the speed of the other wheels as brakes are applied. This situation could cause the car to become unstable and the driver to lose steering control. The system then goes into action to prevent the wheel from locking.

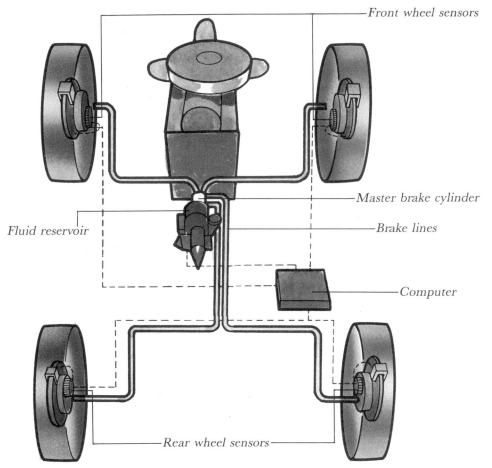

Front wheel sensors

Master brake cylinder

Brake lines

Computer

Fluid reservoir

Rear wheel sensors

Put yourself in this situation: You apply the brakes, but the roadway under one of the wheels is slicker than elsewhere. The rotational speed of that one wheel starts to differ from the speed at which the other wheels are turning. The result could be a skid.

With ABS, the speed at which that wheel turns is being monitored by a pickup ring and sensor at the wheel. The pickup ring is attached to the brake rotor, which turns with the wheel. The sensor is stationary. Thus, as the pickup ring flashes past the sensor, the sensor detects the speed at which the wheel turns. This information is transmitted to a computer.

The computer acts as a switch to activate solenoid valves on the master brake cylinder (see **Hydraulic brakes,** page 45). The solenoid valves oversee the amount of hydraulic pressure that reaches the braking components at the wheels. Each brake line is served by two solenoid valves. During normal operation, one valve is open and one is closed. This maintains equal pressure at all four wheels.

When the computer signals one of the pairs of valves that the wheel it serves is moving more slowly than the others, the open valve closes. This action adjusts the amount of hydraulic pressure reaching the braking components of the wheel so that it turns at a rate equal to the other wheels. The loss of control that could have resulted because of wheel lock is averted.

PARKING METER

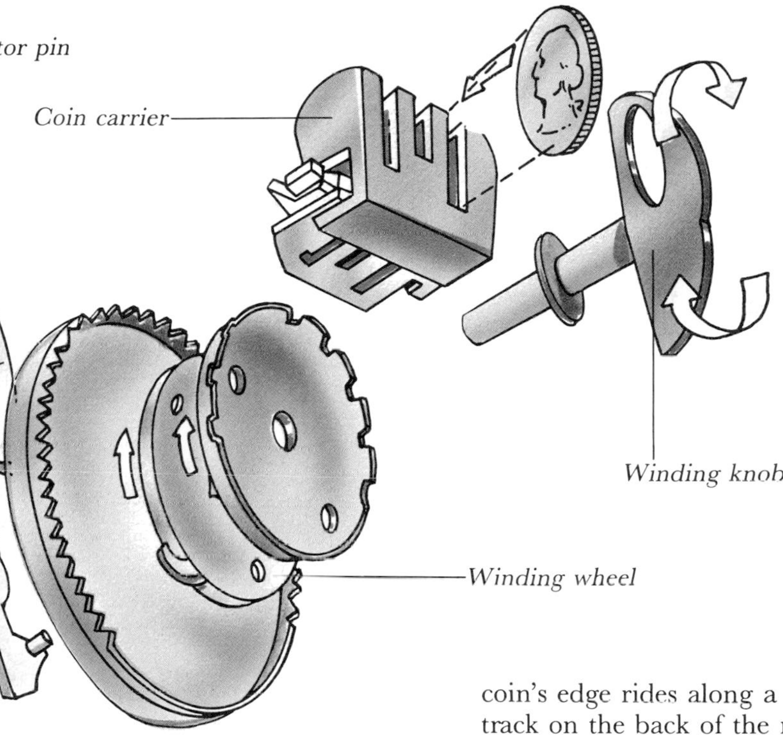

Indicator pin

Coin carrier

Timer mechanism

VIOLATION

Winding knob

Winding wheel

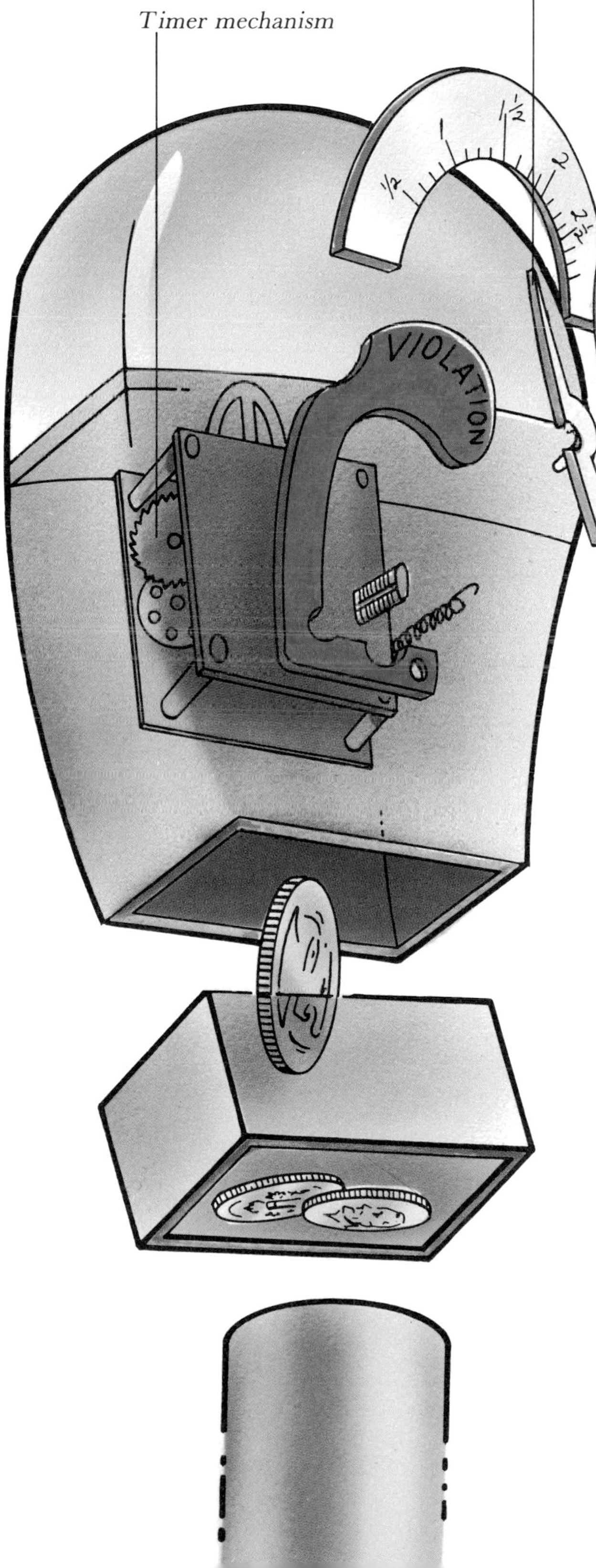

Parking meters are enclosed in a nearly bulletproof steel enclosure with no exposed screws and a lock tough enough to discourage would-be safecrackers. Inside that "street safe," the mechanism does four basic jobs. It detects the type of coin you've inserted. It collects the coins until the meter collector comes around for them. It counts down the amount of time you've bought. And finally, it sends up a signal when your time has run out.

A coin inserted in the meter goes through one of three slots—quarter, dime, or nickel—and into a coin carrier mounted on the shaft of the winding knob. As you turn the knob, sending the coin carrier up and over, the coin's edge rides along a smooth track on the back of the meter's face plate. At a certain point—early for a quarter, much later for a nickel—the coin rides up onto a raised section of the track. The raised section pushes the coin against a mechanism that engages the winding wheel. At this point, your turning motion starts to wind the timer. Since a quarter starts the winding of the timing mechanism earlier than a nickel does, a quarter buys more time.

When the knob reaches the end of its turn, the coin carrier is upside-down and the coin falls into a collection box at the bottom of the meter.

A cam on the shaft with the winding wheel moves the indicator pin, which shows how much time is left on the meter. When the time is up, a spur on the winding wheel catches a pin on the "Violation" flag, pulling the flag up so that it's visible through the glass.

199

RADAR SPEED CONTROL

The radar system used by police in speed traps is a simple form of air traffic control radar (see **Radar,** page 218). Mounted on the police car, or hand-held like a gun, a transmitter sends down the road a continuous beam of microwaves of one frequency, narrow and focused like the beam from a searchlight. Some of the radiation is reflected by moving vehicles back toward the receiver.

If the vehicle is approaching, the reflections arrive at the receiver slightly faster than they do if the vehicle is stationary. This is because each successive wave is reflected from a closer point. The receiver perceives these waves at a higher frequency. If the vehicle is moving away, the opposite is true—the reflecting waves are perceived as a lower frequency. This effect, called the **Doppler shift,** is the same phenomenon that makes the pitch of a train whistle higher while approaching and lower while going away. The radar receiver calculates the speed of the target from the difference in frequency between the transmitted signal and the reflected signal.

Some speed control radars can be operated from a moving patrol car. The radar receiver picks up reflections from the ground, which is moving in relation to the car, along with reflections from moving vehicles. It calculates and subtracts the ground speed—which is the same as the speed of the patrol car—to arrive at the speed of the target vehicle.

Not all the microwaves transmitted by the police speed control radar are reflected; some travel past the target and others are scattered by moisture and dust in the air. These microwaves can be detected by a sensitive receiver, tuned to the microwave frequency of the police radar, in a car several miles down the road from the transmitter. A radar detector is an early warning system that picks up these microwaves even though they may be very weak, amplifies them, and alerts the driver with audio beeps and flashing lights. The detector also has a signal-strength meter that gives an indication of how far away the police radar is.

To defeat radar detectors, patrol officers often wait until a target vehicle is well within range before triggering the beam. Target speed is determined within a fraction of a second. Since the radar is on only briefly, a radar detector in a car that is some distance down the road may miss it completely.

DOPPLER SHIFT

The sound waves are farther apart behind the cab, because the cab is traveling away from them.

This person hears the horn at a lower pitch.

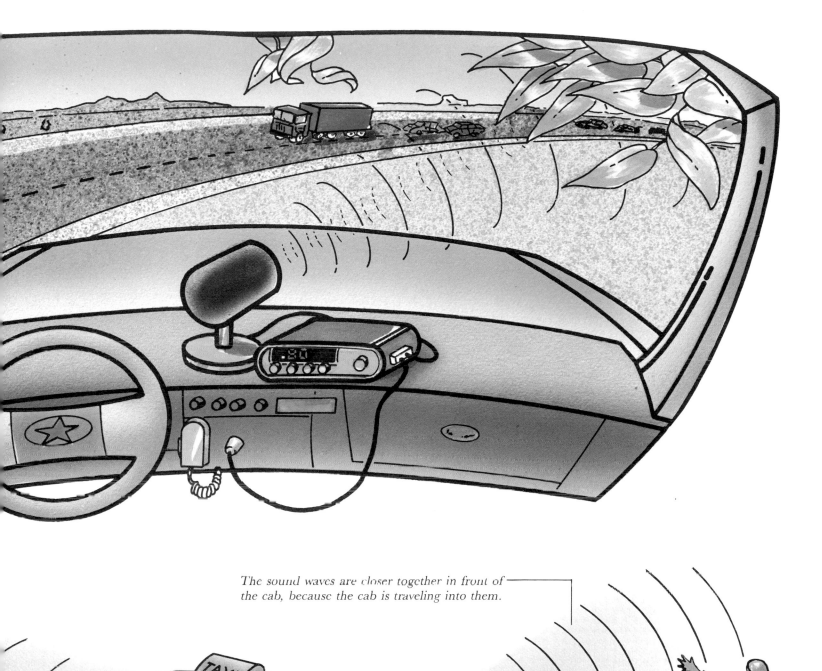

The sound waves are closer together in front of the cab, because the cab is traveling into them.

A taxicab blowing its horn as it travels down the street sends out sound waves in all directions.

This person hears the horn at a higher pitch.

TRAFFIC LIGHT

The basic workings of traffic signals have remained the same for many years; these devices have proven to be very flexible and durable. They can operate automatically or manually, or they can follow orders from remote sensors that react to traffic flow or instructions from a computer. If the signals are disarranged when a traffic officer operates the light manually, they automatically readjust when returned to automatic operation.

The traffic light consists of two parts—the light itself and the control box, which is usually mounted to a nearby pole. Inside the box are switches for each of the lights. The switches are opened and closed by a cam that is controlled by a motor-driven timer. One revolution of the timer drum equals a complete signal cycle. Keys or tabs inserted into the slots in the timer drum determine the percentage of the cycle devoted to each light.

Inside the control box for this traffic light are three switches for the main street's lights and three switches for the side street's lights. Each switch has its own cam to either hold the switch contacts open or allow them to close. The cams rotate together on a camshaft. The profile of the set of cams is such that only certain combinations of lights can be switched on at one time. Both red switches—or worse, both greens—cannot be turned on at the same time.

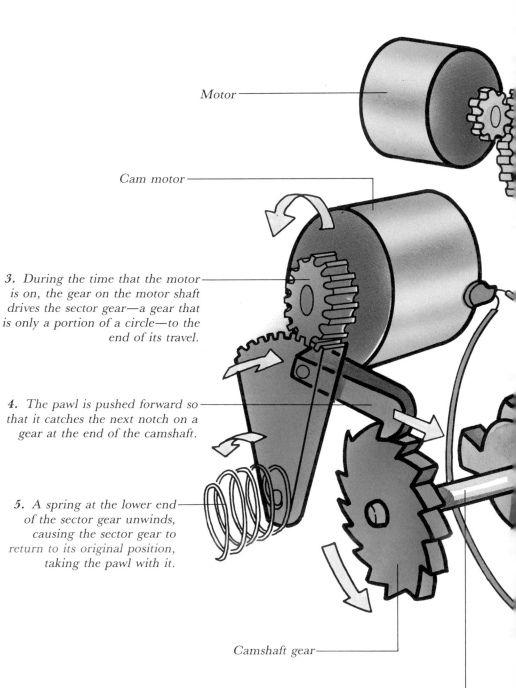

Resynchronizing switch contacts

Motor

Cam motor

3. During the time that the motor is on, the gear on the motor shaft drives the sector gear—a gear that is only a portion of a circle—to the end of its travel.

4. The pawl is pushed forward so that it catches the next notch on a gear at the end of the camshaft.

5. A spring at the lower end of the sector gear unwinds, causing the sector gear to return to its original position, taking the pawl with it.

Camshaft gear

6. The camshaft moves one increment. Two of the light switches are activated, and those lights will stay on until the timer drum brings another of the timing tabs to the camshaft-advance switch.

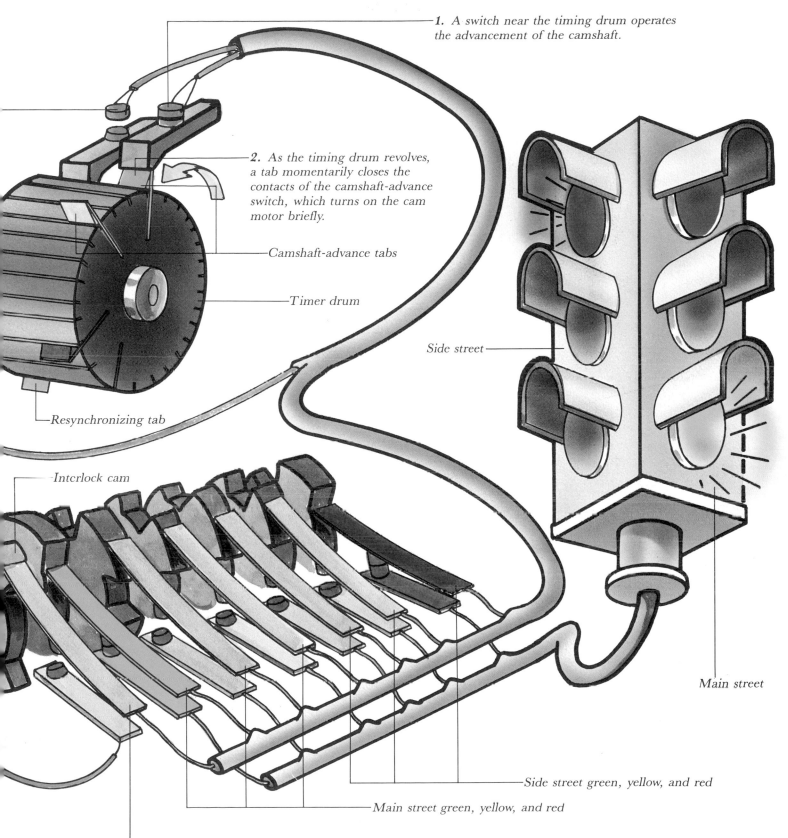

1. A switch near the timing drum operates the advancement of the camshaft.

2. As the timing drum revolves, a tab momentarily closes the contacts of the camshaft-advance switch, which turns on the cam motor briefly.

Camshaft-advance tabs

Timer drum

Resynchronizing tab

Interlock cam

Side street

Main street

Side street green, yellow, and red

Main street green, yellow, and red

7. At a certain point in the cycle, one of the cams opens the interlock switch. The timing tabs then have no effect until the resynchronizing tab, set at this point, closes the contacts of the resynchronizing switch and energizes the cam motor again. In this way, the timing cycle is reprogrammed, which compensates for mixups and power failures.

BREATH ANALYZER

Drivers under the influence of alcohol are a leading factor in traffic accidents. In order to convict drunk drivers, police officers needed a sobriety test more accurate than asking the driver to walk a straight line. The breath analyzer was developed to give police officers a portable tool for determining whether a driver is intoxicated. The actual amount of alcohol in the bloodstream that makes a person unfit to drive has been debated, but the legal maximum is generally 0.10 percent.

Several methods of testing have been developed. The portable ones test the amount of alcohol in the breath, relying on a direct correlation between that and the amount of alcohol in the bloodstream. This correlation is valid, because some of the alcohol dissolved in the bloodstream can cross into the lung and enter its air spaces as alcohol vapor.

BALLOON BREATH ANALYZER

A commonly used breath analyzer consists of an inflatable bag attached to a transparent tube of chemicals. The suspect blows into a mouthpiece attached to the tube. The mixture of chemicals in the tube—sulphuric acid and potassium dichromate—is orange-yellow in color to begin with. Alcohol in the breath combines with the chemicals to form acetic acid, and at the same time the potassium dichromate breaks down to form potassium and chromium sulphates. The color of the chemicals changes to blue-green. The extent of the change is proportional to the amount of alcohol in the breath.

FUEL-CELL BREATH ANALYZER

The inflatable-bag type of breath analyzer was limited in accuracy, yet police officers needed a valid test that could be performed without taking the driver to the police station. An electronic device that uses a fuel filter is portable, durable, and more accurate than the inflatable bag.

When oxygen in the air is mixed with alcohol and directed to a fuel cell, the fuel cell generates an electric current. The strength of the current is porportional to the amount of alcohol present. To use this type of breath analyzer, the suspect blows into a tube, and a measured amount of breath is drawn into the fuel cell. Colored lights indicate the result of the test. A green light shows a concentration of alcohol below the legal limit, a yellow light shows an alcohol level close to the legal limit, and a red light indicates a failure—the subject is not legally fit to drive.

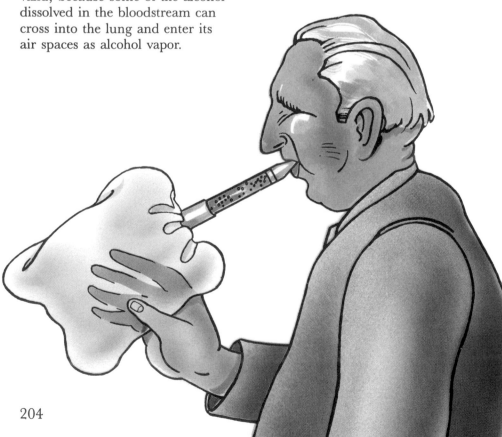

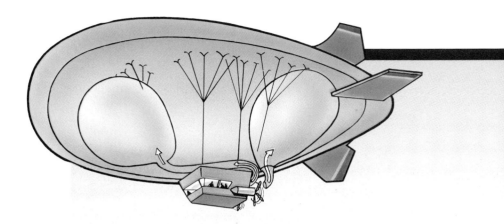

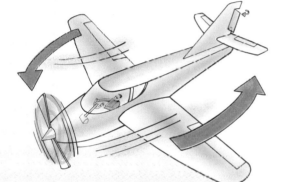

Flight

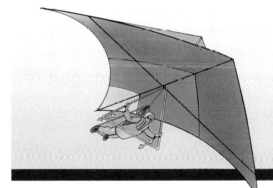

BOOMERANG

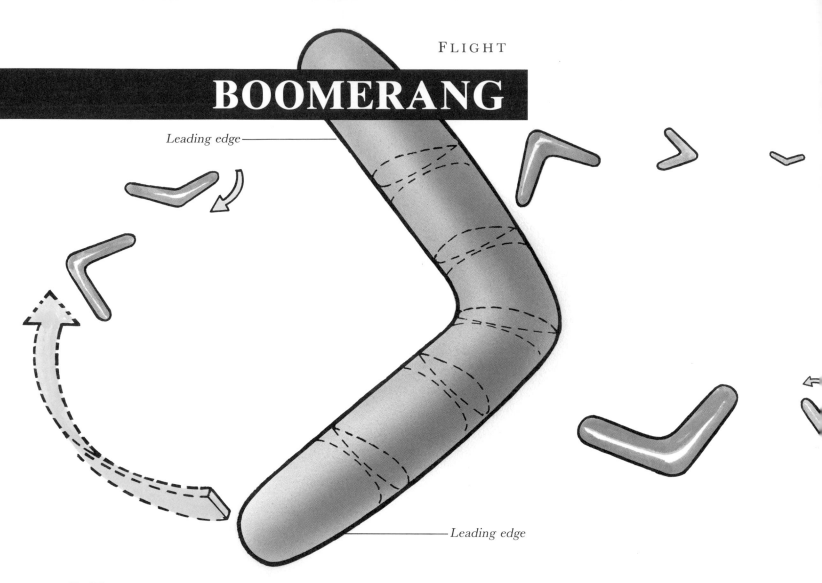

Leading edge

Leading edge

Not all boomerangs return to the thrower; whether they do or not depends upon their construction. The nonreturning type was used, not only by native Australians but also by prehistoric peoples in other areas, as a weapon or tool. The returning type is a fascinating sporting device that goes through complex aerodynamic contortions in its flight.

A boomerang returns because of the shape of its two arms and because it spins. Each arm is shaped like a classic **airfoil** (see **Airplane,** page 212). This shape has a rounded edge—the leading edge—and a sharper edge that

trails. The leading edge of one arm faces toward the inside of the boomerang's curve and the leading edge of the other arm faces toward the outside. A boomerang is thrown vertically and turns over and over in its flight; this configuration causes both leading edges to cut into the air as the boomerang rotates.

Just as in the case of an airplane wing, the airfoil shape of the boomerang's arms creates lift. But since the boomerang is vertical when thrown, the lift is in a horizontal direction, not a vertical one. Furthermore, the lift is always greater on the arm that is uppermost, because its forward travel through the air is faster than the lower arm's. The upper arm moves through the air at the speed of the boomerang plus the

speed at which it is rotating, whereas the lower arm moves through the air at the speed of the boomerang minus the speed at which it is rotating. Therefore the speed of the air passing the top arm is greater, and more lift is generated.

The effect of this difference in lift is that the boomerang is rotating and also traveling in a large circle. It's the same effect you can see in a spinning top. Gravity pulls the top down, causing the rotating top to wobble in a circle around its vertical axis. The boomerang's rotation is on a horizontal axis, and the extra pressure on the top arm causes it to circle back to the thrower.

To throw a boomerang, grasp one of the arms so that the flat bottom is against your palm. The leading edge of the other arm is then in a position to cut through the air as the boomerang rotates. (Left-handed boomerangs are the mirror image of right-handed ones.) Lift the boomerang behind your head and throw it out with a snapping motion. The snapping motion gives the boomerang its spin.

The boomerang travels, head over heels, in a circle back to you. By the time it has returned, its rotation is nearly horizontal.

FRISBEE

Almost anything flat and round will fly if you throw it so that it spins. That's because of **gyroscopic force,** which causes anything spinning to remain stable in its orientation (see **Whirling wheels,** page 43).

A Frisbee flying disk flies much farther than an ordinary flat, round item because its curved top generates lift. A flick of the wrist spins the disk and sends it forward at the same time. The forward motion gives it the lift, in the way that the forward motion of a wing lifts an airplane. The gyroscopic effect of the spin keeps the disk flying steadily.

If thrown in a vertical plane, a Frisbee will return to you the way a boomerang does. You can make curved throws by releasing the disk at a slight angle. As long as it has enough forward momentum, the disk climbs in the direction of the angle. Once its speed drops off, the disk heads back down, but because the gyroscopic force keeps it at one angle, it goes down in the same orientation it went up.

AIRSHIP

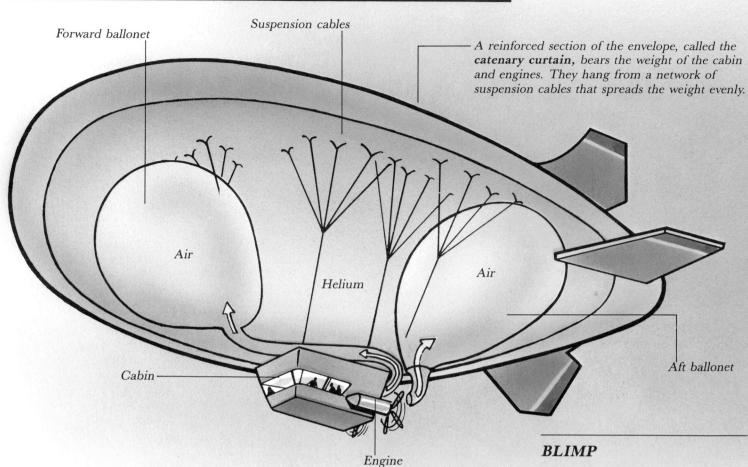

Forward ballonet

Suspension cables

A reinforced section of the envelope, called the **catenary curtain,** bears the weight of the cabin and engines. They hang from a network of suspension cables that spreads the weight evenly.

Air

Helium

Air

Cabin

Engine

Aft ballonet

Man's first controllable flight took place in the 1850s—in an airship. The slow, bulbous airship beat the airplane into commercial service; it carried passengers across the Atlantic during the 1920s and 1930s.

The airship's head start on airplanes was probably due to the airship's simplicity. Airships are big bags of lighter-than-air gas with movable control fins at the rear, motor-driven propellers, and a cab for passengers to ride in. Dirigibles are rigid airships that enclose several gas bags inside a light metal framework; blimps are nonrigid and hold their shape because of the pressure of the gas inside.

Unlike the early airplanes, airships were reasonably foolproof. They could get off the ground as long as they didn't leak. And they stayed aloft even with engine failure. But early airships used flammable hydrogen rather than helium as the lighter-than-air gas, and eventually the airplane proved faster. After the explosion of the German dirigible *Hindenburg* in 1937, the great age of the airships was past. Today, blimps are used chiefly as advertising tools and camera platforms for bird's-eye views of sports events.

BLIMP

A blimp rises because its sealed fabric "envelope" is filled with helium, which is lighter than air. The helium is slightly pressurized, pushing out the walls of the envelope so that the blimp retains its shape. The envelope contains air compartments called **ballonets,** which help the blimp adjust to changes in atmospheric pressure. The pilot uses electric blowers, or vents that catch the breeze from the propeller, to expand or contract the ballonets. This keeps the blimp's pressure constant as the temperature and atmospheric pressure vary.

Control fins at the rear of the blimp work like the horizontal and vertical stabilizers of an airplane. Hinged surfaces at the rear of the fins allow the blimp to be steered. Using controls located in the cabin, the pilot can move these surfaces to send the blimp up, down, or to either side.

GLIDER

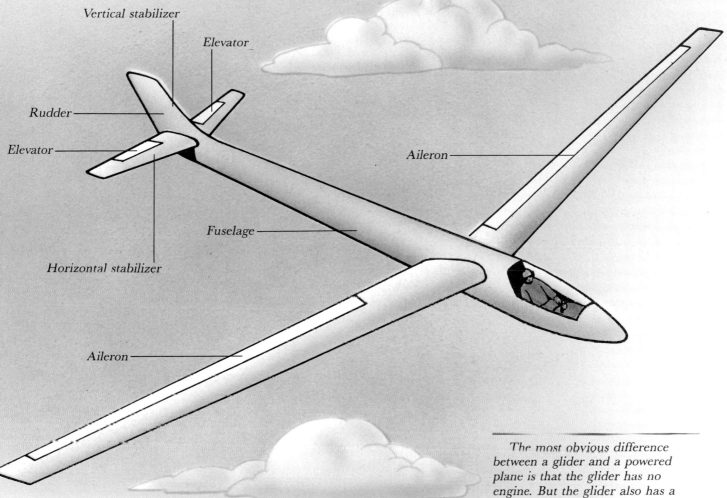

Vertical stabilizer

Elevator

Rudder

Elevator

Aileron

Fuselage

Horizontal stabilizer

Aileron

A glider is a plane that uses the Earth's gravity and the sun's energy to propel it through the skies. It has no engine, throttle, or fuel tank.

All the glider needs to take off is a way to get it moving fast. You can tow it behind a car or powered plane, or you can attach it to a high-speed winch. Once the glider is airborne, the pilot begins looking for hot-air **thermals,** which are air currents rising up from the Earth's surface. Thermals consist of warm air, which is less dense than cold air and therefore rises.

Thermals can be found over warm areas of land and under cloud formations. Another way to get the glider to rise is to take advantage of wind blowing up a nearby hill.

The glider pilot finds air currents by keeping a close eye on the **variometer,** an indicator that tells how fast the glider is climbing or descending. Experienced pilots know how to look for natural signs of thermals—certain cloud patterns, for example, or birds riding the currents without flapping their wings. Gliders have climbed up to 46,000 feet above sea level.

The most obvious difference between a glider and a powered plane is that the glider has no engine. But the glider also has a longer wingspan for extra lift, along with a thin, lightweight fuselage for minimal drag.

*Like an airplane, the glider has horizontal and vertical stabilizers at the tail and a rudder, elevator, and ailerons for control. (See **Airplane,** page 212.) The glider pilot usually has a stick to manipulate these control surfaces. Federal regulations for gliders require very little instrumentation—a typical glider has only an altimeter, an air speed indicator, a compass, and a variometer.*

Most gliders hold a pilot and one passenger seated behind the pilot. A large canopy, free of obstructions, gives the flier a spectacular view.

HANG GLIDER

If you increase the size of a kite so that it's big enough to lift a person off the ground, you have a hang glider. NASA developed the hang glider as a recovery vehicle for spacecraft. Sports enthusiasts added a control bar and a harness to support a pilot. The result is an inexpensive, easy-to-carry glider with no moving parts. Hang gliders can take off and land in just a few feet of space, and they don't require a pilot's license. These reasons make hang gliders a popular sport aircraft.

The pilot's weight, which hangs from the hang glider's center of gravity, is used as the controlling force. As long as the pilot is relaxed, the hang glider flies straight and level. To turn left, the pilot leans to the left, which presses the pilot's weight down on the control bar and pushes the left wing downward. To go into a dive, the pilot leans forward, pushing the hang glider's nose down. Leaning backward pulls the nose up, but since no engine provides a thrust for climbing, this action slows the craft. To reduce air resistance, the pilot adopts a prone position while flying.

The pilot takes off by running forward into the wind until the combined speed of the pilot and the wind is enough to get the glider airborne. Hang gliders must rely on gravity to maintain their forward momentum, so the pilot needs to launch from a hill or cliff. With a little luck, the pilot soon finds a column of rising air and circles in the column to gain altitude.

The basic structure of a hang glider is similar to that of a kite. A fabric or plastic sheet covering is stretched over a frame made of lightweight aluminum tubing. A typical hang glider is 20 feet long with a 30-foot wingspan. The whole assembly folds into a 45- to 70-pound package that can be carried on an automobile rooftop rack and stored in a garage.

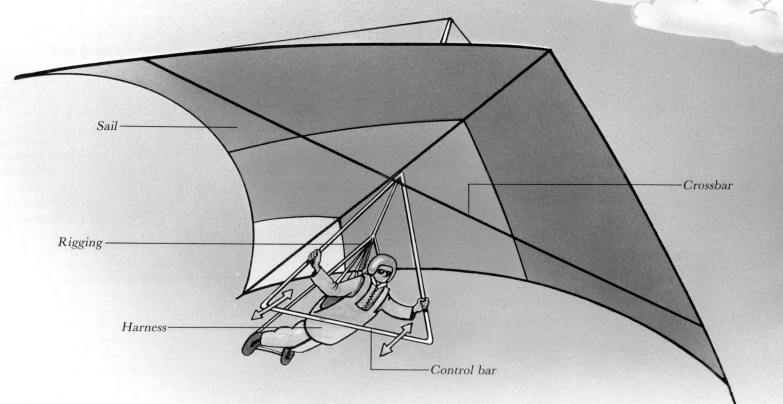

Sail

Rigging

Harness

Crossbar

Control bar

PARACHUTE

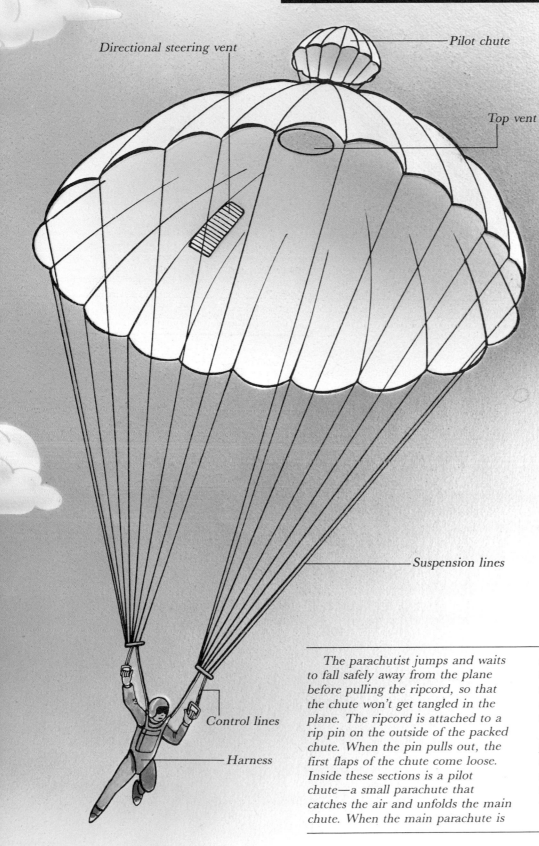

Directional steering vent

Pilot chute

Top vent

Suspension lines

Control lines

Harness

Gravity pulls a falling object to the Earth. A parachute counteracts the force of gravity by providing a large surface to increase air resistance. The pressure of the air trapped inside the canopy slows the fall. A little air needs to escape through a top vent in order to keep the parachute stable; if the air could escape only from the side of the canopy, it would rock to and fro. Most parachutes incorporate a couple of side vents that can be opened and closed with control lines. These allow the parachutist to steer the chute by directing escaping air away from the direction the parachutist wants to go.

Before they leave the ground, parachutists carefully fold their chutes onto a harness that straps to their backs. Unless the chute is folded properly, the suspension lines can tangle, which would prevent the parachute from opening fully. Sport parachutists wear a tightly packed emergency chute strapped to their stomachs. Emergency chutes must be repacked periodically by an expert parachute maker to ensure that they will not fail.

The parachutist jumps and waits to fall safely away from the plane before pulling the ripcord, so that the chute won't get tangled in the plane. The ripcord is attached to a rip pin on the outside of the packed chute. When the pin pulls out, the first flaps of the chute come loose. Inside these sections is a pilot chute—a small parachute that catches the air and unfolds the main chute. When the main parachute is filled with air, the parachutist's speed quickly drops from about 150 feet per second to just a few feet per second. This creates a breathtaking shock.

On reaching the ground, the parachutist uses quick-releasing hooks to release the chute immediately, to prevent being dragged across the ground by the wind.

AIRPLANE

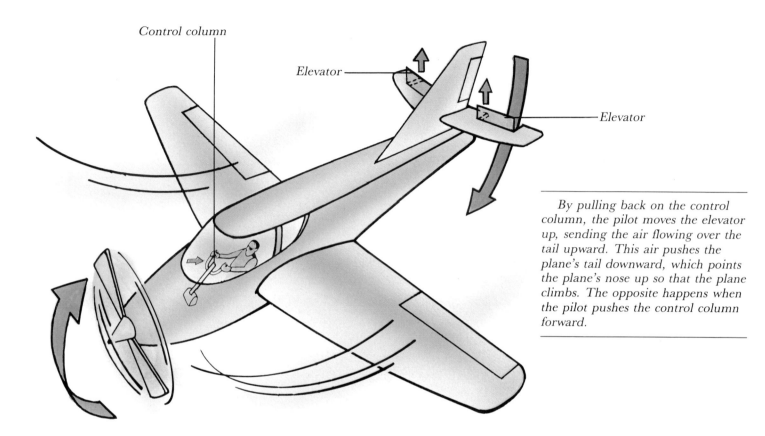

Control column

Elevator

Elevator

By pulling back on the control column, the pilot moves the elevator up, sending the air flowing over the tail upward. This air pushes the plane's tail downward, which points the plane's nose up so that the plane climbs. The opposite happens when the pilot pushes the control column forward.

A cross section of a bird's wing, a boomerang, a paper airplane, and a Stealth bomber are all the same shape: relatively flat on the bottom, with an asymmetrical curve on top. Whether it's part of a sparrow or a jet plane, aerospace engineers call this shape an **airfoil.** It's the shape used in airplane wings.

When an airfoil moves forward, air passing under it flows straight. Air passing over the airfoil must travel a longer distance, around the curve. Spreading this air over the longer distance decreases the air's pressure above the airfoil. The

higher pressure of the air under the airfoil tends to press the airfoil up. And as the airfoil moves up, so does the plane.

The airfoil's curved top is shaped so that the front—the leading edge—is thicker than the back. Air passing over the thickest section jumps over the part of the wing directly behind the airfoil's peak. This jump leaves very little air, and very low air pressure, behind the peak. Because that area is near the center of the wing, the lift force balances out. If the curve were symmetrical, the lift force would be focused toward the rear of the wing.

The pilot controls the airplane by directing air in the direction opposite to the way he or she wants the plane to move. This is accomplished through the control surfaces: the **rudder,** the **elevator,** and the **ailerons.** Through a control column, the pilot works the ailerons and the elevator. A pair of pedals pulls the rudder. But none of this equipment works unless the plane is moving through air. The job of the engine is to keep the plane moving fast enough so that it will fly.

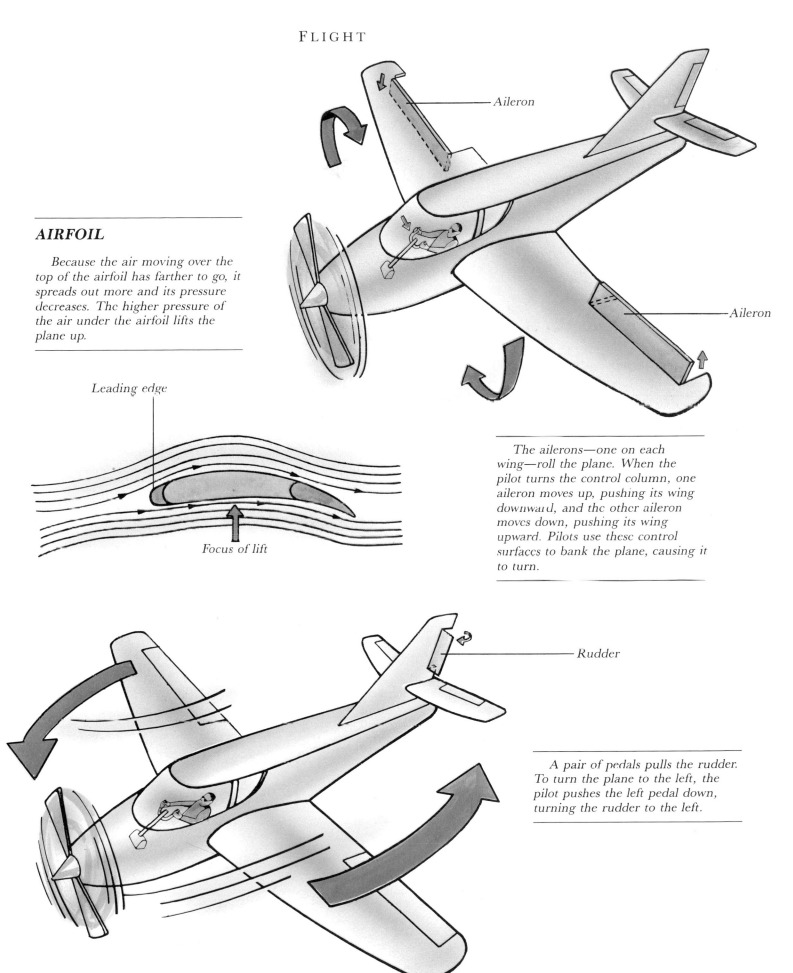

AIRFOIL

Because the air moving over the top of the airfoil has farther to go, it spreads out more and its pressure decreases. The higher pressure of the air under the airfoil lifts the plane up.

Leading edge

Focus of lift

Aileron

Aileron

The ailerons—one on each wing—roll the plane. When the pilot turns the control column, one aileron moves up, pushing its wing downward, and the other aileron moves down, pushing its wing upward. Pilots use these control surfaces to bank the plane, causing it to turn.

Rudder

A pair of pedals pulls the rudder. To turn the plane to the left, the pilot pushes the left pedal down, turning the rudder to the left.

213

JET ENGINE

A jet engine works on the mechanical principle that an action has an equal, opposite reaction. To produce the action, air is scooped into the engine and compressed until it is very hot (see **Compression,** page 12). That air is used to burn a fuel mixture, and the exhaust is expelled from the rear of the engine much faster than the air went in the front. The reaction from the exhaust impels the plane forward.

The compressor in a turbojet engine is an **axial compressor.** It has alternating sets of rotating and stationary blades radiating from a shaft. The action of each set of blades compresses the air slightly, so that by the time the air has moved through the blade system it is hot enough to ignite the fuel.

The fuel combustion results in a hot mass of gases that expands and rushes out of the combustion chamber past a **turbine.** A turbine is a wheel with blades on it that can be turned by either gas or liquid. In the jet engine, the movement of the gases turns the turbine, which turns the shaft of the compressor.

The gases shoot out the end of the engine, creating a powerful thrust that moves the plane forward. A nozzle at the rear of the engine increases the pressure of the escaping gases.

To slow down on landing, the jet reverses its thrust. The **thrust reversers,** hinged sections at the rear of the engine, swing together to block the exhaust, forcing it into the opposite direction to slow the plane.

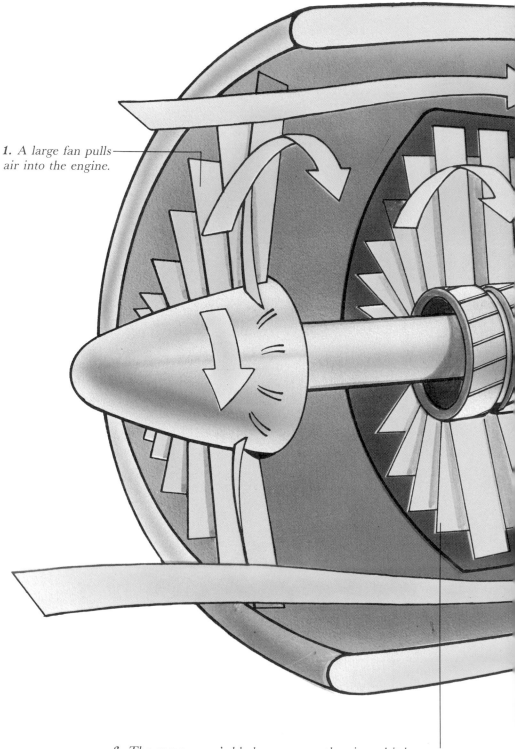

1. A large fan pulls air into the engine.

3. The compressor's blades compress the air and it heats up.

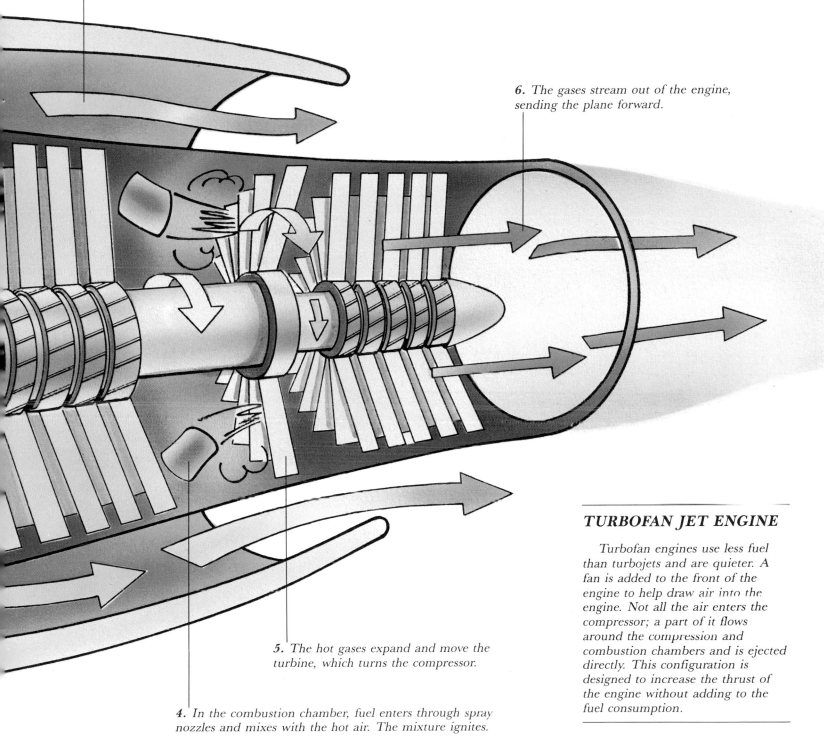

2. *Part of the air bypasses the compressor.*

6. *The gases stream out of the engine, sending the plane forward.*

5. *The hot gases expand and move the turbine, which turns the compressor.*

4. *In the combustion chamber, fuel enters through spray nozzles and mixes with the hot air. The mixture ignites.*

TURBOFAN JET ENGINE

Turbofan engines use less fuel than turbojets and are quieter. A fan is added to the front of the engine to help draw air into the engine. Not all the air enters the compressor; a part of it flows around the compression and combustion chambers and is ejected directly. This configuration is designed to increase the thrust of the engine without adding to the fuel consumption.

HELICOPTER

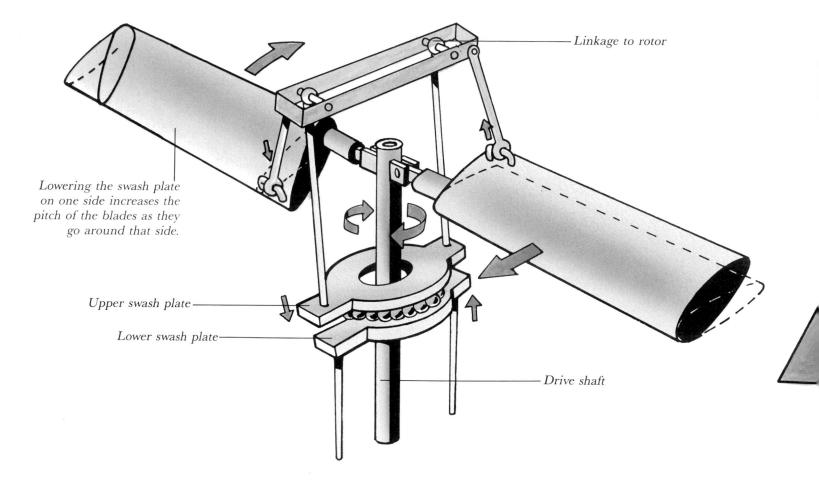

Linkage to rotor

Lowering the swash plate on one side increases the pitch of the blades as they go around that side.

Upper swash plate

Lower swash plate

Drive shaft

Helicopters can take off vertically and can hover in the air. Most airplanes can't, because the wings of airplanes are fixed to their bodies. They must be moving forward very fast in order to create enough lift at the wings to rise (see **Airplane, page 212**). Helicopters' rotor blades are also the airfoil shape, and by rotating the blades through the air, enough lift is generated to get the helicopter off the ground.

As the helicopter's blades rotate, a blade moving forward, into the flow of air, develops much more lift than a blade moving downwind does (see **Boomerang, page 206**). If not

corrected, this would create an unstable ride. To compensate, the blades are hinged, which allows them to flap up and down in response to the variation in lift.

The pilot steers the helicopter by varying the pitch, or tilt, of the blades. The greater the pitch of an airfoil, the more lift it has. By increasing the pitch of the blades as they pass the rear of the helicopter and decreasing pitch as they swing around the front, the lift is greater toward the back of the helicopter. The whole rotor then tilts forward, and the helicopter moves forward.

*The rotor head is designed to change the pitch of the blades in varying degrees as they rotate. The key to its operation is the **swash plate**, a pair of round metal plates separated by bearings and with a hole in the middle for the drive shaft. The upper swash plate rotates along with the blades; the lower swash plate does not rotate.*

One edge of each rotor is connected through a linkage to the upper swash plate. The lower swash plate is linked to the pilot's controls. When the pilot raises one side of the lower swash plate to tilt the upper swash plate, the linkages to the rotors ride up and down as they go around the drive shaft, changing the pitch of each rotor blade automatically.

216

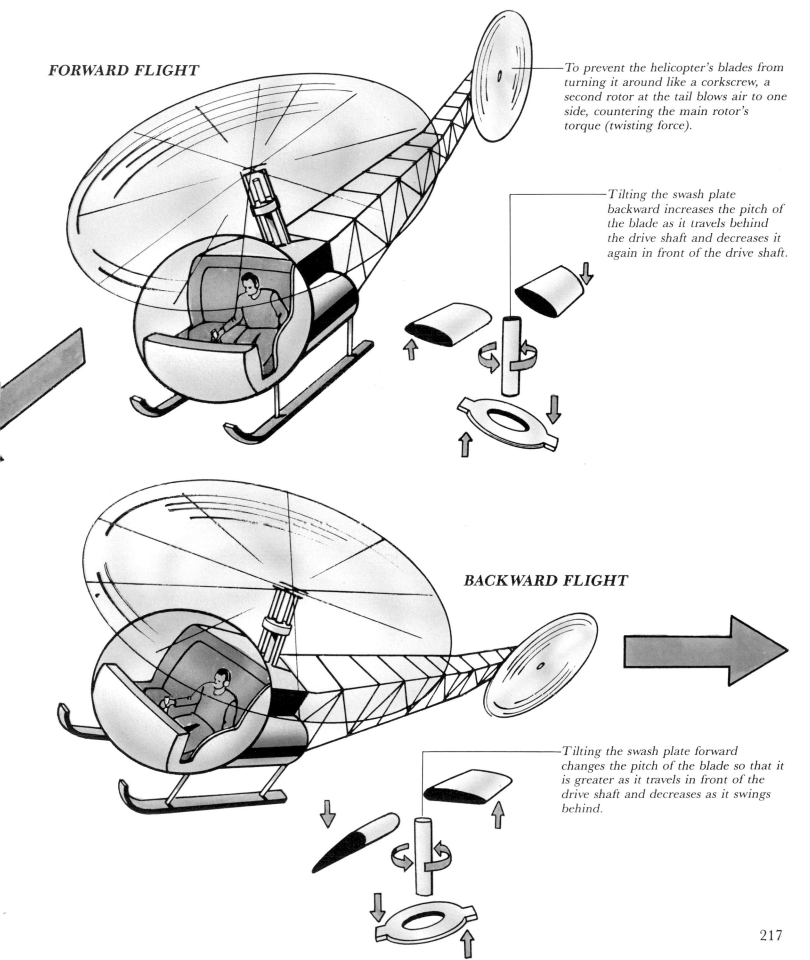

FORWARD FLIGHT

To prevent the helicopter's blades from turning it around like a corkscrew, a second rotor at the tail blows air to one side, countering the main rotor's torque (twisting force).

Tilting the swash plate backward increases the pitch of the blade as it travels behind the drive shaft and decreases it again in front of the drive shaft.

BACKWARD FLIGHT

Tilting the swash plate forward changes the pitch of the blade so that it is greater as it travels in front of the drive shaft and decreases as it swings behind.

217

RADAR

Radar was developed during World War II to track enemy planes and ships, but today radar systems have many forms and applications, including air traffic control and satellite tracking.

Radar is an acronym for RAdio Detecting And Ranging. Short pulses of microwaves are sent out by a powerful radio transmitter. When the waves strike an object, a portion of them are reflected back to the source as an echo. (See **Radio waves and microwaves,** page 52.) The pulses are separated by intervals long enough to allow for the returning echo. Since the speed of radio waves is known—it is the same as the speed of light—the time it takes for the signal to return indicates how far away the object is. The bearing of the object is the direction from which the echo returns.

Because the radar signal is sent out intermittently, the same antenna can be used for both transmitting and receiving signals. The antenna reflectors are dish-shaped. Microwave pulses bounce off the screen and are focused into a strong beam that can travel hundreds of miles. When the transmission is turned off to wait for a returning echo, the echo pulses bounce off the screen into the receiver.

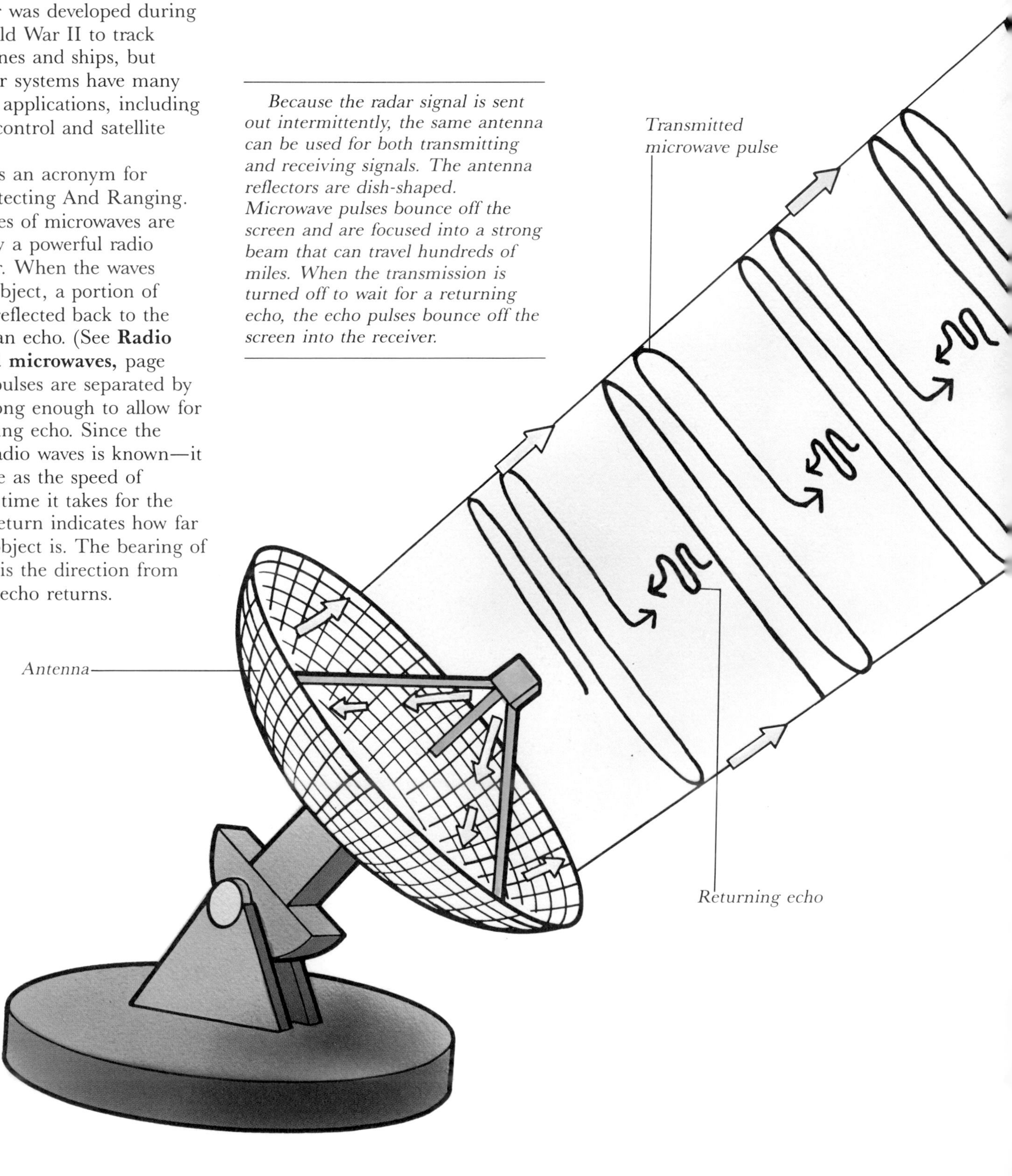

Transmitted microwave pulse

Antenna

Returning echo

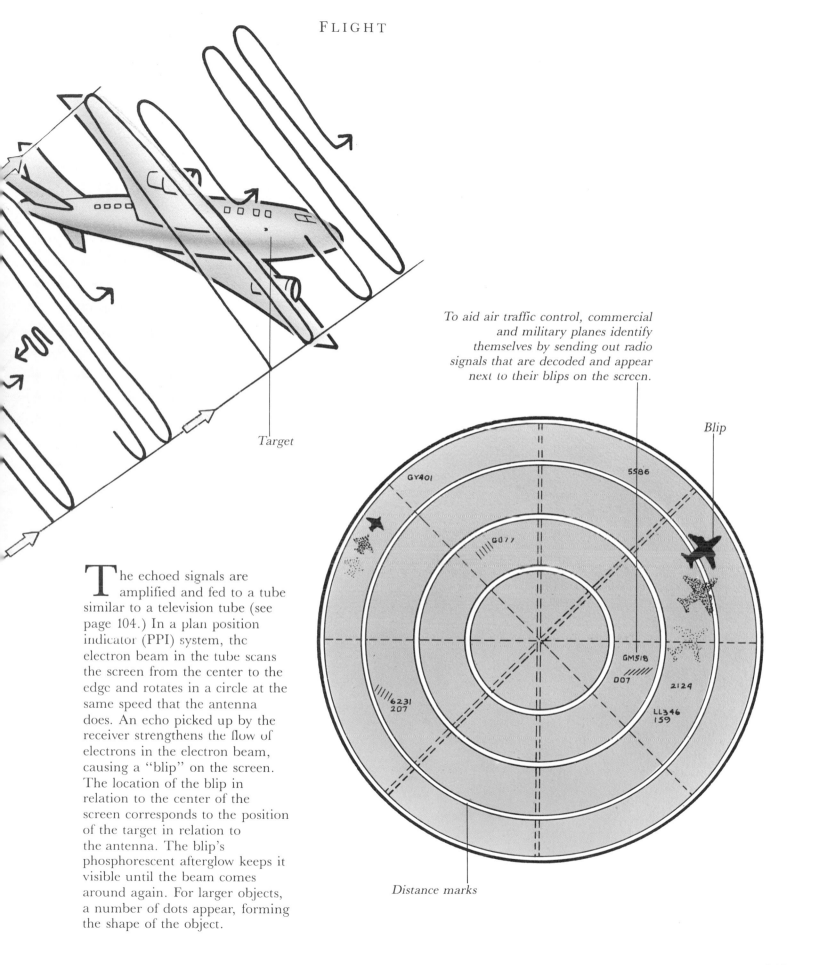

Target

To aid air traffic control, commercial and military planes identify themselves by sending out radio signals that are decoded and appear next to their blips on the screen.

Blip

GY401

5586

G077

GM518

D07

2124

6231
207

LL346
159

Distance marks

The echoed signals are amplified and fed to a tube similar to a television tube (see page 104.) In a plan position indicator (PPI) system, the electron beam in the tube scans the screen from the center to the edge and rotates in a circle at the same speed that the antenna does. An echo picked up by the receiver strengthens the flow of electrons in the electron beam, causing a "blip" on the screen. The location of the blip in relation to the center of the screen corresponds to the position of the target in relation to the antenna. The blip's phosphorescent afterglow keeps it visible until the beam comes around again. For larger objects, a number of dots appear, forming the shape of the object.

AUTOPILOT

The autopilot system of an aircraft depends chiefly on motor-driven gyroscopes (see page 43) for its information. Since the axis of a spinning gyroscope remains stable regardless of the position of the airplane, a gyroscope can indicate whether the plane is pitching or rolling or if it is off course. Some gyroscopes are part of the plane's navigation equipment. Planes use a gyromagnetic compass, which uses a gyroscope to stabilize the magnetic compass. Another gyroscope is part of the artificial horizon, which has a horizontal indicator bar that can be compared with the axis of the gyroscope. If the two are even, the plane is level.

When the plane's attitude changes in respect to that of the gyroscopes, a signal is sent to the plane's autopilot system. The signal is amplified and sent to mechanisms that operate the airplane's **control surfaces** (see **Airplane,** page 212) just as the pilot would operate them manually. The autopilot also receives information from the plane's accelerometer (which senses increasing or decreasing speed) and altimeter (a radar device that detects the plane's distance from the ground). As a fail-safe measure, the plane has three separate systems working simultaneously.

The autopilot can actually keep a plane on a level course far more accurately than a human pilot could, and by relieving the pilot of routine activity it keeps the pilot alert for situations in which the judgment of a human is necessary.

INSTRUMENT LANDING SYSTEM

Using the instrument landing system (ILS), a plane can land in practically zero visibility. From the airport, two antennas transmit radio beams in almost the same direction. Receivers in the plane direct instruments to guide the plane midway between the two beams, which is a path directly to the runway. Two other beams, one above the other, govern the slope of the plane's approach.

Glide-path beams indicate the proper altitude for approach.

Localizer beams show the correct direction for approach.

Marker beacons indicate the distance to the runway.

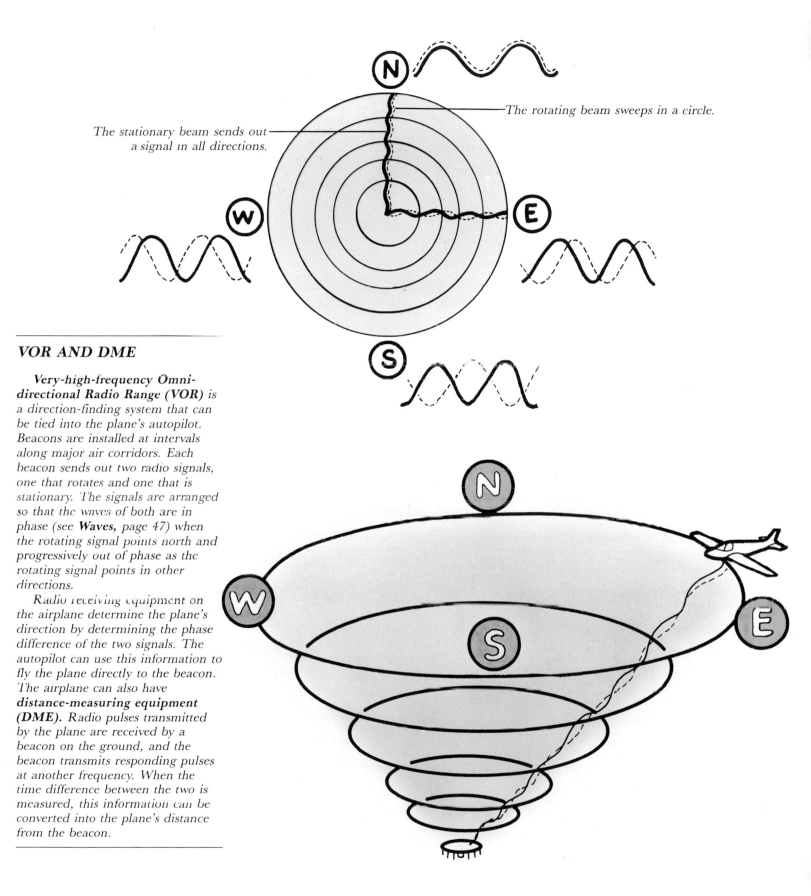

The stationary beam sends out a signal in all directions.

The rotating beam sweeps in a circle.

VOR AND DME

Very-high-frequency Omni-directional Radio Range (VOR) is a direction-finding system that can be tied into the plane's autopilot. Beacons are installed at intervals along major air corridors. Each beacon sends out two radio signals, one that rotates and one that is stationary. The signals are arranged so that the waves of both are in phase (see **Waves,** page 47) when the rotating signal points north and progressively out of phase as the rotating signal points in other directions.

Radio receiving equipment on the airplane determine the plane's direction by determining the phase difference of the two signals. The autopilot can use this information to fly the plane directly to the beacon. The airplane can also have **distance-measuring equipment (DME).** Radio pulses transmitted by the plane are received by a beacon on the ground, and the beacon transmits responding pulses at another frequency. When the time difference between the two is measured, this information can be converted into the plane's distance from the beacon.

SUPERSONIC FLIGHT

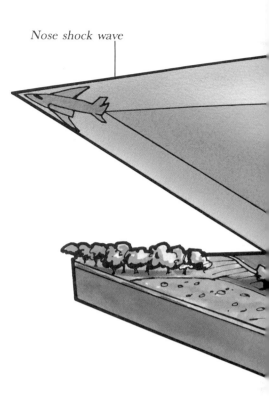

Nose shock wave

In the 1940s, before Chuck Yeager had flown faster than sound, supersonic flight was called "breaking the sound barrier." It was thought that a plane flying faster than its own sound waves would break up in the air. In a sense, a "sound barrier" actually exists, because a plane traveling at the speed of sound creates a wall of sound waves right in front of it.

Imagine a boat traveling faster than the water's waves. It creates a V-shaped pattern of crests on the water's surface that streams out from the bow. This is called a **bow wave.** A plane traveling faster than the speed of sound also creates a bow wave, but it is three-dimensional, with the crest of the wave streaming from the plane's nose and surrounding the plane in a cone shape. This shape is called the **Mach cone,** after the Austrian physicist Ernst Mach. **Mach numbers** are the ratio of the speed of the plane to the speed of sound in the surrounding atmosphere; Mach 1 is the speed of sound at sea level (760 miles per hour).

Once a plane has exceeded Mach 1, it is traveling faster than its own sound waves and is in complete silence.

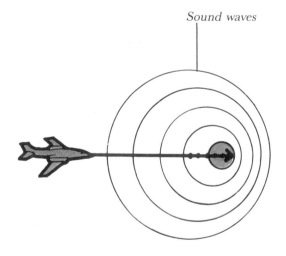

Sound waves

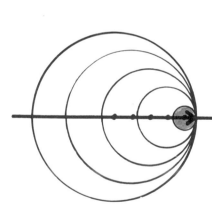

Mach cone

*A plane traveling under the speed of sound is pushing sound pressure waves around it in all directions. (See **Sound,** page 48.) The sound waves going ahead of the plane push the air out of the plane's way before the plane gets to the waves.*

A plane traveling exactly as fast as the speed of sound encounters a bunched-up group of waves where the air has not moved aside. Because of this air compression, the plane is subject to increased pressure and turbulence.

Once the plane has gone past the "sound barrier" and is traveling faster than the sound waves, it enters a zone of total silence where none of its sound waves can catch up with it.

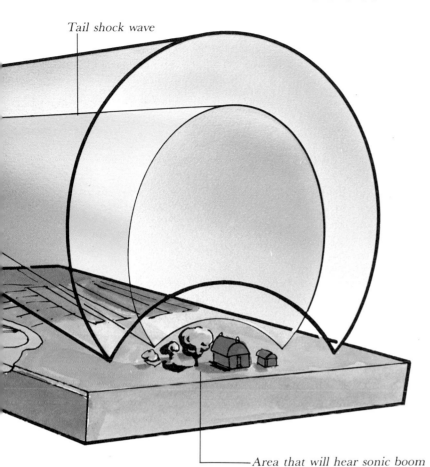

Tail shock wave

Area that will hear sonic boom

At supersonic speeds, the bow wave becomes a shock wave. The compression of the air in the wall of bunched-up sound waves heats the air. Because sound travels faster in warmer air, subsequent sound waves from the trailing part of the plane are traveling faster than the previous waves. These faster waves catch up with the first crest of waves, producing an extremely intense wave front. This causes the sonic boom associated with supersonic flight. The pressure of this shock wave is lessened by the time it arrives on the ground, but nonetheless it can cause considerable damage.

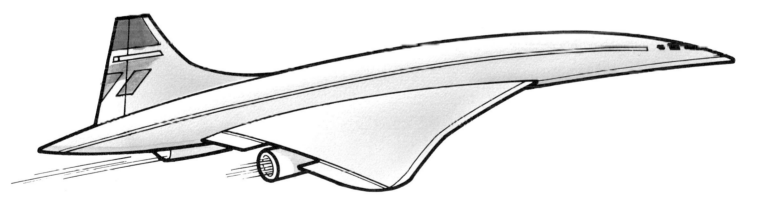

Because the shock waves created at speeds greater than Mach 1 cause considerable drag, supersonic planes are designed differently from subsonic planes. The airfoil shape that gives the ordinary plane its lift (see **Airplane,** page 212) can develop excessive compression at the thicker leading edge. A new shape, with a sharp leading edge and more symmetrical contours, produces less drag and enough lift because of the plane's great speed. The whole plane is proportioned to fit inside the Mach cone it generates. The wings are swept back in a delta shape. At speeds faster than Mach 3, the increased friction with the air produces so much heat that the usual construction materials become softened. Therefore, stainless steel and titanium are used.

AIRPORT SECURITY SYSTEMS

To protect air passengers from terrorism and sabotage, airports use devices that search the passengers and their luggage for guns or other ominous objects. You remove keys and other metal objects you are carrying and walk through metal detectors, which measure the change in a magnetic field when metal passes through it. Older metal detectors could not distinguish different types of metals, and guns made of some metals could slip past them. New, more sensitive detectors not only indicate these metals, but can also distinguish among certain types of narcotics and explosives, and they even identify the make and caliber of a hidden gun.

To do this, a detector probes a package or person with electric and magnetic fields at harmless levels. All types of materials that conduct electricity register a "signature," a unique graphic form that can be plotted. These signatures can be stored in a computer memory. When the detector finds a signature that matches, it is cause for suspicion.

To search baggage, X-ray machines show shadows of the metal objects inside the suitcases the same way an X-ray machine can show the bones in our bodies (see page 286). The original idea was to search for guns, which would indicate a potential hijacker. But plastic explosives, which have been used to blow up airplanes, do not show up well on X-ray pictures. At a few airports, a new type of bomb detector is being used that detects materials having a high concentration of nitrogen—such as plastic explosives. This detector, a thermal neutron activation (TNA) machine, bombards the suitcase with neutrons, which are subatomic particles. The neutrons enable the machine to detect the presence of a certain level of nitrogen.

A baggage X-ray machine exposes the luggage to X rays that are detected by diodes, and an image of the inside of the suitcase is registered on a video screen. Metal is a dense material that absorbs a lot of X rays, so that metal objects show up as a shadow. Plastics, such as plastic explosives, do not absorb so many X rays and create an indistinct or practically invisible shadow.

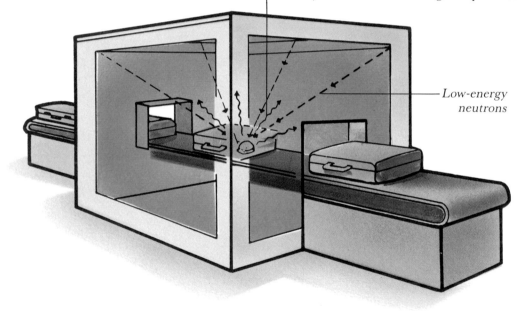

Gamma rays are emitted if nitrogen is present.

Low-energy neutrons

BAGGAGE BOMB DETECTOR

In a thermal neutron activation (TNA) detector, the luggage is bombarded with low-energy neutrons. The neutrons cause nitrogen to emit gamma rays, which are electromagnetic rays of a particular frequency (see page 50). Detectors then sense the gamma rays. If a high enough level of nitrogen is detected in a suitcase, operators will be able to spot it.

The machine is set at a level of sensitivity that discourages false alarms. Some materials—certain types of fabrics and large plastic items—can show densities of nitrogen high enough that the objects can be mistaken for explosives.

FLIGHT RECORDER

An air accident, no matter how destructive or how remote, has at least two witnesses: the plane's flight data recorder and cockpit voice recorder. These devices became mandatory on commercial aircraft decades ago, and they have helped investigators piece together the causes of hundreds of air accidents since then.

Cockpit voice recorders are simply tape recorders (see page 78) that record the voices of the crew through a microphone. The flight data recorder, or FDR, is like a tape recorder in that it receives signals, amplifies them, and puts them on magnetic tape using an electromagnetic head. The signals come from a variety of sensors that monitor such factors as air speed, climb rate, engine thrust, altitude, and heading. Federal regulations now require that FDRs measure at least 17 different performance statistics.

The signals travel first to the flight data acquisition unit (FDAU), which converts the wide variety of signals to a digital signal—a series of ones and zeros—that a computer can read (see **Binary code,** page 155). From the FDAU, the signals go to tape. The tape records for 25 hours and then begins recording over old information.

To meet Federal regulations, these units must be able to survive a 2,000°F fire for 30 minutes and be able to withstand the impact of a 500-pound steel rod dropped on their cases. Aircraft manufacturers go one step further by isolating the units on shock-damping mounts in the plane's tail.

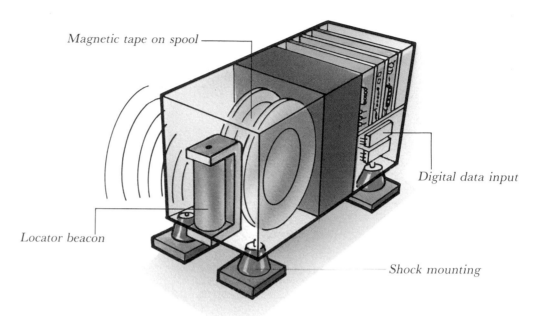

Magnetic tape on spool

Locator beacon

Digital data input

Shock mounting

The rugged unit for the FDR houses electronic processors for the the digital data, a motor and heads for the recorder, and a spool of tape to record on. If a plane crashes, the investigators really need only the tape spool, so it has more fire protection than the rest of the FDR.

These recorders are popularly known as "black boxes," but they are actually bright orange or yellow and covered with strips of reflective tape to make them easy to find. In case of a crash in the ocean, each flight data recorder and cockpit voice recorder includes a locator beacon, which is a small tube that emits a loud, high-frequency ping if it gets wet.

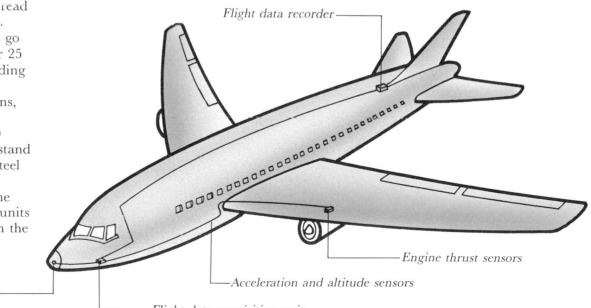

Flight data recorder

Air-speed sensors

Flight data acquisition unit

Acceleration and altitude sensors

Engine thrust sensors

FLIGHT SIMULATOR

Projectors above the cockpit throw their images onto a curved screen. This image is reflected by a curved mirror that surrounds the simulator cockpit. When the pilot looks out, he or she sees a reflection of the computer-controlled image. This gives the simulation a feeling of depth and perspective.

After slipping a coin into a video arcade simulator, you can feel as if you are skidding a Ferrari sports car around a corner or chasing enemy jets at supersonic speeds. Professional flight simulators that help train pilots do the same kind of thing—they make you feel like you are flying. But the detailed realism they attain puts even the best arcade graphics to shame.

Sophisticated projectors are linked to powerful computers. Three and sometimes as many as five projectors are in constant operation, with computers changing the perspective seen by the pilot in the cockpit. As the pilot simulates taxiing the plane onto the runway and throttling up to begin takeoff, he or she experiences exactly the same sights, sounds, and vibrations that would accompany an actual takeoff. Once "in the air," roadways and buildings seem to become tinier and tinier as hydraulic jacks pitch the simulator upward. Computers, responding to the pilot's pressure on the controls, direct this illusion with the help of screens and mirrors outside the cockpit window.

Inside the simulator, pilots practice landing, turning, and controlling the plane. With this kind of experience, novice pilots can become expert flyers before they ever step into the cockpit of a real aircraft. Their training lets them know what to expect and how the aircraft "feels" while it's

in the air. Flight simulators are so realistic that some pilots even get airsick in simulated flight.

Besides saving money and time, flight simulators can help save lives. Such emergency flying situations as engine loss, severe turbulence, and system failures can be simulated in the cockpit, giving the pilots-in-training a chance to encounter life-and-death situations while in training. Flight simulation can give pilots valuable experience in how to deal with real emergency situations.

Although they imitate flight down to the smallest detail, the exteriors of flight simulators look nothing like airplanes. The whole structure stands atop a platform that is supported by hydraulic jacks. When the pilot inside the simulator uses the controls, computers adjust the jacks automatically to reproduce the feeling of turning, rising, and diving in an airplane.

The cockpits of professional flight simulators are exact replicas of the aircraft that they imitate. If the pilot is training on a 747, for example, the simulator cockpit reproduces every dial, gauge, switch, and lever that the real plane has. The seats and even the seat belts are the same as those used on real 747s.

Pilot cockpit

Hydraulic jacks

On larger simulators, instructors sit behind the pilot surveying a computer console. From here the instructor can unexpectedly add emergency situations or stop the simulation, correct the student, and continue the training.

ROCKET ENGINE

A rocket is one of the simplest engines in the world. A "bottle rocket" or firework rocket is a tiny example of the same principles that lift multimillion-dollar satellites and spacecraft into orbit around the Earth. If you take a cylinder with one open end, pack it with materials that burn fiercely, and ignite them with a fuse, a furious reaction takes place. The hot gases produced by combustion rush out of the open end at great speed. The rocket reacts by lifting up into the air. (See **Jet engine**, page 214.) Except for valves and pumps that control the rate of fuel consumption, rocket engines have no moving parts.

Solid-fuel rockets are closely related to firework rockets—both use a solid chemical mixture as a propellant, and, once ignited, both kinds of rockets will "burn" until all the propellant is gone. The primary difference between firework rockets and the elaborate space rockets that took the Apollo astronauts to the moon is the fuel, or propellant, used to achieve thrust. Before the introduction of the space shuttle to America's aerospace fleet, only liquid-propelled rockets were used to lift astronauts and satellites into space. These rockets were giant-size, multi-stage devices that slowly blended two or more liquids together to produce the explosive thrust of lift-off.

Today, NASA has nearly abandoned its large, multi-stage liquid-fuel rockets. The newest launch designs combine liquid-fuel rockets with solid-fuel rockets. (See **Shuttle launcher**, page 230). Experience has shown that loading liquid-fuel rockets with explosive fuel mixtures is a dangerous undertaking.

In fact, the primary drawback to liquid-fuel rockets is their danger. The combustion process that lifts thousands of tons of machinery off the ground is actually a kind of controlled explosion, which can easily become an out-of-control explosion. Solid-fuel rockets are likely to be more stable and less prone to accidental combustion.

The nozzle area of the rocket is all-important. It must be tough to withstand the intense heat and pressure of rocket gases escaping out the rear. At the same time, it must be able to swivel, so that the direction of the rocket can be controlled.

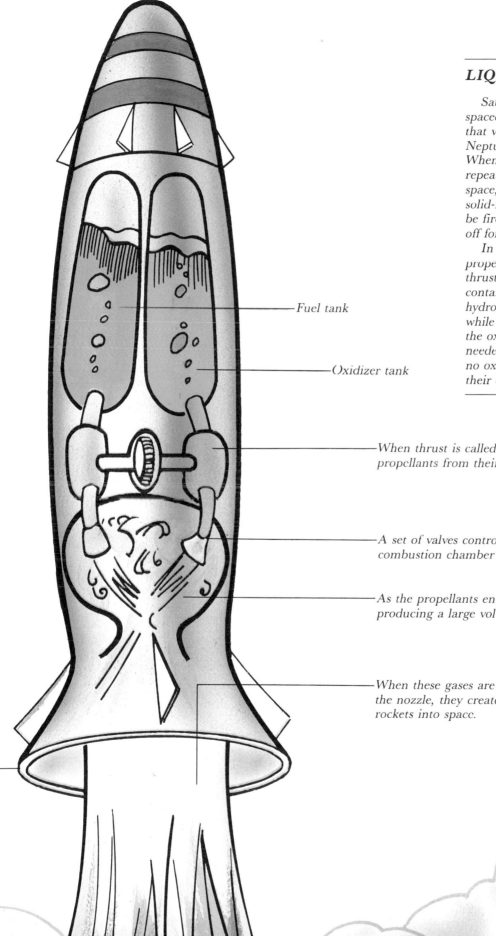

Fuel tank

Oxidizer tank

LIQUID-FUEL ROCKETS

Satellites and interplanetary spacecraft such as the Voyager probe that visited Jupiter, Saturn, and Neptune use liquid-fuel rockets. When engines must be fired repeatedly, often for maneuvering in space, liquid fuels are best. Unlike solid-fuel rockets, these rockets can be fired for a short time and shut off for use later.

In liquid-fuel rockets, the propellants that create explosive thrust must be kept in separate containers. In most cases, liquid hydrogen occupies the fuel tank, while liquid oxygen (LOX) occupies the oxidizer tank. Since oxygen is needed for combustion and there is no oxygen in space, rockets bring their own along.

When thrust is called for, pumps draw the propellants from their storage tanks.

A set of valves controls the amount of fuel entering the combustion chamber near the bottom of the rocket.

As the propellants enter the chamber, they ignite, producing a large volume of hot gases.

When these gases are forced out of the rocket through the nozzle, they create thrust—the force that hurls rockets into space.

SHUTTLE LAUNCHER

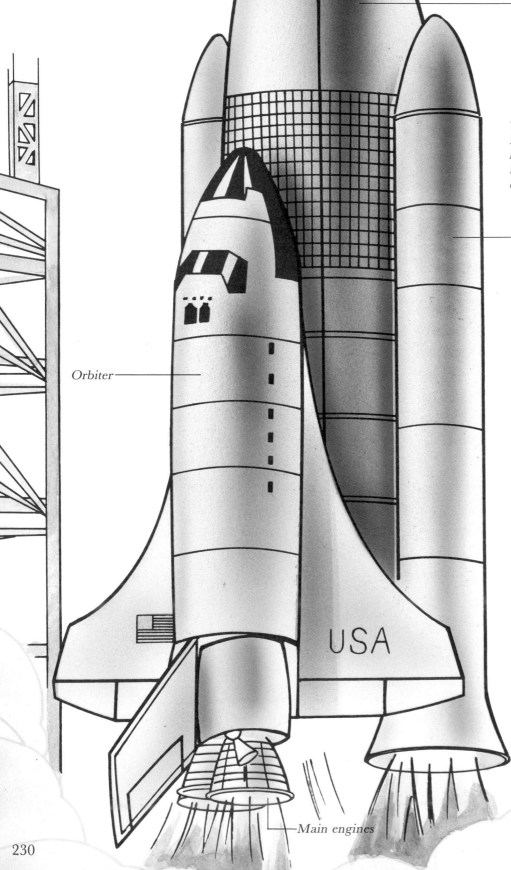

The center external tank carries LOX in one chamber and liquid hydrogen in another chamber. This is the fuel for the three main engines at the rear of the orbiter.

Twin solid-fuel rocket boosters flank the big external tank in the center.

Orbiter

USA

Main engines

Five principal engines make up the space shuttle launch system—two solid-fuel rocket boosters and three main engines on the orbiter. The orbiter's main engines are liquid-fuel rockets that are exposed to severe temperature extremes. Liquid hydrogen, at –423°F, is the second coldest liquid on Earth. When it combines with liquid oxygen (LOX) in the engine combustion chamber, the temperature rises to 6,000°F, higher than the boiling point of iron. The power released by the main engines at full throttle is equivalent to the output of 23 Hoover Dams.

The two solid-fuel rocket boosters generate a combined thrust of 5.3 million pounds, equivalent to 44 million horsepower. It would take 14,700 six-axle diesel locomotives to produce 44 million horsepower.

The space shuttle launch system is called a "hybrid" rocket system. Both liquid-fuel rockets and solid-fuel rockets are used to propel the craft into orbit. The entire flight system is a four-piece unit, with the orbiter composing just one of the pieces. The other three pieces—two solid-fuel rocket boosters and a large external fuel tank—launch the orbiter into space.

As the countdown clock reaches zero, the three main engines of the orbiter begin gulping fuel from the external tank, sending out a storm of flame and smoke. About four seconds later, the solid-fuel boosters kick in. With a deafening roar, the shuttle lifts off from its launching pad.

The first eight minutes of flight are the most critical. Two minutes after lift-off, when the shuttle is about 30 miles high, the twin solid-fuel boosters are nearly burned out. At this point, tiny explosions separate them from the external tank. As they drop toward Earth, parachutes pop out of their nose cones. The solid-fuel boosters drift gently into the sea to be picked up by tugboats and towed home. They will be refitted and used for another shuttle flight.

At the eight-minute mark, about 70 miles above the planet, the orbiter separates from the external fuel tank—the large center tank that it rides "piggyback" into space. Unlike the solid-fuel boosters and the orbiter itself, the external tank is not used again. As it falls back to Earth, it breaks up and tumbles into the ocean.

On the shuttle program's 25th mission, January 28, 1986, problems with the solid-fuel rocket boosters triggered an explosion of the external tank. The shuttle *Challenger* was destroyed and seven crew members were killed. Experts say that, since then, design changes in the solid-fuel rocket boosters have eliminated the danger. Nonetheless, engineers are studying the possibility of adding emergency escape capsules to the next generation of orbiters.

SPACE SHUTTLE

Since 1981, the space shuttle has become NASA's primary craft for space exploration and research. This machine is like three machines blended together: The vehicle takes off like a rocket, orbits the Earth like a spacecraft, and lands like an airplane.

The crew and payload ride in the orbiter section of the space shuttle system. Its sleek shape helps it stream through the air on its way to space. On its way home, the orbiter enters the Earth's atmosphere at a speed of more than 16,000 miles per hour. By the time it lands, the craft's speed drops to only 200 miles per hour.

Satellites to be put into orbit are stored in the payload section of the orbiter. Once the shuttle is in orbit, the crew pilots the space shuttle to an exact location and places the satellite into orbit. Usually, a specially trained technician operates a robot arm in the payload bay, gently picking the satellite up and carefully placing it in orbit. The robot arm can also be used to pluck faulty satellites from orbit for a trip back to Earth.

The shuttle crew eats, works, and sleeps in their airtight compartments near the front of the spacecraft. Up to seven astronauts and mission specialists can be accommodated in the crew's quarters. Unlike earlier astronauts, shuttle astronauts have no need for airtight space suits. As long as they stay inside the crew compartments, they can work in shirtsleeves. However, if they want to leave this area, they must wear space suits. Crew members can put on space suits and leave the orbiter to fix satellites or examine the ship for damage.

The space shuttle is the first reusable spacecraft to be built. Each ship in the shuttle fleet is designed to last for 100 missions. Experts predict that the space shuttle will allow astronauts to build and maintain a fully operational space station.

The upper section, or flight deck, is the control cockpit for the shuttle's pilot and commander. Below this deck are the sleeping quarters and the galley.

The surface of the orbiter is covered with more that 25,000 individual tiles that are specially treated to withstand the intense heat caused when the orbiter reenters Earth's atmosphere.

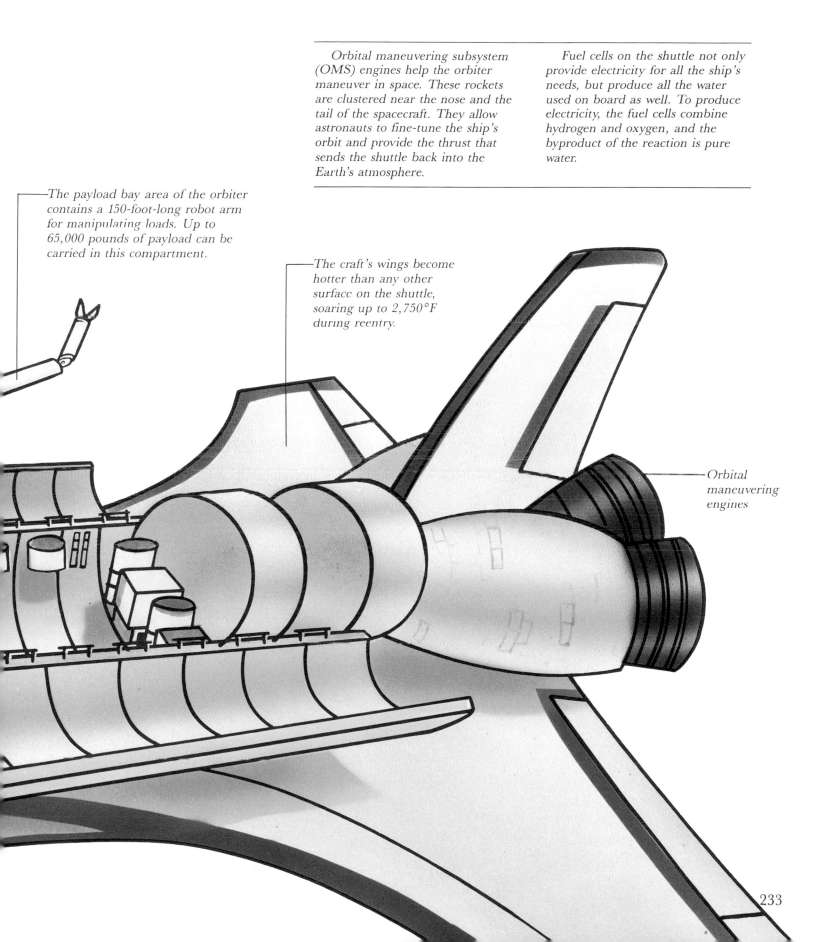

Orbital maneuvering subsystem (OMS) engines help the orbiter maneuver in space. These rockets are clustered near the nose and the tail of the spacecraft. They allow astronauts to fine-tune the ship's orbit and provide the thrust that sends the shuttle back into the Earth's atmosphere.

Fuel cells on the shuttle not only provide electricity for all the ship's needs, but produce all the water used on board as well. To produce electricity, the fuel cells combine hydrogen and oxygen, and the byproduct of the reaction is pure water.

The payload bay area of the orbiter contains a 150-foot-long robot arm for manipulating loads. Up to 65,000 pounds of payload can be carried in this compartment.

The craft's wings become hotter than any other surface on the shuttle, soaring up to 2,750°F during reentry.

Orbital maneuvering engines

GUIDED MISSILES

Since jet fighter planes travel so fast, bullets from machine guns find it difficult to catch them. Modern guided missiles, however, can stay on target even if the enemy plane attempts evasive maneuvers.

Air-to-air missiles use a variety of guidance systems, but the most common is probably infrared homing, or heat-seeking. This system searches for heat, specifically the exhaust from a jet or piston engine. The missile is launched by a rocket engine (see page 228). On pulling away from the plane, the missile is already pointed in roughly the right direction. To locate the target precisely, heat-seeking missiles use a **goniometer,** a sensitive direction-finding device that connects to thermal sensors in the missile's nose. When the goniometer detects heat, it sends power to move flaps on the missile's fins so that it is steered in the direction of the heat.

Ground-to-air (antiaircraft) missiles, which also have airplanes as targets, frequently use radar guidance (see **Radar,** page 218). Heat-seeking sensors don't work well at long distance, and since ground-to-air missiles protect surface facilities, they try to strike attacking aircraft while they are a long way off.

Inside the missile is a radio receiver and an autopilot mechanism (see **Autopilot,** page 220). A targeting system on the ground leads the antiaircraft missile to its prey. One radar tracks the enemy aircraft; another follows the missile's path. The radars feed information into a computer that projects the path of the enemy aircraft and sends

SIDEWINDER INFRARED HOMING MISSILE

MINUTEMAN ICBM

course information over a radio to the missile. The missile's autopilot mechanism controls its course.

Ground-to-ground missiles range in size from antitank missiles small enough to be carried by a soldier all the way to intercontinental ballistic missiles (ICBMs) that stand several stories tall. The ICBM's guidance system is neither a radar nor a heat-seeking system. The missile carries an internal guidance system—gyroscopes and **accelerometers** (devices that sense the missile's speed). If the instruments detect a course error, they send electrical power to move the rocket nozzle to compensate for the error. The system is accurate to within a few yards and cannot be jammed. Part of the journey of an ICBM is in the outer reaches of the Earth's atmosphere, where air resistance is not present to slow the missile down.

On the Water

BUOYANCY

An object floating in water—from a cork to a supertanker—stays afloat because of the principle of water displacement. As it sits in the water, the object is taking up space that would normally be filled with water. It is displacing water, making space for itself by pushing water out of the way. But the water reacts to displacement by creating a force of its own called **buoyancy.** The more the object settles into the water, the more buoyancy pushes back, keeping it afloat. So long as the object displaces enough water to provide the necessary buoyancy to support its weight, it floats.

Things weigh less under water than they do above the water's surface. An object totally submerged in a liquid is buoyed up by a force equal to the weight of the liquid displaced by the object. A five-pound metal block the size and shape of a pint jar will displace one pint of water, which weighs one pound. Under water, the block will weigh only four pounds. A floating object sinks into the water just enough so that the amount of water displaced is the same weight as the object.

Objects that sink must be heavy for their size. Objects that float must be light for their size. To predict whether an object will float, a measure is needed that includes both weight and volume. The measure that is commonly used is **specific gravity,** or the ratio of the weight of an object to the weight of an equal volume of water. Anything with a specific gravity of less than one will float; anything with a specific gravity greater than one will sink.

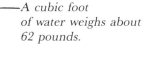
A cubic foot of water weighs about 62 pounds.

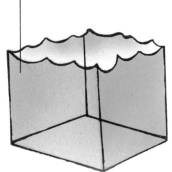

A cube of steel measuring one foot on each edge—a cubic foot of steel—weighs about 486 pounds. So the cube sinks.

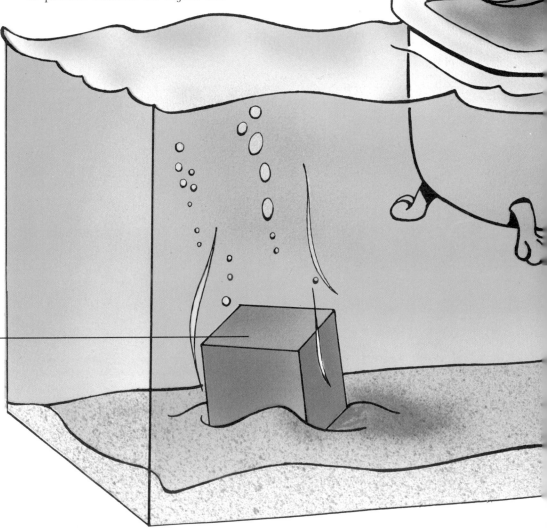

If that cube of steel is used to create a vessel 4 feet long, 2 feet wide, and 3 feet deep, the vessel still weighs 486 pounds on dry land. But even with a 100-pound boy inside, the vessel floats.

The vessel has a volume of 24 cubic feet (4 times 2 times 3). But it displaces only about 9½ cubic feet of water (586 pounds of boat and boy divided by 62 pounds per cubic foot of water).

Boats are often made of materials with specific gravities greater than one—boats have successfully been built of extremely heavy materials including steel and concrete. But boats are hollow. The air inside them normally assures that the overall specific gravity of the boat is much less than one.

If a boat springs a leak, however, the light air contained in its structure is replaced by water. Its overall specific gravity increases until it becomes greater than one, and the boat sinks. Raising a sunken ship consists of lowering its specific gravity until it floats again. This can be accomplished by sealing underwater leaks and pumping air into the vessel, or even by forcing low-density objects, such as ping pong balls, into the wreck.

SAILBOAT

The earliest boats were propelled by oars, but it didn't take long for prehistoric sailors to recognize that the wind was a potential source of propulsion. Although the first unquestionable evidence of boats with sails comes from Egypt more than 4,000 years ago, the sail was probably a well-known device as long as 6,000 years ago.

Exactly how the sail functions to propel the boat is quite complex. The surface of the sail catches the wind, which moves the craft along. Because the sail billows out in a curve, it behaves in the same way an airplane wing does (see **Airplane,** page 212). Lower air pressure develops on the outward-curving side of the sail, moving the boat in that direction.

Nearly all modern sailboats use triangular-shaped sails. This shape is far more versatile than a square sail—because all three sides are supported, the sail doesn't flutter along its leading edge the way a square sail would.

All but the smallest yachts use two sails. The **jib** or **foresail** is stationed in the front, and a larger **mainsail** extends behind the mast. The mainsail is secured at the bottom to a **boom,** which can swing about to position the sail in the best way to catch the wind.

When the sailor wants the boat to go in a direction other than the direction the wind is blowing, a **tacking** maneuver is used. The sailor sails at an angle to the wind and then cuts across in the opposite direction, following a zigzag course to its destination.

The keel—a weighted fin extending downward from the hull—makes an effective counterbalance to the force of the wind. Without a keel, a sailboat would skid in whatever direction the wind blew. But the resistance created by a boat's keel keeps the craft from skidding, giving it "traction" in the water. Furthermore, its weight can help prevent the boat from capsizing.

A boat's rudder is what turns the boat. As the craft moves through water, water flows past the curves of the ship's hull and keel. The rudder deflects the flow's current, moving the stern (rear of the ship). As a result, the whole vessel turns.

Jib

Mainsail

Boom

Stern

Rudder

Bow

Keel

238

Foremast

Mainmast

Mizzenmast

Aftermast

Rigging

CLIPPER SHIP

The clippers were the biggest, the tallest, and the last of the great sailing ships. Their heyday was between 1820 and 1880. Before they died out, ships such as the clippers achieved the status of the best and fastest sailing ships in the world.

Its sleek shape and its thick forest of masts distinguished the clipper ship from other sailing ships of the 1800s. Between three and six masts, with a total of up to 35 sails of varying sizes and shapes, helped the clipper to take maximum advantage of the wind.

Some of the larger clipper ships needed crews of 60 people to handle all the sails. Ships this large could carry up to 4,500 tons of goods. At top speed, a clipper ship could exceed 20 knots, crossing the Atlantic Ocean in as few as 12 days.

STEAMSHIPS

American inventor Robert Fulton's first steam-powered vessel, the *Clermont,* was an immediate success in 1807. The *Clermont* used a steam engine designed by James Watt to drive a pair of large, side-mounted paddle wheels. The paddles spun through the water, propelling the craft forward.

In principle, the workings of a steam engine are simple enough. Coal is used to fire a furnace hot enough to boil water. The steam from the boiling water is sprayed into a cylinder. When it expands, it forces the piston to move, creating mechanical power from heat energy.

The benefits of the steam engine were clear: No longer would the speed of a ship be at the mercy of capricious and contrary winds. With an engine on board, the ship could maintain constant speeds in dead-calm water or when the water was rough. Engine power also gave skippers better control of their vessels when approaching or leaving docks and when navigating tricky near-shore waters.

Early steam-driven paddle wheelers set the stage for modern ships. Engines had become proven power plants for ocean-going vessels.

Stabilizer fins

Propeller

Rudder

◀ PADDLE WHEELER

Steam-driven paddle-wheel boats were poor performers on the open ocean, but they were ideally suited for shallow waters. The rigors of the Mississippi River led to a new development, the rear-mounted paddle wheel. Earlier designs placed the paddles on the side of the ship. But rear-mounted paddles gave the craft a shallower draft, or depth in the water.

The design of Mississippi River paddle-wheel steamships was well suited to navigating in the shifting shallows of the great river. The ship's hull was long, somewhat narrow, and nearly flat, with many layers of passenger accommodations stacked on top. This design gave the ship a shallower draft, but it also made the paddle wheeler top-heavy and difficult to control.

In the case of "side-wheelers"— paddle boats with their paddles mounted on the side of the ship— the pistons spun a drive shaft that powered the movement of the paddles. Mississippi riverboats had large, rear-mounted paddles that were usually linked directly to the steam engine.

PASSENGER SHIP

Most passenger ships use steam-driven turbines for power. Water is heated by diesel-fuel boilers, and the steam spins the turbines, converting the energy of steam into mechanical energy. The turbine powers a drive shaft, which in turn spins one or more propellers at the stern of the ship.

The propellers bear a remarkable resemblance to airplane propellers and perform the same function—to drive fluid (in the case of a plane, air acts as a fluid) past the rear of the ship, thus propelling the ship forward. A rudder directs the flow of water at the stern of the ship, giving the pilot control over the vessel's direction.

Because of the tremendous size attained by some passenger ships, they tend to roll and pitch when the sea becomes rough. Stabilizer fins projecting from either side of the ship's hull help keep the vessel steady. These fins are hinged like the flaps of an airplane wing to counteract the rolling motion of the waves, and they are usually controlled automatically by gyroscopes that sense the ship's movement. Stabilizer fins can cut roll by as much as 90 percent.

CANAL LOCKS

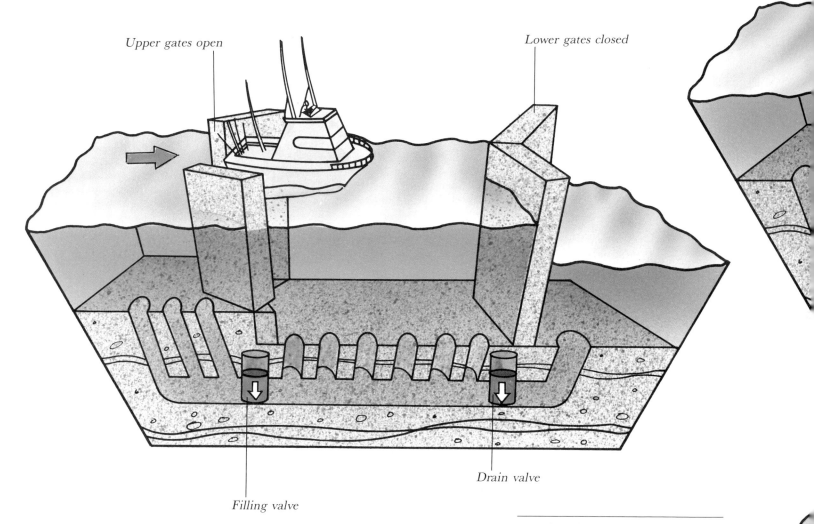

Upper gates open

Lower gates closed

Drain valve

Filling valve

Canals frequently connect two bodies of water that have different water levels. The Panama Canal, for instance, connects the Atlantic and Pacific Oceans; the Pacific Ocean has a mean water level several feet higher than that of the Atlantic. If the two were connected directly, a roaring torrent would result as water flowed toward the Atlantic. Boats would find it difficult to get through the canal.

The solution is to break the canal into sections, each with its own water level. The sections are connected by locks—which are like stairways for boats. A lock is an enclosure with watertight gates at each end. A system of valves and water passages changes the water level in the lock so that the boat can rise or fall to the next water level.

A boat is traveling from the higher water level to the lower water level. With the lock gates at the lower end closed and those at the upper end open, the water level in the lock is the same as the water level in the high-water section of the canal. The gates are angled toward the high-water section, so that the water pressure helps keep them shut. The upper gates can open easily, because the water pressure inside and outside the lock is the same.

A filling valve at the higher end of the lock has let water into the lock. A drain valve at the lower end is closed. The boat travels through the upper gates and stops. The gates close behind it.

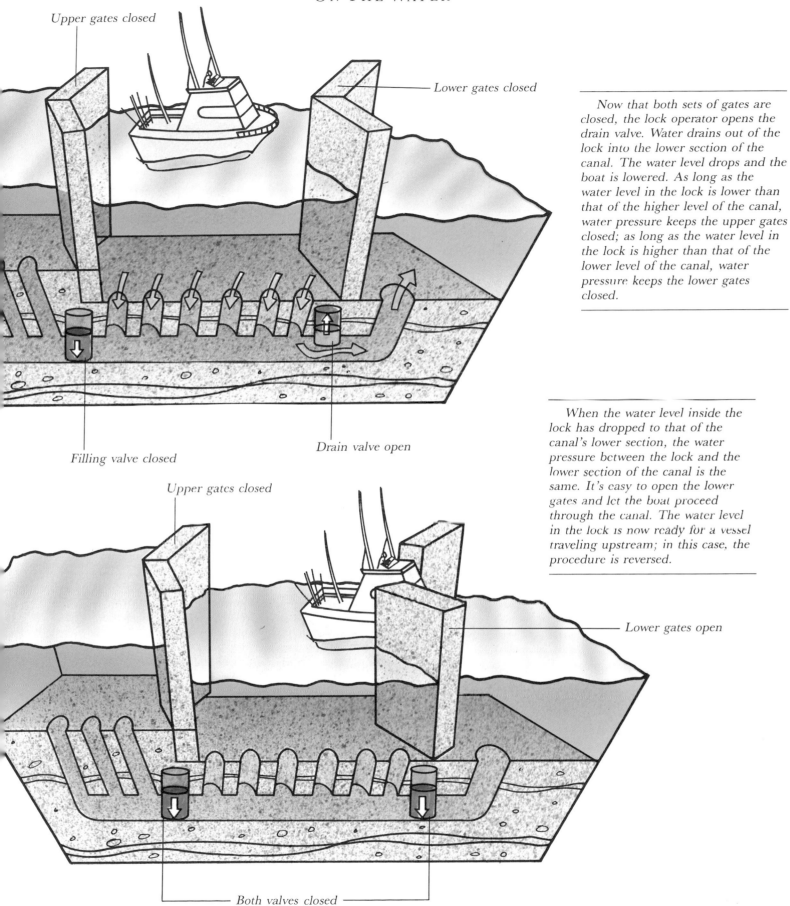

Upper gates closed

Lower gates closed

Now that both sets of gates are closed, the lock operator opens the drain valve. Water drains out of the lock into the lower section of the canal. The water level drops and the boat is lowered. As long as the water level in the lock is lower than that of the higher level of the canal, water pressure keeps the upper gates closed; as long as the water level in the lock is higher than that of the lower level of the canal, water pressure keeps the lower gates closed.

Filling valve closed

Drain valve open

Upper gates closed

When the water level inside the lock has dropped to that of the canal's lower section, the water pressure between the lock and the lower section of the canal is the same. It's easy to open the lower gates and let the boat proceed through the canal. The water level in the lock is now ready for a vessel traveling upstream; in this case, the procedure is reversed.

Lower gates open

Both valves closed

HYDROFOIL

Anything that floats will move through the water if it is pushed. The harder it is pushed, the faster it will move. As you approach a speed—referred to by boat designers as **hull speed**—the amount of push required to make the boat go faster increases at a rapid rate. Hull speed defines a practical limit to how fast any vessel can travel if it depends on buoyancy to keep afloat. Surprisingly, hull speed depends only upon the boat's length. Sleeker lines and smoother hulls may allow the boat to reach that speed with less effort, but they have no effect on what that speed is.

Hull speed exists because any vessel traveling through the water makes waves. Most boats produce one wave where the hull enters the water at the front—the **bow wave**—and one where the hull leaves the water at the stern—the **stern wave.**

Both the bow and the stern waves follow the familiar crest-and-trough pattern (see **Waves,** page 46). When the boat is moving very slowly, the waves are short and not very high, and they have little effect on one another. As the speed increases, they become longer and longer. Eventually the boat is traveling so fast that the trough, or low point, of the bow wave is trying to depress the same water that the crest of the stern wave is trying to lift. Since water can't be in two places at once, the two waves opposing each other create a great deal of turbulence. The result is a powerful drag, which acts like a brake on the boat. Applying more power does no good; it only increases the turbulence.

It's possible to make a boat exceed its hull speed by a large margin, without requiring a huge increase in power, by using **hydrodynamic lift** (see **Water skis,** page 254). If the boat starts to lift partway out of the water as it speeds up, it can literally climb over the hull-speed limitation. The trough of the bow wave will suddenly move back to clear the crest of the stern wave as the boat's bow leaves the water. This type of hydrodynamic lift is called **planing.**

But a hydrofoil greatly overcomes the drag of the water because its hull rises out of the water to soar through the air. The hydrofoil is not limited by hull speed—hydrofoils can achieve speeds between 45 and 75 miles per hour.

Although hydrofoils are not a new idea—Alexander Graham Bell built a successful one in 1918 that was clocked at 60 knots (about 69 miles per hour)—they were not developed into useful vessels until the 1950s.

*The hydrofoil gets its lift from small submerged "wings" that act in the water just the same way an airfoil acts (see **Airplane,** page 212). Since water is 600 times denser than air, a relatively small foil can lift a large boat.*

At low speeds, hydrofoils move through the water as conventional craft do. As they speed up, the lift from the foils causes them to take off from the surface and ride with the boat's weight completely supported by the foils.

The power for a hydrofoil usually comes from an engine driving a propeller on a shaft long enough to keep the propeller submerged when the hull is "flying." Some newer hydrofoils use a water-jet engine, which avoids the drag of the propeller. These engines take in water and thrust it out behind the ship with great force, which propels the ship forward.

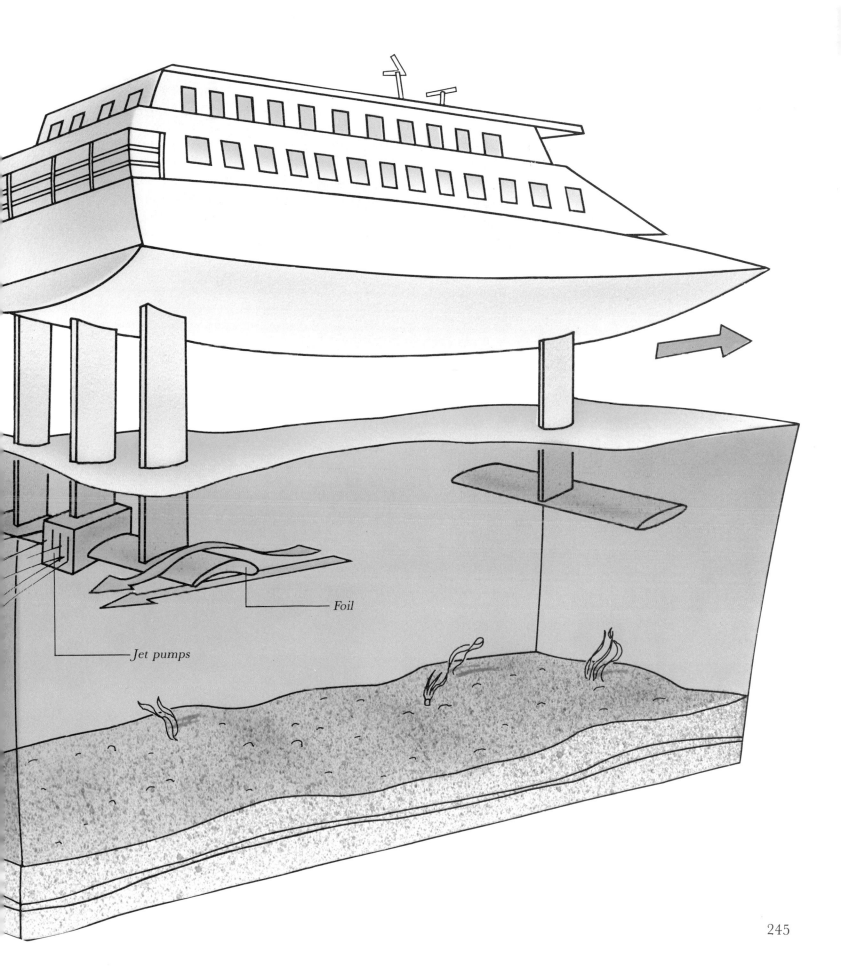

Foil

Jet pumps

SUBMARINE

A one-person submarine powered by hand-cranked propellers was built by an American in 1776; it was put in use during the Revolutionary War to attempt to attach explosives to British ships. But the use of submarines as a practical military vessel awaited three developments: the rechargeable battery, the electric motor, and the diesel-powered generator.

Nonnuclear submarines use an electric motor powered by large storage batteries, which are recharged on the surface by a diesel-powered generator. Until the development of nuclear energy, electric power was the only form of power available that didn't require oxygen while operating, and while submerged, a submarine is cut off from additional oxygen. Rechargeable storage batteries gave the vessel a wide range of travel. The generator powered by a diesel motor provided a way of periodically recharging the batteries using a compact source of energy.

In Germany, during World War II, the snorkel was developed. This breathing tube permitted diesel engines to operate while the submarine was submerged, which greatly reduced the submarine's vulnerability to detection.

Nuclear submarines need refueling so seldom that they can stay submerged for months at a time. They do not even have to surface for a new supply of breathing air: Steam-powered generators provide electricity to manufacture oxygen from the surrounding sea water.

The Turtle, *invented by American David Bushnell in 1776, was probably the first submarine used as an offensive weapon. A drill in the top of the craft was supposed to drill a hole in the hull of a British ship so that a charge of gunpowder could be attached, but the copper sheathing on the ship's hull proved too difficult to get through.*

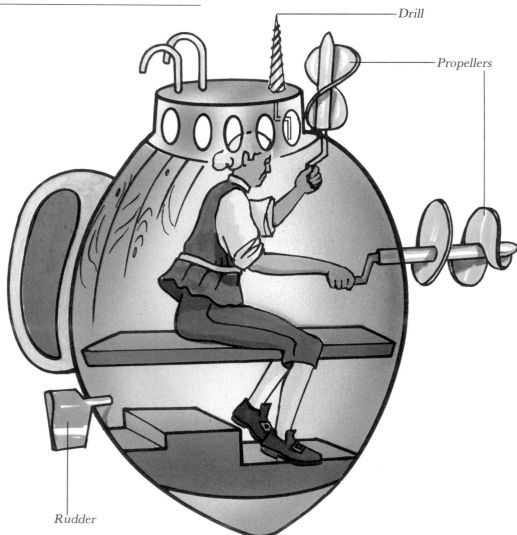

Propeller

Diving rudders

Directional rudder

Drill

Propellers

Rudder

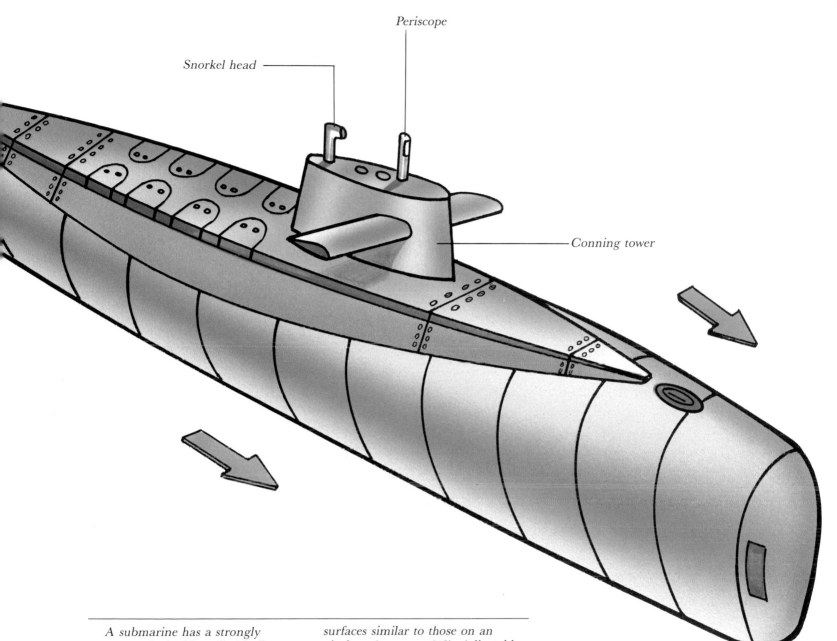

Periscope

Snorkel head

Conning tower

A submarine has a strongly constructed, double-walled hull to withstand great water pressure. Between the hull's double walls are ballast chambers to allow the submarine to raise or lower itself (see **Buoyancy,** page 236). Water is allowed into the ballast chambers until the submarine has a specific gravity greater than water, and it submerges. To rise, compressed air is blown into the ballast chambers, which pushes the water out and makes the submarine more buoyant.

The submarine is helped in its underwater movements by control surfaces similar to those on an airplane (see page 212). Adjustable fins or planes steer it vertically as well as horizontally. One or more propellers move it through the water.

A raised **conning tower** mounted on the hull provides a way to survey the submarine's surroundings. The conning tower houses periscopes, radar equipment, and radio antennas; it also serves as a raised command center. Modern instrumentation such as sonar has lessened the conning tower's importance.

SUBMERSIBLE

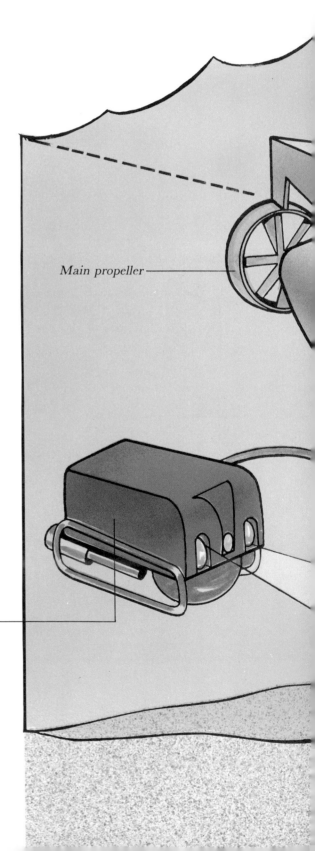

A submersible is a deep-diving submarine designed for exploration and investigation rather than travel. Submersibles have been used to investigate marine conditions at depths of two miles or more. They are also used to recover the contents of sunken ships; in 1987, a French submersible was used to recover valuable artifacts from the sunken *Titanic*.

Like a submarine (see page 246), a submersible will sink unless its ballast tanks are filled with air. A submarine, however, is designed for a high underwater speed. A submersible performs delicate, precise tasks while submerged. The fins or planes that a submarine uses to turn work only while the submarine is in forward motion. A submersible uses thrusters—auxiliary propellers mounted in various directions. These allow fine adjustments in the submersible's position even if the vessel is not moving forward.

Submersibles are usually much smaller than submarines. Although a fighting submarine must carry months' worth of life-support material for a fighting crew, a submersible's workday is usually just that—one day. After the day is over, the submersible returns to the surface.

The cabin of a submersible, which holds the crew, is often a metal sphere—a shape that is unsurpassed for resisting high pressures. It is built strong enough to withstand the extreme pressures encountered during deep dives. A lighter shell surrounding the cabin contains storage batteries, a main motor, auxiliary maneuvering thruster motors, ballast, and flotation material.

The crew can see out through portholes in the cabin and by means of closed-circuit TV. Additional lights and a TV camera can be contained in a remote-controlled vehicle.

A submersible usually has one or more sets of manipulators—jointed mechanical arms equipped with grippers that the crew can direct from the cabin to collect samples, debris, or treasures from the ocean floor.

Main propeller

Remote-controlled vehicle

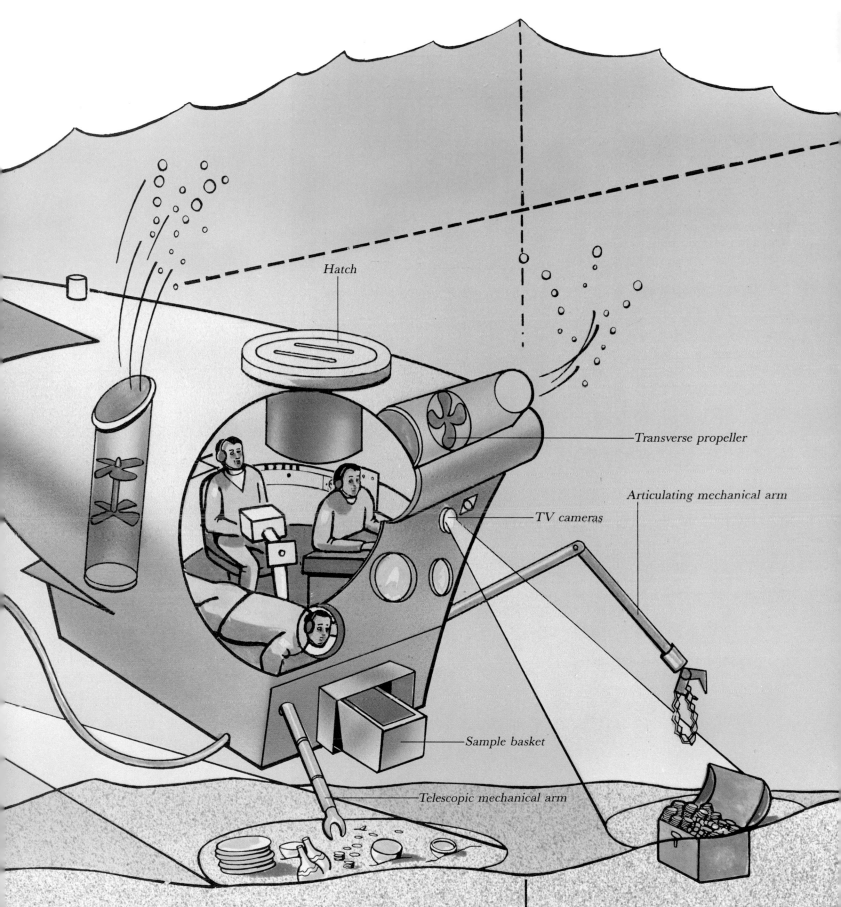

Hatch

Transverse propeller

Articulating mechanical arm

TV cameras

Sample basket

Telescopic mechanical arm

249

OFFSHORE OIL DRILLING PLATFORM

Petroleum is almost invariably located under the earth; you have to drill to get at it. Since some wells are as much as five miles deep, the difficulty of controlling that kind of operation has been compared to that of a dentist working on a patient located at the opposite end of a football field.

Drilling for oil is almost always done by a rotary method. A rotating turntable, powered by an engine, securely holds a series of pipes that are connected to the drill bit. The turntable's rotation turns the drill bit, and the pipes follow the drill bit into the deepening hole.

Offshore oil drilling uses techniques that are much the same as those used on land. But when the earth and rock being drilled through are under the sea, it's essential to have a precisely located, stable platform above the waves.

The design of an offshore drilling platform depends greatly on the depth of the water. The earliest offshore drilling was done from converted ships that were anchored in place. Since even the heaviest anchors can be dragged away from their original location, this technique is now used only in predictably calm waters. In shallow water, drilling can be done from a pier extending out from the shore.

Stand-alone, fixed platforms, supported by pilings set in the ocean floor, can be built any distance from the shore and are used to depths of around 100 feet. They are usually raised well above the water level to avoid buffeting by waves. These fixed platforms are large enough so that the drilling crew can live on them for a long time.

In water that is too deep to permit sinking support piles for a fixed platform, a jack-up platform can be used. This is a self-contained oil drilling rig, with all necessary support functions, mounted on a floating base. The entire unit is towed to the desired location, and the legs are jacked down until they stand on the sea bottom. The jacking process is continued until the base is raised high enough above the sea's surface to be clear of even high waves. Jack-up platforms can be used in waters as deep as 300 feet. If the water were deeper, the longer jack-up legs would buckle under the platform's weight.

SEMISUBMERSIBLE RIG

The most recent development in offshore drilling platforms is the semisubmersible rig. It consists of a number of large hulls carrying a permanent framework for the drilling platform, which is 100 to 150 feet above the hulls. The hulls are ballasted so that they will be stable well below the water's surface. A large semisubmersible rig can be 200 feet by 250 feet; its hulls can be up to 90 feet below the surface and its platform 50 feet above the surface. These rigs are rugged. A large one can withstand 85-foot waves and winds of 140 miles per hour.

A semisubmersible is massive enough to permit multiple anchors to secure it in place. Many of them, however, use propulsion units to move them automatically into position over a radio beacon on the ocean floor.

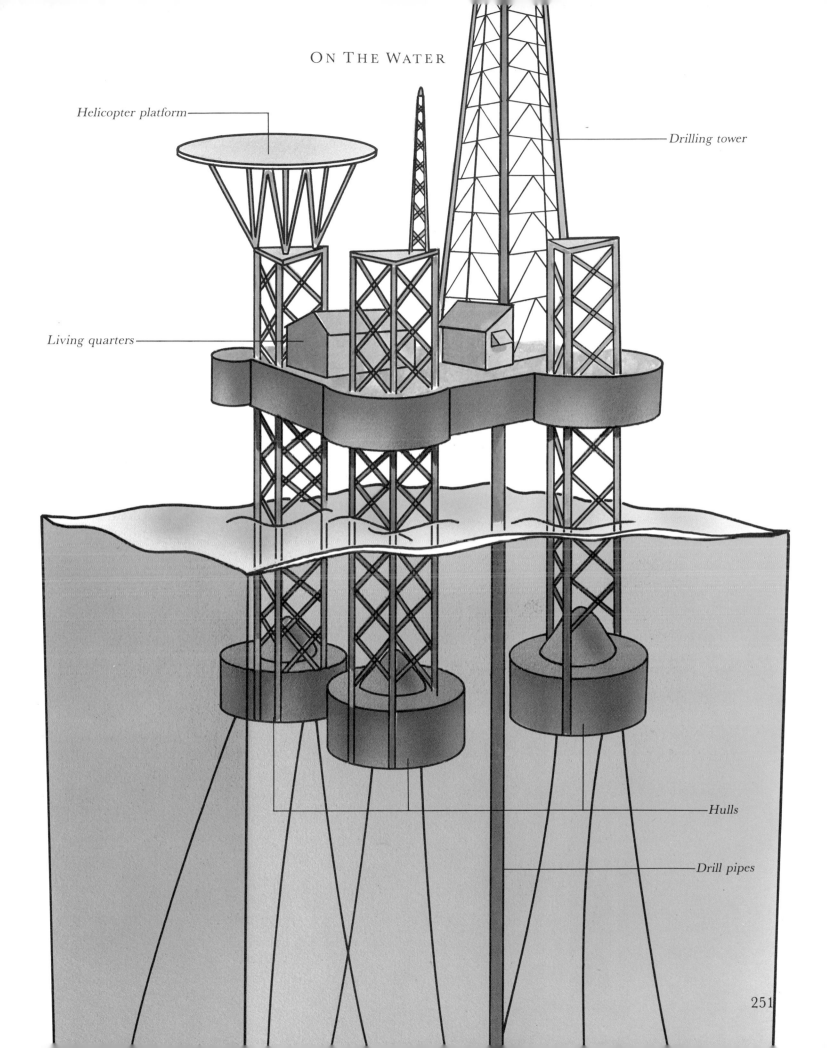

Helicopter platform

Drilling tower

Living quarters

Hulls

Drill pipes

251

BRIDGES

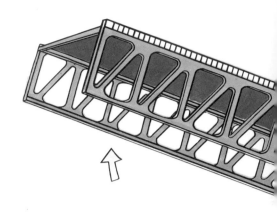

The simplest bridge—a plank suspended from the two shores of a stream—is a beam bridge. Like all bridges, it is subject to two kinds of stress, not only from the traffic across it but also from the weight of the bridge itself. The downward force on the plank's supports on the banks creates the first kind of stress—compression, which pushes a material together. The second is tension, which pulls a material apart. A heavy load in the middle of the beam bends the beam. The top of the beam is compressed, and the bottom of it stretches apart. If the water under the bridge is shallow enough, one or more piers can be installed to support the center of the bridge. This converts one long beam bridge to what are essentially several short ones. Many waterways, however, are not shallow enough or calm enough to allow the building of extra piers. Other bridge designs are needed.

TRUSS BRIDGE

Long spans can be strengthened with a truss, which is a framework made up of straight pieces that does not depend on the stiffness of its fasteners to maintain its shape and strength. A triangle retains its shape when pressure is exerted on one of its corners. The two legs attached to that corner are compressed; the third leg is under tension. A truss is composed of triangular sections, and a truss bridge is very strong for its weight. Trussed bridge sections can be either above or below the bridge's roadway, or both.

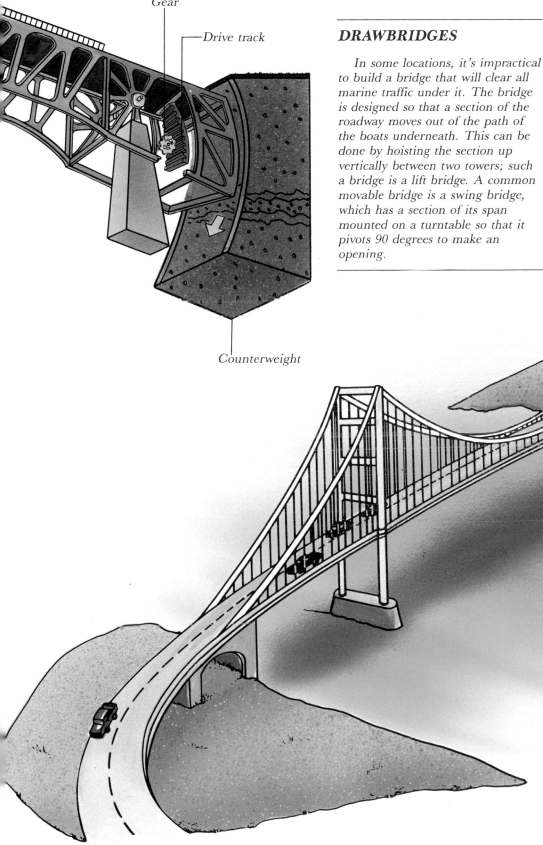

Gear

Drive track

Counterweight

DRAWBRIDGES

In some locations, it's impractical to build a bridge that will clear all marine traffic under it. The bridge is designed so that a section of the roadway moves out of the path of the boats underneath. This can be done by hoisting the section up vertically between two towers; such a bridge is a lift bridge. A common movable bridge is a swing bridge, which has a section of its span mounted on a turntable so that it pivots 90 degrees to make an opening.

The most familiar type of drawbridge is the **bascule** bridge, in which the roadway, or leaf, swings upward. The leaf rotates on an axle parallel to the bank of the waterway and is balanced by a massive counterweight buried beneath the ground. The counterweight and the leaf are so carefully balanced that a small motor can raise the bridge. The motor rotates a gear that forces a drive track downward. A wide waterway can have a double-leaf bascule bridge.

SUSPENSION BRIDGE

A suspension bridge hangs the roadway section from overhead supports by means of cables. The most common design is a pair of suspension cables, one over each side of the roadway, supported by two towers. The roadway, which can be stiffened by trusses or by a girder mounted beneath it, is hung from the suspension cables by smaller cables. The bridge's load is carried by the suspension cables, which are under tension. The stress on the towers is compression.

WATER SKIS

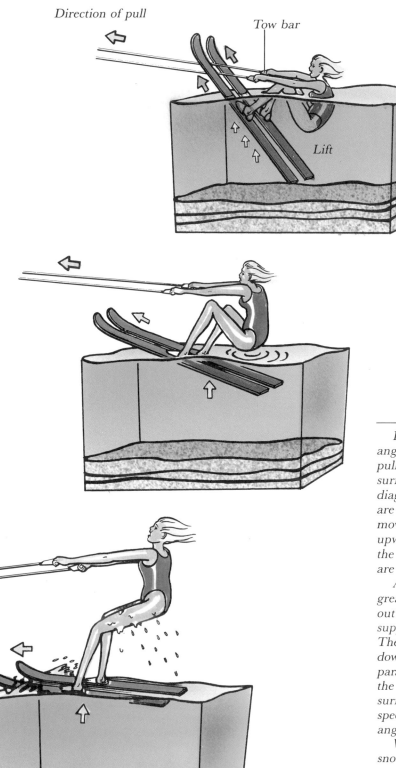

Direction of pull

Tow bar

Lift

If you've skipped a flat stone across the surface of a pond, you've encountered one form of **hydrodynamic lift.** The word "hydrodynamic" comes from the Latin words meaning water and motion. To keep a stone skipping, you need a stone with a flat shape, and you need to throw it with sufficient speed and at the right angle between the stone and the water's surface. If any of these requirements is missing, the stone sinks.

Water skis work in much the same way. Skis are able to float on their own, but they can't support a skier unless the skis are moving. Their flat shape, the speed provided by the tow boat, and the angle the skier causes them to make with the water allow them to skim along the water's surface.

If partially submerged skis at an angle to the water's surface are pulled in a direction parallel to the surface, the skis slide forward in the diagonal direction in which the skis are pointed. This results in a movement that is both forward and upward. The upward component of the motion is lift. The faster the skis are pulled, the greater the lift.

As the skis pick up speed, a greater portion of each ski is lifted out of the water, where it's not supported by hydrodynamic lift. The front portion of the skis tilt downward, bringing the skis more parallel with the surface. Eventually, the skis will be almost parallel to the surface. The extra lift due to high speed makes up for the smaller angle of the skis and the water.

Water skis have turned-up tips, as snow skis do, to prevent their catching in the water. Most of them have a shallow fin to help keep them going in a straight line.

SURFBOARD

The surfboard is unique among watercraft—it is propelled through the water by the force of gravity. A native Hawaiian sport, surfing was described by Captain James Cook in 1778, but despite glowing reports by such mainland reporters as Mark Twain and Jack London, surfing didn't begin to take hold outside of Hawaii until almost 150 years later.

Early surfboards were long, heavy, solid wooden boards. A typical board from the early 1900s might have been 10 or 12 feet long, absolutely flat, and more than 200 pounds in weight. These boards were made obsolete when, in 1928, a lighter-weight hollow board was introduced that proved to be much faster and easier to control. These new boards were flat-bottomed, however, and were almost as likely to skid sideways as they were to go straight ahead. The simple addition of a fixed fin, or **skeg,** attached to the bottom of the board made surfboards hold their course better.

The development of plastics allowed such design sophistication as turned-up tips and contoured bottoms. By the late 1950s, strong urethane foams became available, and the fiberglass-covered foam board 5 to 8 feet long and weighing 20 to 25 pounds became widely accepted.

A surfboard stays afloat because it is sufficiently buoyant to support itself and its rider—heavier riders require larger boards. Surfing is similar to snow skiing; instead of sliding down a snow-covered hill, the surfboarder is floating downhill on the face of a wave. The trick is to adjust your speed by steering the board so that you slide across the face of the wave for as long as possible.

A surfer turns by shifting his or her weight, forcing the board to pivot on its skeg. To steer to the right, you lean to the right, shifting your weight backward at the same time. Then the nose of the board, which has little weight on it, swerves to the right, forced by the curvature of the board's bottom. The tail, guided by the skeg, continues to go straight.

If the board is aimed too high on the wave front, the board won't move fast enough to stay in front of the wave. To remedy that, the surfer steers the nose of the board farther downhill. If the nose is pointed too far down, the board will outrun the wave. The cure is to cut speed by steering uphill.

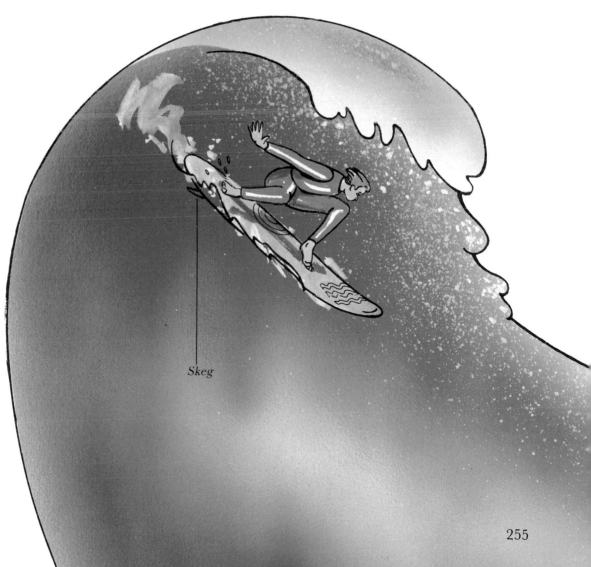

Skeg

255

SAILBOARD

A sailboard is like a surfboard with a sail attached to it. The use of surfboards is limited to areas where there is surf, but a sailboard can be used on lakes, ponds, and rivers—any body of water that has some wind. Sailboarding was developed in the late 1960s and received official recognition in 1984 when sailboarding became one of the yachting events in the Summer Olympics.

The sailboard has a lightweight hull with a downward fin, or **skeg,** at the rear, that helps keep the board moving through the water in a straight line. A larger second fin, the **daggerboard,** extends down through a slot in the hull at the center; it also helps keeps the board from moving sideways.

The sail is supported by a mast that slips into a sleeve sewn into the sail. A **wishbone boom** attaches to the mast and extends completely around the sail. The wishbone boom performs the dual functions of stretching out the sail and providing a hand grip for the sailboarder. A transparent window in the sail lets you see where you're going.

Unlike the mast on a sailboat, the mast on a sailboard doesn't stand up by itself. It's attached to the hull by means of a universal joint (one that allows movement in any direction) and is kept upright by the sailboarder. If you let go of the wishbone boom, the entire mast, boom, and sail assembly falls into the water. An **uphaul line** lets the sailor pull it back up into position.

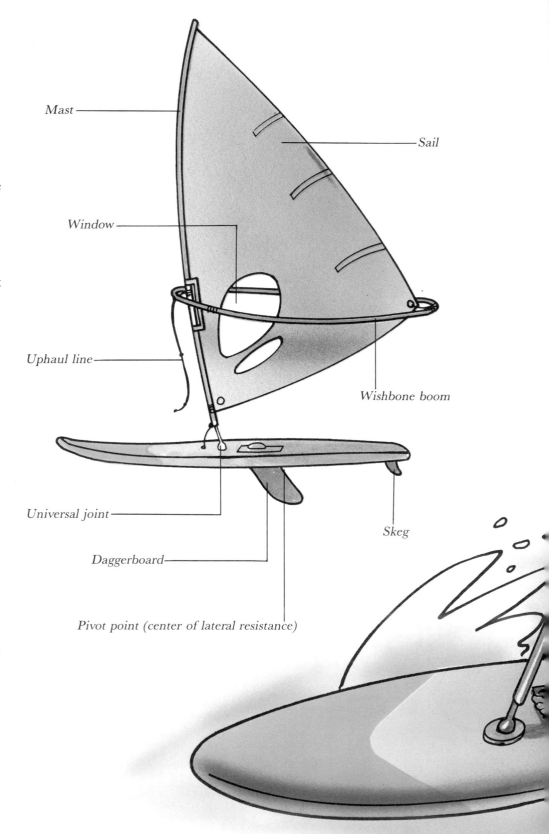

Mast

Sail

Window

Uphaul line

Wishbone boom

Universal joint

Skeg

Daggerboard

Pivot point (center of lateral resistance)

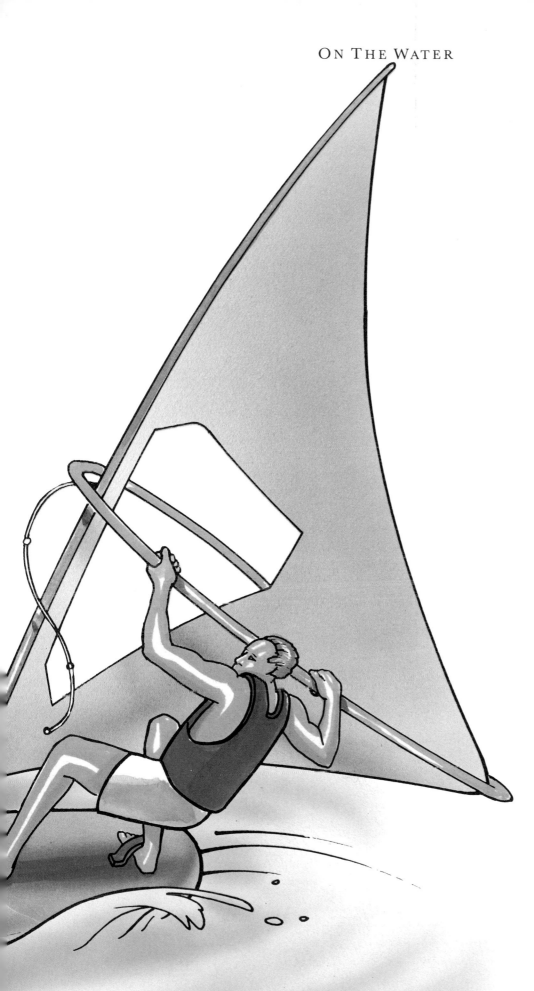

The most striking feature of a sailboard is its apparent lack of any rudder or other provision for steering. The sailboarder relies on the flexible connection between mast and hull, tilting the mast forward or aft to make the turn.

The hull of a sailboard acts as if it pivoted on a point somewhere behind the daggerboard—an imaginary point called the **center of lateral resistance**, which is the center of the sideways force acting on the sailboard. If you tilt the mast forward, the wind's force on the sail pushes more strongly on the bow of the sailboard than on the stern, because more of the sail is toward the bow. The sailboard pivots on its center of lateral resistance until the board is pointing farther downwind. Similarly, if you tilt the mast backward, the wind pushes on the stern section, pointing the board more upwind.

Changing direction so that the wind comes from the other side of the board involves either **tacking** or **jibing.** For tacking, or turning with the wind in front of you, the sailboarder tilts the mast backward to head into the wind and then steps around the mast to the other side before tilting it upright again. To jibe, or turn with the wind behind you, the sailboarder stands behind the mast, tilting it forward to head the board downwind. The sailor then allows the sail to swing 180 degrees in front of the mast before raising the mast and grasping the opposite side of the wishbone boom.

SCUBA DIVING

Until the early 1940s, the only underwater breathing gear available consisted of a rigid diving helmet sealed to a pressurized suit. Air was fed through a hose from a vessel on the surface. In 1943, two Frenchmen—Jacques-Yves Cousteau and Emile Gagnan— invented a compressed-air regulating system that allowed divers to swim about freely, carrying their air supply with them. This new development was scuba, an acronym for Self-Contained Underwater Breathing Apparatus.

The two fundamental elements of a scuba are the air tank and the regulator. The most popular air tanks are made of aluminum. A typical tank uses a pressure of 3,000 pounds per square inch to compress 80 cubic feet of air. That amount will last an average person 80 minutes in a dive that is just below the surface. Deeper dives require a greater rate of air flow; at a depth of 33 feet, the same size tank will last about half that time.

The regulator is an intermediate stage between the tank and the diver, so that the diver breathes air, not at 3,000 pounds per square inch directly, but at the same pressure as the surrounding water. It has a mouthpiece, an exhaust valve for exhaled air, and an inlet valve connected to the tank. The inlet valve is controlled by a lever that rests on a flexible rubber diaphragm.

When the diver inhales, this diaphragm is sucked inward. This moves the lever that opens the inlet valve, and air is admitted from the tank. When the diver exhales, the diaphragm returns to its original position, closing the inlet valve. At the same time, the pressure from exhalation forces the exhaust valve open. The spent air is released as a stream of bubbles in the water.

This design is a single-stage regulator, which has the disadvantage that the air flow from the tank turns on and off with uncomfortable abruptness. A modern two-stage regulator first reduces the tank pressure to about 100 pounds per square inch. This first-stage regulator is usually mounted on the tank.

INHALING

Diaphragm
Air from tank
Inlet valve open
Movable lever
Mouthpiece
Exhaust valve closed

EXHALING

Exhaust valve open
Inlet valve closed

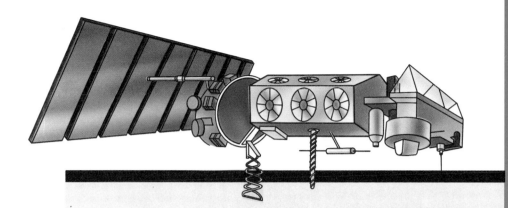

Interacting with Nature

LIGHTNING ROD

Even though close to 45,000 strokes of lightning zap our planet each day, the processes that produce lightning are poorly understood. Scientists do know that lightning is generated by atmospheric friction. A small-scale example of lightning-causing friction can be produced by a furry sweater: When you rub a sweater against itself and then pull it apart, sparks of static electricity seem to jump out.

In a very basic way, the same thing takes place when a lightning bolt is developing. When the wind inside a thundercloud grows strong, it produces the same kind of friction that is produced when you rub a soft sweater against itself. The friction makes the cloud become polarized; that is, positive electrical charges float to the top of the thundercloud while negative charges settle at the bottom of the cloud. Positive and negative charges gravitate toward each other, but the fury of wind keeps the cloud charges apart. When the accumulated electric charge is great enough to overcome the insulating effects of air, a lightning strike can result.

When lightning strikes, a stream of negative charges—a **leader**—extends down from the cloud toward the Earth, which is a conductor of electricity. As the leader nears the ground, the strong electric current it carries increases the electric field between the leader and the ground. The electric field becomes intense enough to cause electrical discharges, which rise from the ground—or trees or buildings—toward the cloud. One of the streamers from the ground makes contact with the leader, completing the electric circuit. The cloud then sends a return stroke of very high current through the pathway to the ground.

Benjamin Franklin invented the lightning rod in 1752, and the design has not changed much since that time. It is a metal rod, usually copper, that is connected by cables to another rod buried in the ground. The rod is often higher than the structure it is protecting under the theory that lightning strikes taller objects. Because it is metal, it conducts the electricity of the lightning into the ground.

Metal-framed structures, in general, have no need for lightning rods. The frame itself provides a suitable path for lightning to disperse harmlessly into the ground. Wooden buildings and trees, which do not conduct electricity, become intensely heated if struck by lightning and can be damaged by fire. Older, larger houses are sometimes fitted with elaborate systems of lightning rods.

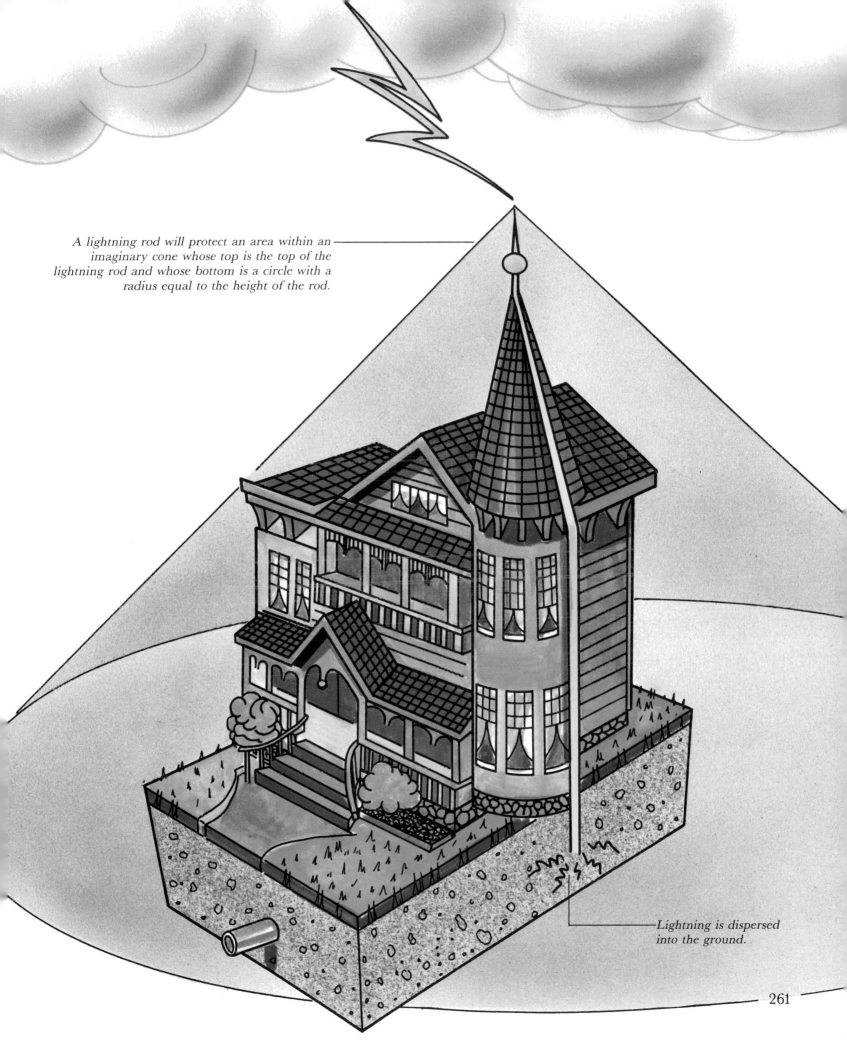

A lightning rod will protect an area within an imaginary cone whose top is the top of the lightning rod and whose bottom is a circle with a radius equal to the height of the rod.

Lightning is dispersed into the ground.

261

THERMOMETER

Most people can easily tell the difference between 90-degree weather and 20-degree weather. Changing temperatures can make us change our moods. They can also have a profound effect on other physical substances. Temperature changes influence the volume of liquids and solids and the electrical resistance of metals. When these changes are closely observed and calibrated, they can be used to tell the temperature.

The most common thermometers—found hanging outside a window, under a patient's tongue, or inside a roasting turkey—are called liquid-in-glass thermometers. They rely on a very basic physical principle. As the temperature goes up, the volume of a liquid expands and the liquid inside the thermometer rises. When the temperature goes down, the volume of a liquid contracts and the liquid inside the thermometer falls. Calibrations on the side of the glass tell you what the temperature is when the liquid reaches the volume that's indicated.

Another type of thermometer is characterized by a needle pointing to a temperature reading. These devices are called bimetallic thermometers. The bimetallic thermometer has a spring mechanism, usually made of a strip of brass layered on top of a strip of steel. Because the lengths of solids change in response to temperature and because the spring itself is made of two different metals, the metals expand and contract by differing amounts, causing the spring to bend. A needle at the end of the coil points to a calibration that shows the temperature.

The most high-tech of common thermometers are electrical thermometers, devices designed to measure electrical resistance (see page 19). Temperature changes cause a change in the electrical resistance of metals, and these changes are measured and expressed as temperature. Platinum resistance thermometers are used by scientists because of their high accuracy and their wide range of measurements—from –434°F to 1167°F.

There are several ways to calibrate temperature changes. In the U.S., the Fahrenheit system is used. In this system, 32° is the temperature at which water freezes; 212° is the temperature at which water boils. In Europe and much of the world, the Celsius scale is used. In the Celsius system, water freezes at 0° and boils at 100°. Scientists around the world use the Kelvin scale of measurement, in which water freezes at 273 kelvins and boils at 373 kelvins.

In a liquid-in-glass thermometer, either mercury or alcohol is used. A change in temperature warms or cools the sealed glass tube holding the liquid. The glass conducts the temperature to the liquid, which responds by either rising or falling. Since mercury freezes solid at –39°F, alcohol thermometers are used in northern latitudes where severely cold weather is common.

At the heart of the bimetallic thermometer is a spiral-shaped spring composed of two different types of metal sandwiched together. All metals expand when heated and contract when cooled; different metals do this to different degrees. If you fasten ribbons of two different metals together, this difference in expansion rate makes the composite strip coil more tightly if you cool it or makes it expand to a looser coil if you heat it. A needle gauge at the end of the spring appears on the face of the thermometer, indicating the temperature.

BAROMETER

We move through the air surrounding us with such ease that it is easy to forget that air is a physical substance with its own weight. Moreover, the weight of air changes constantly, and these changes, as well as the weight of the air itself, can be measured by a barometer.

The easiest way to understand how a barometer works is to visualize the atmosphere as a vast, planet-wide blanket that extends at least five miles into the sky. On a mountaintop two miles high in the sky, only three miles of air is piled atop the mountain peak. But at sea level, a full five miles of atmosphere weighs down on the surface. This "air weight" is referred to as air pressure; the air pressure at sea level is greater than is the air pressure on the mountaintop.

A barometer measures these differences in air pressure and goes one step beyond. Because changes in air pressure signal changes in the weather, the barometer's needle points to the kind of weather that can be expected in the very near future. It can predict anything from hurricane conditions to cloudless, sunny skies. Low air pressure signals damp, cloudy conditions, whereas high pressure conditions are the forerunners of sunny skies. A barometer "on the rise" refers to increasingly drier conditions; a "falling" barometer forecasts increasingly damp conditions.

Unfortunately, there is more to weather than a simple rising or falling of air pressure. The temperature of a weather front, its speed, its direction, and the weather system it is displacing or confronting all influence the weather. Nonetheless, the humble barometer does a remarkably good job at predicting the weather.

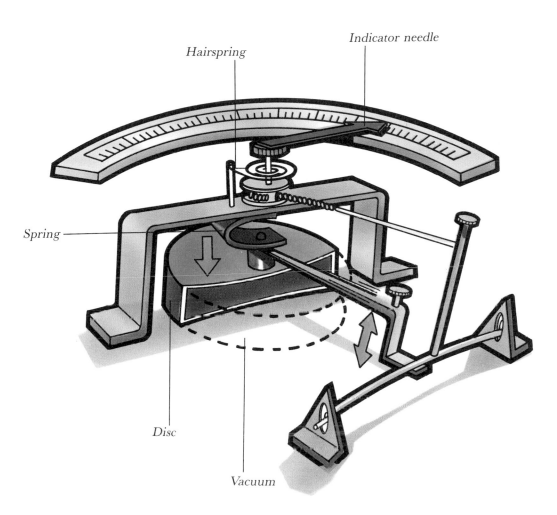

Hairspring

Indicator needle

Spring

Disc

Vacuum

The key mechanism inside a barometer is a vacuum chamber—a carefully sealed, hollow disc made of thin metal. All the air is drawn out of the disc, leaving a near-vacuum inside. Air from the atmosphere presses on the disc in varying degrees depending on the air pressure. The disc yields to the pressure, collapsing or expanding slightly.

The rest of the mechanism is a series of clever springs and levers that amplify the tiny rises and falls of the metal disc into movements big enough to move an indicator needle. A spring makes contact with the vacuum disc, rising and falling when the metal vacuum chamber does. This in turn moves an arm that activates a lever that either tightens or loosens a chain wrapped around the base of the indicator needle. The indicator needle on the face of the barometer moves when the chain pulls tighter or goes slack. A hairspring at the base of the indicator needle keeps a small amount of tension in the lever system to ensure concise translation of the movement of the vacuum chamber to the movement of the needle.

SEISMOGRAPH

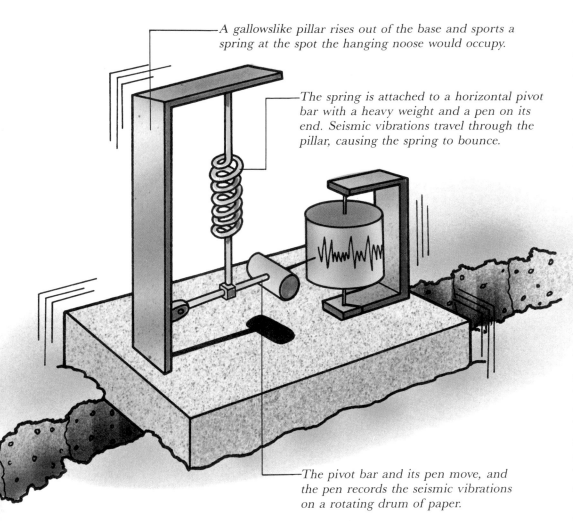

A gallowslike pillar rises out of the base and sports a spring at the spot the hanging noose would occupy.

The spring is attached to a horizontal pivot bar with a heavy weight and a pen on its end. Seismic vibrations travel through the pillar, causing the spring to bounce.

The pivot bar and its pen move, and the pen records the seismic vibrations on a rotating drum of paper.

VERTICAL SEISMOGRAPH

The mechanical seismograph is a device designed to measure vibration in the ground. Although at times the vibrations it records are those of trucks or low-flying airplanes, seismographs are fair and accurate recorders of earthquake-caused vibrations.

This vertical seismograph embodies the basic mechanics used in more sophisticated seismographs. To accurately record earthquakes, the seismograph is firmly linked to solid rock surface beneath the soil by means of a slab of concrete. Vibrations from deep beneath the ground activate a pen, which records the vibrations on paper.

Electronic seismographs have supplanted the mechanical seismograph. Not only do they need less maintenance, but electronic seismographs can be dovetailed with radio devices that automatically broadcast a report on any seismic activity in the area. When solar panels are fitted to the units as an energy source, electronic seismographs make excellent field tools that can be left in remote locations.

The Chinese are credited with the invention of the first seismograph. This device featured four metal balls delicately balanced in the mouths of four dragons that were aligned on the four geographic compass points. When the Earth vibrated as a result of an earthquake, one of the balls would fall out of the dragon's mouth, indicating the direction from which the vibration had come. Like the early Chinese models, modern seismographs are earthquake detection devices, but they are also delicate recording devices for recording seismic vibrations.

Although the ground beneath our feet seems solid enough, it is in a constant state of motion. Indeed, the entire surface of our planet is like a loosely fitting jigsaw puzzle with individual pieces drifting, pulling, and bumping into the pieces surrounding them. The long-term changes wrought by continental drift are not readily visible. But the slight shudders or slips along the edges of the plates of the Earth's crust are highly visible and can result in disastrous earthquakes.

When an earthquake occurs, it releases pent-up stress between plates. The energy of the released stress radiates out in circular waves like the concentric ripples produced by casting a pebble into a calm pool. These radiating waves are so energetic that they are actually Earth-shaking. They become less energetic as they begin to spread out farther and farther from the **epicenter,** or source of the earthquake. Yet they still retain a profile of the quake that produced them. For this reason, a seismograph in Oklahoma can accurately detect and record an earthquake taking place thousands of miles away.

WEATHER BALLOON

Weather balloons are key tools in helping meteorologists explore the upper atmosphere. When filled with a gas that is lighter than air, weather balloons float scientific instruments through the atmosphere. As the balloon rises, it carries aloft an atmospheric data-collecting package called a **radiosonde**. This is a miniature radio transmitter that collects data from an increasingly thin atmosphere and radios the information to weather stations on the ground.

Radiosondes are usually sent aloft on expandable balloons. These balloons are made of a thick, flexible rubber. As the balloon rises higher and higher into the sky, air pressure drops, and the gas inside the balloon begins to expand. When it takes off from the ground, a weather balloon is usually five feet in diameter. By the time it reaches 100,000 feet, the balloon has expanded to a diameter of more than 20 feet. Then it pops.

Engineers have wisely added small parachutes to the radiosonde so that, when the balloon finally breaks, the sensitive devices have at least a chance of reaching the ground intact. More than 700 weather balloons fitted with radiosondes are released from weather stations around the globe twice each day, at noon and at midnight Greenwich Mean Time. They are part of an international effort to keep an eye on the weather.

To explore the atmosphere above 100,000 feet, a different kind of weather balloon is needed. A zero-pressure balloon is similar to an expandable balloon, but when it lifts off, it is only half-filled with gas. As it rises, the gas expands to fill the balloon. This method lets the balloon rise much higher, because the expanding gas has more room and fills, rather than bursts, the balloon. Zero-pressure balloons can attain heights of up to 30 miles above the ground and can stay aloft for months. The instruments on board are usually too expensive to waste, so radio commands bring the balloon and its freight safely back to the ground.

Some weather balloons, especially very large, high-flying zero-pressure balloons, have been mistaken for UFOs. Tricks of reflected light and the translucent materials used for the balloons can make them look like glittering saucers or spheres.

Wind velocity at the upper reaches of the atmosphere can be measured by weather balloons. Since they are simply large, helium-filled balloons, they drift in the wind when sent aloft. Homing devices in the radiosonde pinpoint the balloon's location. Ground stations plot the balloon's changing position to see how fast the wind is blowing.

Expandable weather balloons rise at a rate of 1000 feet per minute. It takes about 90 minutes for the balloon to reach an altitude where it pops. Experts estimate that about one out of every three radiosondes launched makes it safely to the ground to be found and returned.

In contrast to the shape of a recreational hot-air balloon, like that of an upside-down pear, the shape of an expandable weather balloon is nearly spherical. The shape helps evenly distribute the gas, which expands as the balloon rises.

Sensors in the radiosonde measure such atmospheric conditions as temperature, relative humidity, and air pressure. A radio transmitter inside the device continually relays the data to stations on the ground during the flight.

WEATHER SATELLITE

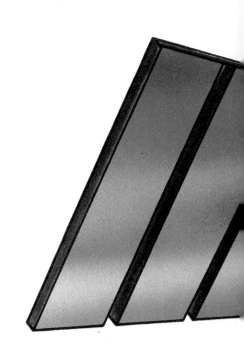

When the first astronauts orbited the Earth and brought home pictures, scientists were especially excited by the pictures of clouds. They had expected to see a chaos of clouds and weather fronts bumping into each other at random around the world. Instead they saw a vaguely ordered pattern of circulation. Meteorologists realized that they could study cloud formations for weather prediction if they had their own cameras aboard orbiting satellites.

Weather satellites are simply orbiting cameras that photograph the movement of weather systems. The photographic image is converted into a digital code and radioed to stations on the ground. A computer unscrambles the signal and displays it as a picture. The weather-satellite pictures you see on the evening weather report are pieced together from the pictures taken by these satellites.

What a satellite sees depends on how high it is. For polar-orbiting satellites, the altitude varies between 500 and 900 miles. Cloud formations photographed from this height are especially good for determining local weather conditions. It's also the ideal altitude for tracking hurricanes and other severe storms.

Other weather satellites orbit the Earth at a far higher altitude, 22,279 miles above the equator. At this height, the speed of the satellite's orbit matches the speed of the rotating Earth below; thus the satellite remains above one point on the ground. For this reason they are called **geostationary** weather satellites. Images from geostationary satellites are broad enough to span one quarter of the globe. They show weather patterns and formations that will soon supersede local weather conditions.

In weather reports, maps that show the weather conditions in the continental United States rely on satellite images for the general picture and turn to reports from weather balloons and surface weather stations for details about temperature and sky conditions in various parts of the country.

TIROS-N

TIROS-N is the latest incarnation of the hardy TIROS series of weather satellites, which have been watching the weather since the early 1960s. A pair of them operate in a polar orbit. In a 24-hour period, each satellite orbits the Earth 14 times and transmits 1,000 pictures to ground stations.

The **solar panels** that power the satellite always point toward the sun. They are adjusted by a small motor on a boom that supports the panels.

The primary instrument on board TIROS-N is a **radiometer,** a sophisticated electronic camera that takes pictures of cloud cover. The radiometer also doubles as an infrared sensor and recorder for detecting heat from the ground.

The satellite itself—a boxlike structure about 12 feet long and 6 feet wide—could easily fit into the back of a pickup truck.

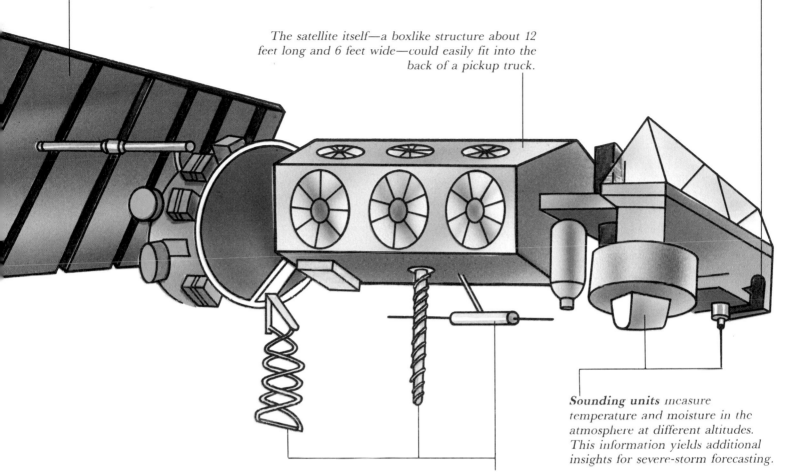

Sounding units measure temperature and moisture in the atmosphere at different altitudes. This information yields additional insights for severe-storm forecasting.

Three different antennas broadcast pictures to ground stations. TIROS-N can broadcast live pictures directly to the ground, or it can store pictures in a tape system for later relaying to ground stations.

HYDROELECTRIC POWER

Water is usually thought of as a soft, benign liquid. But step under a raging waterfall, and the hard facts are clear: The weight of falling water can be crushing. At least 2,000 years ago, this observation led to a variety of inventions. These early inventions used the weight of falling water to spin paddle wheels that were perched under a waterfall or half-submerged in the currents of a swift river. The paddles caught the current, spun the shaft they were attached to, and provided mechanical energy for pumps, millstones, and a host of other devices.

The weight of falling water is still an excellent source of energy. Hydroelectric power plants, earmarked by towering dams, use the force of falling water to generate electricity. In most of these plants, water flashes past the blades of turbine generators inside the dam structure, spinning the blades. The spinning motion, in turn, spins a generator that produces electricity. When the water has finished its work, it continues to flow downriver.

In contrast to electric plants that use coal, uranium, or oil to produce electricity, hydroelectric power does not consume natural resources. In addition, it does not pollute the atmosphere. However, hydroelectric power does have some ecological drawbacks. When a river is dammed, thousands of upstream acres are submerged under an artificial lake. Since the water to power the turbines is drawn from the base of the dam, it is cold and oxygen-poor. Once it exits the power station and reenters the river, that portion of the river has water that is colder and has less oxygen than it formerly had. This creates an environment for a different kind of fish. Finally, fish such as salmon that migrate upstream to spawn find the dam a formidable obstacle. But with careful planning, hydroelectric plants can be designed to minimize these problems.

Newer designs in hydroelectric power generators attempt to take advantage of the power of moving water without blocking it with a dam. These generators, largely experimental, resemble jet airplane engines with a turbine blade recessed into the unit. They are submerged into a river or the ocean, with several units strung along in a line across a strong current. A network of cables brings the electric current to stations on the shore.

The traditional hydroelectric power station creates an artificial lake, or a reservoir, behind it when the dam is laid in place. This ensures a constant supply of water regardless of seasonal fluctuations in the river's flow.

*When falling water shoots through the sluices (called **penstocks**), the turbine blades spin. The motion is used to power a generator (see page 23) to produce a flow of electric current. The current flows into a transformer that boosts its voltage. Finally, the current flows into high-tension wires, traveling to areas where it is needed.*

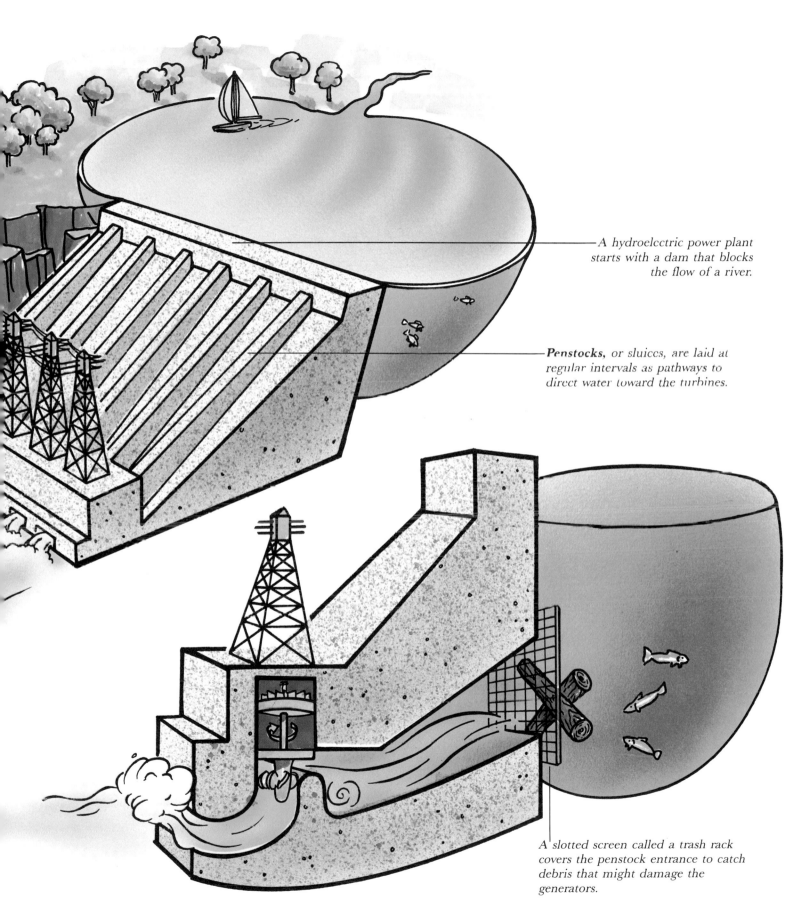

A hydroelectric power plant starts with a dam that blocks the flow of a river.

Penstocks, or sluices, are laid at regular intervals as pathways to direct water toward the turbines.

A slotted screen called a trash rack covers the penstock entrance to catch debris that might damage the generators.

SOLAR ENERGY

Double-glazed glass

Any object left in the sun, be it a hammer, a car, or even a human being, becomes warmer. This simple observation has led to ways of producing heat with solar energy. Passive solar collectors are solar energy systems that have no moving parts or mechanical devices and therefore need no maintenance. Active solar collectors are so called because pumps and motors help them do their work. Both active solar collectors and passive solar collectors share the same goal: They collect and store the sun's natural heat for use in home heating or for hot water.

Although passive solar collectors can attain elaborate heights of engineering refinement, they are very simple in concept. Even a thick concrete wall facing south can be considered a passive solar collector: When the sun's radiant energy strikes it, the wall begins to heat up. When the sun sets, the heat stored within the masonry is released.

The trick is to release the heat where and when you want it. There are literally scores of passive solar designs. One novel and especially successful design is called a roof pond. Atop a flat steel roof lie four clear, water-filled plastic bags that cover the roof's entire surface. During the day, covers are removed, and the sun's heat is absorbed by the water. When cooler nightfall approaches, the covers are replaced, and the heat energy collected during the day radiates into the house's living areas.

Active solar collectors, which are aided by pumps and motors, are especially well suited for manufacturing hot water. Many

possible designs exist, some of which employ mirrors to direct and focus the sun's heat. Two experimental power stations now in operation use fields of mirrors that pinpoint their collective heat on a "power tower" fitted with a network of pipes. The heat becomes so intense that water inside the pipes turns to steam. The steam, in turn, spins the blades of a turbine generator to produce electricity.

A more modest system can collect enough heat to help supply hot water for homes. The solar collectors sometimes seen on rooftops are like car radiators. Antifreeze fluid flows through a meandering network of tubes, which are heated by the sun. Pumps circulate the fluid, which has collected heat energy, through the system. The hot pipes heat water in a storage tank.

Some active solar collectors use air instead of liquid to collect the sun's heat. Fans blow cool air into a "hot box" solar collector mounted on the roof. When the air becomes hot, it is drawn into a storage bin tightly packed with stones. The warm air heats the stones. When heat is needed, vents are opened and heat from the stones is circulated through ducts by electrical fans.

Under ideal conditions, solar power can produce 100 percent of a home's heating and hot water needs. But the best performance level attained by a combination of passive and active solar systems commonly yields about 75 percent of a home's heating needs and about 50 percent of its hot water.

PASSIVE SOLAR COLLECTOR: TROMBE WALL

French scientist Felix Trombe invented a passive system that went one step beyond the simple wall collecting and storing heat. In front of a concrete wall painted black, he placed a layer of double-glazed glass, leaving an air gap between the glass and the wall.

Sun shines through the glass and strikes the wall, heating it up. In the meantime, air from the interior enters the heated gap between the wall and the glass through air vents at the base of the wall. The air heats up and rises past the warm concrete and glass, exiting from vents near the top of the Trombe wall. The heated air enters the living space, warming it. At the same time, more air is drawn into the vents at the bottom of the wall. Without fans or motors, a Trombe wall creates its own air circulation throughout the room. In addition, the wall itself absorbs heat and radiates it directly into the living space.

Slight alterations to Professor Trombe's original design can improve the system's effectiveness. Louvered shades can help deflect unwanted heat from the absorbing surface. At the same time, fans installed at the intake and exhaust vents at the bottom and the top of the Trombe wall help collect and distribute air throughout the room.

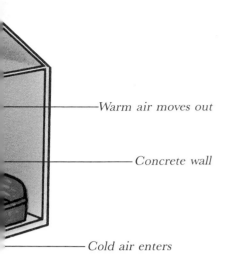

—Warm air moves out

—Concrete wall

—Cold air enters

ACTIVE SOLAR COLLECTOR: SOLAR WATER HEATER

To heat water, a solar collector is set in place, usually on the roof of the house. The collectors themselves are usually black, to absorb as much solar radiation as possible, and are carefully sealed and insulated. Hollow copper tubes filled with antifreeze run through the glass-covered collectors.

A pump sets the heat-collecting fluid in motion, circulating it through the solar collector. During its journey, the fluid heats up. The heated fluid enters the heat exchanger, a device that extracts heat energy from the fluid. Then the fluid recirculates to the rooftop collectors.

Water warmed by the heat exchanger is dumped into a storage tank and, from there, into the domestic hot water unit. Additional energy may be needed to bring the water up to a higher temperature. In many cases, solar water heaters augment, rather than supplant, existing water heater systems.

The same mechanics that produce hot water can also be used to warm rooms. Water circulates through the hot solar collectors, but instead of being dumped into a storage tank, the water circulates through baseboard radiators situated throughout the house, warming the pipes that in turn warm the room.

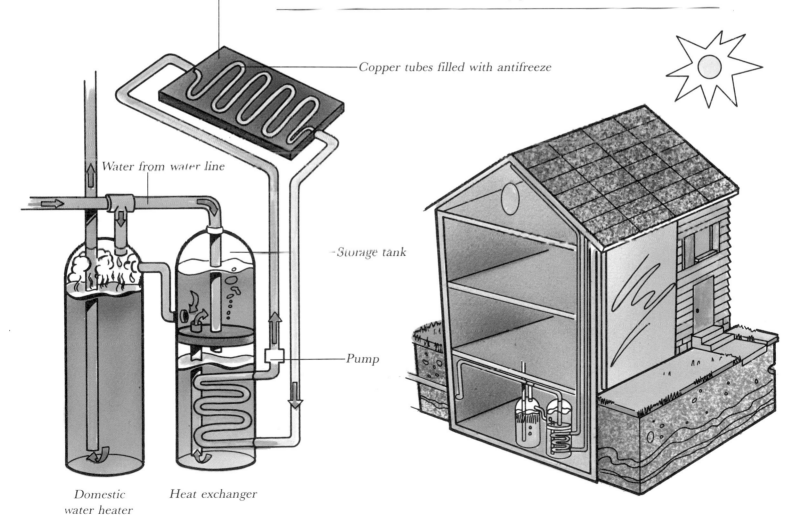

Solar collectors

Copper tubes filled with antifreeze

Water from water line

Storage tank

Pump

Domestic water heater

Heat exchanger

PHOTOVOLTAIC ENERGY

Sunlight

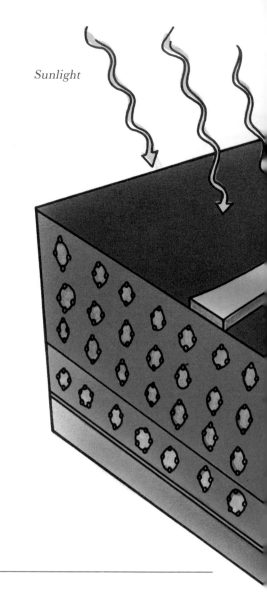

Although the sun has been bathing our planet with energy for millions of years, only recently have scientists learned how to "capture" wavelengths of light to produce electrical current. Photovoltaic cells generate electricity from the sun's radiant energy.

Perhaps the most amazing aspect of photovoltaic electricity is that no moving parts are used to produce electricity. Instead, carefully trimmed sheets of silica crystals are layered together and framed in cells. When sunlight falls on the cells, it excites their molecular structures, creating an electric current. The current is collected and trafficked into a battery for storage and later use.

Photovoltaic, or solar, cells were first developed as exotic and expensive power sources for satellites. Improvements in manufacturing and materials have brought their price down so that today their use is far more commonplace. Isolated research stations, without nearby power lines, can use arrays of photovoltaic cells to generate electrical power. Desert water pumps, emergency phones, or even air conditioners can be linked to arrays of solar cells for power.

Some solar-powered devices, such as hand-held photovoltaic calculators, don't use storage batteries but operate as soon as light falls on the cells. Natural sunlight is unnecessary because artificial light produces wavelengths that will excite the cells to make electricity. But if the lights fail or if you move the calculator into the shadows, it shuts itself off and your calculations are lost.

The one serious drawback to photovoltaic cells is that if the sun isn't shining, no electricity is produced. For this reason, photovoltaic arrays can help supplement electricity needs, but they will probably continue to be subordinate to more conventional electricity sources.

SOLAR CELL

The essential key to turning sunlight into an energy source for electricity is the solar cell. These cells are crystals of a semiconducting material such as silicon. Semiconductors will conduct electricity, but only under special conditions. Sunlight causes the silicon's molecular structure to become excited. In its excited state, it conducts electricity.

The crystals are "doped" (see **Transistors,** page 28) so that one side carries a slight positive charge and the other side a slight negative charge. They are cut into thin sheets to form individual cells.

When sunlight falls on the cell, it permeates the cell with light energy. The light energy striking the negative layer bumps free electrons through a wire as an electric current. The electrons return to the cell's positive layer, where they can combine again with the positively charged silicon. Meanwhile, the positively charged "holes" vacated by the free electrons migrate to the positive layer. (See **Transistors,** page 28.)

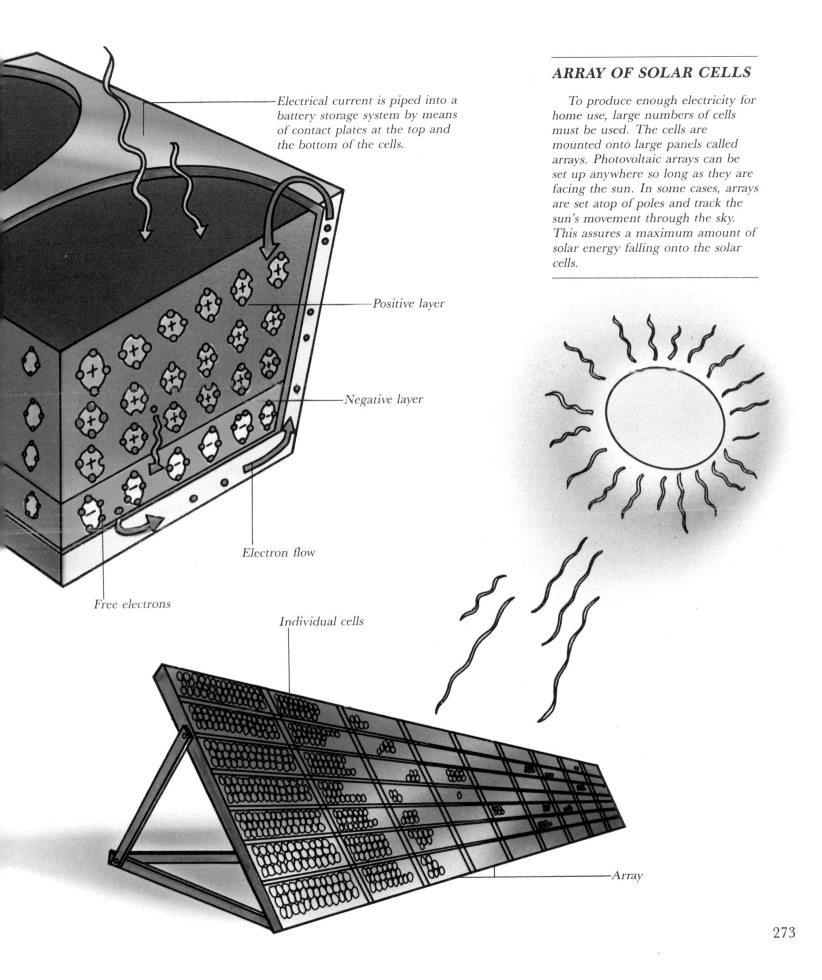

Electrical current is piped into a battery storage system by means of contact plates at the top and the bottom of the cells.

ARRAY OF SOLAR CELLS

To produce enough electricity for home use, large numbers of cells must be used. The cells are mounted onto large panels called arrays. Photovoltaic arrays can be set up anywhere so long as they are facing the sun. In some cases, arrays are set atop of poles and track the sun's movement through the sky. This assures a maximum amount of solar energy falling onto the solar cells.

Positive layer

Negative layer

Electron flow

Free electrons

Individual cells

Array

273

NUCLEAR REACTOR

The enormous amount of energy produced by a nuclear fission reaction (see **Energy from atoms,** page 15) can be converted to electricity. A nuclear reactor is an installation that generates energy from a self-sustaining, controlled nuclear fission reaction.

To be self-sustaining, the reaction must be a chain reaction in which at least one neutron produced by a fission causes another atom to split. A sufficient quantity of atomic fuel such as uranium 235 or plutonium 239 is needed to ensure that the neutrons will find a target. That quantity is called a critical mass. The critical mass can be reduced by surrounding the fuel with a

material that reflects neutrons back into the fuel rather than allowing them to escape.

A second consideration is the speed of the neutrons. Fast-moving neutrons are less effective in splitting nuclei than slower-moving ones. To slow down the neutrons, the fuel is surrounded by a **moderator**—a material that forces the neutrons to bounce off the nuclei of the moderator without the moderator absorbing too many of them. Sometimes water, which doubles as a coolant, is used as a moderator.

A coolant or heat-transfer medium is necessary to remove

the energy from the reactor core and transfer it to a place where it can be used. In addition, the coolant keeps reactor temperatures low enough to prevent the reactor core from melting.

If the nuclear reaction proceeded unchecked, the result could be a melted reactor core. Movable **control rods** made of a material that absorbs neutrons are inserted at precise depths into the reactor core. The absorbed neutrons do not cause a fission and the reaction is kept under control. Inserting the control rods completely absorbs enough neutrons to stop the reaction; removing them could result in an uncontrolled reaction.

Reinforced-concrete secondary containment building

Pressurizer

Boiler

Steam

Generator

Turbine

Control rods

Reactor core

Pump

Heat exchanger

Reactor vessel

The potential of nuclear reactors for producing energy is very great—a pound of uranium 235, if all of its atoms could be used, would produce as much energy as the burning of 1,500 tons of coal. Nonetheless, their use is controversial. The fission process produces large amounts of radiation, an energy form that can be harmful to animal life. For this reason, the reactor vessel is designed to contain the radiation and is usually enclosed in a thick concrete casing, which acts as a secondary containment. Any waste products, excess coolant, or spent fuel must also be contained in radiation-absorbing materials and containers.

A common type of nuclear reactor is a pressurized-water reactor. Specially prepared water is pumped into the core, where it is pressurized to keep it liquid at a higher temperature. This superheated water is passed through a heat exchanger, where it boils a secondary, noncontaminated water supply into steam. The steam powers a turbine that turns a generator (see page 23) to make electricity.

The used steam that leaves the turbine is pumped through a condenser to return it to liquid form and a demineralizer to remove any metallic salts it may have picked up. Then the water goes back to the heat exchanger.

The molten sodium coolant permits operation at very high temperatures—over 1,100°F—which improves the efficiency of the reactor.

A molten metal such as sodium is commonly used as a coolant, since it neither absorbs nor slows down neutrons.

Because of the high operating temperature, a two-step heat exchanger is normally used.

In the secondary loop, molten sodium is used as a heat-transfer fluid. Heat from the reactor's coolant is transferred to make superheated steam used to drive the generator's turbines.

◀ NUCLEAR BREEDER REACTOR

*Fuel availability is a fundamental problem with conventional nuclear reactors. The uranium 235 used for fuel is an **isotope,** or variant form, of uranium; more than 99 percent of the uranium that exists in nature is uranium 238, which is not fissionable. It's possible, though, to create fissionable elements that are not normally available in nature. Plutonium 239 is a fissionable element that can be obtained from the more readily available uranium 238 by bombarding the uranium with fast-moving neutrons. That's the idea behind a breeder reactor.*

No moderator is used that would slow down the neutrons. Because they use fast-moving neutrons, breeder reactors contain a larger mass of fissionable material. They also contain a supply of uranium 238. The fast-moving neutrons that do not split the nuclei of the fissionable material strike the atoms of the uranium 238. This reaction leads to the creation of plutonium 239. An on-line breeder reactor almost invariably creates fissionable plutonium faster than it consumes the original fuel.

NUCLEAR WEAPONS

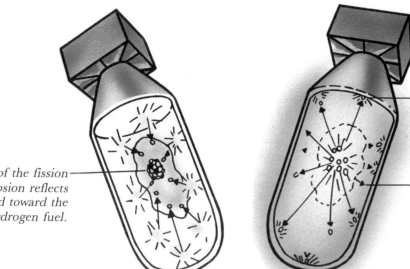

The force of the fission bomb's explosion reflects inward toward the hydrogen fuel.

Neutrons strike the uranium container, splitting the uranium atoms. The fission adds to the destruction.

Fission of the hydrogen atoms releases high-energy neutrons.

Many of the basic concerns relating to peaceful uses of nuclear energy—control of the reaction and shielding from radiation—don't apply to explosions, although bombs must be carefully designed so that they are safe to transport. A bomb's reaction is meant to be sudden, violent, and uncontrolled.

A bomb like the first one that was dropped on Hiroshima in World War II has an explosive force of 13 kilotons—releasing the same amount of energy that would be obtained from exploding 13,000 tons of TNT, but with less than 100 pounds of uranium fuel. Since then, bombs that are up to a thousand times more powerful have been tested. This awesome power, combined with the compact size that permits a nuclear explosive to be carried by aircraft, missiles, or artillery shells, has drastically changed the way in which warfare can be considered.

The Hiroshima bomb was a fission bomb, which is usually called an atomic bomb. (See

Energy from atoms, page 15, and **Nuclear reactor,** page 274.) A fission bomb can be of two types: a gun-type assembly or an implosion-type assembly. A gun-type bomb contains two masses of uranium 235, each of less-than-critical mass but having critical mass when they are combined. To explode the bomb, the two masses are forced together very rapidly by a chemical explosive in a gunlike device. After neutrons are introduced, a chain reaction is produced that grows progressively more violent until the heat buildup within the outer shell of the bomb is great enough to cause an explosion. In an implosion-type bomb, a subcritical mass of plutonium 239 is surrounded by a chemical explosive. The detonation of the explosive compresses the plutonium 239 so that it forms a supercritical mass. Neutrons are introduced to initiate an explosive chain reaction.

THERMONUCLEAR BOMB

Thermonuclear bombs are so called because of the high temperatures required to set one off. They are sometimes called hydrogen bombs, after one component of the fuel that is used. They use fusion rather than fission and can be much more powerful than fission bombs.

Instead of splitting heavy atoms, fusion reactions combine light atoms (see **Energy from atoms,** page 15). Isotopes of hydrogen or other light elements are used as fuels. Such a reaction requires extremely high pressure and temperature to initiate. At this time, the only method with enough energy to force the atoms together is to use a fission bomb.

In addition to the detonator—the fission bomb—a thermonuclear bomb contains a separate component containing the fusion fuel. The container is uranium. When the fission bomb explodes, the container reflects the force of the explosion inward to the fuel. The heat and energy produced squeezes the fuel's atoms together. An enormous amount of energy is released, along with high-energy neutrons. The neutrons strike the uranium and the uranium's atoms split, amplifying the explosion even further.

Health and Personal Care

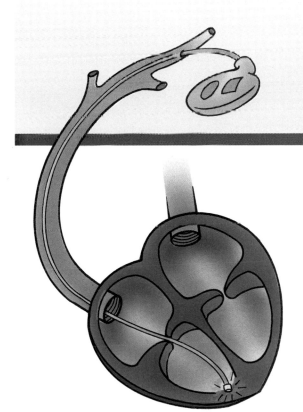

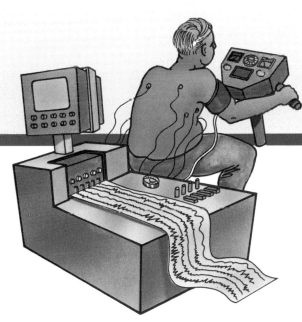

RUNNING SHOE

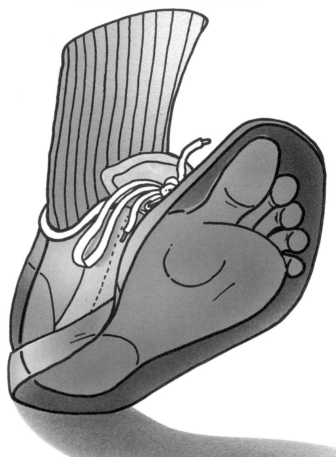

Twenty-five years ago, if you wanted footwear to run, jog, or play tennis or basketball in, you bought a simple pair of sneakers. Today, you can buy specialized shoes for almost every form of exercise. The rubber-soled canvas-topped sneaker has evolved into a running shoe that's a high-tech support system for your feet.

Running is no longer restricted to the young, the fit, and the athletic. More people are running, and they run on inhospitable pavements and roadways. The human foot was never designed to run on concrete. Increased interest in running has generated medical and scientific research on the mechanics of running and the causes of running injuries. As a result, proper footwear has been designed to reduce the risk of injury and pain.

THE MECHANICS OF RUNNING

Nearly every runner starts a stride on the outside edge of the foot that strikes the ground. Most hit heel first, but some runners strike farther forward on the foot. The impact of this first strike is roughly twice the runner's weight.

As the runner continues to shift his or her weight to this foot, the foot rolls inward, flattening out as it rolls. This is called **pronation.** *During pronation, the slight flattening of the arch and the inward roll of the foot probably absorb the shock that would otherwise be transmitted to the ankle, knee, and hip joints. The stress on the foot is reduced during this stage.*

Heel counter

Midsole

Sole

Finally, the foot continues its roll onto the ball of the foot and slightly beyond, in a motion called **supination.** At this point, the joints lock, leaving the foot rigid and ready to push off for the next stride. This third stage applies the greatest force to the foot—at distance-running speeds, about three times the runner's weight.

RUNNING SHOE

The running shoe's few basic functions are extremely important to the runner. It provides cushioning to help absorb the impact of your foot striking an unyielding road surface.

Next, a shoe provides support, or stability. A stiff plastic **heel counter** cups the heel and keeps it from shifting laterally while you run. The midsole is shaped so that it helps keep your foot lined up in the shoe. Midsole design is a compromise between support and cushioning— the more cushioning, the less stability, and vice versa.

The type of **lasting,** or attachment between the upper and the sole of the shoe, is one factor that determines stability. A **board-lasted** shoe, which has the upper mounted to a full-length fiber board, provides the greatest stability; a **slip-lasted** shoe, with the upper attached directly to the sole in one piece, provides the least. A **combination-lasted** shoe, fastened to a board at the heel and sewn directly around the forefoot, has an intermediate range of flexibility.

A running shoe can also compensate for problems in the runner's feet or stride. Runners with unusually high arches tend to **underpronate**—their feet don't flatten out enough to provide adequate shock absorption. They need running shoes that are well cushioned and flexible.

Overpronation is usually associated with flat, extremely flexible feet. An overpronater's foot continues to roll past the point of resting flat on the ground. Overpronaters need a shoe with good support—a board or combination last with a firm midsole and stiff counter.

ELECTRIC RAZOR

The average man has about 14,000 whiskers on his face, and every day they grow 0.02 inch. Each whisker is attached at the root and grows out of a hole in the skin's surface. Pressing down on the skin surrounding the whisker makes it poke farther out of its hole, and it can then be sliced off closer to the skin. Electric razors are designed to take advantage of this.

The electric razor has a very thin metal guard that is perforated with closely spaced tiny holes to trap the whiskers extending through. A vibrating cutter is spring-loaded to hold it close to the guard as it moves back and forth to shear off the whiskers.

ELECTRIC RAZOR

*A vibrating electric razor has a tiny **induction motor** (see page 27) that works on alternating current. The current through a coil produces a magnetic field, and because the current alternates, the magnetic field changes polarity. The armature within the magnetic field is spring-loaded. The combination of the changing magnetic field and the action of the spring makes the armature move back and forth very rapidly. Through a linkage, the blade is connected to the armature, so that it moves back and forth under the perforated guard.*

Armature

Blades

Coil

HAIR DRYER

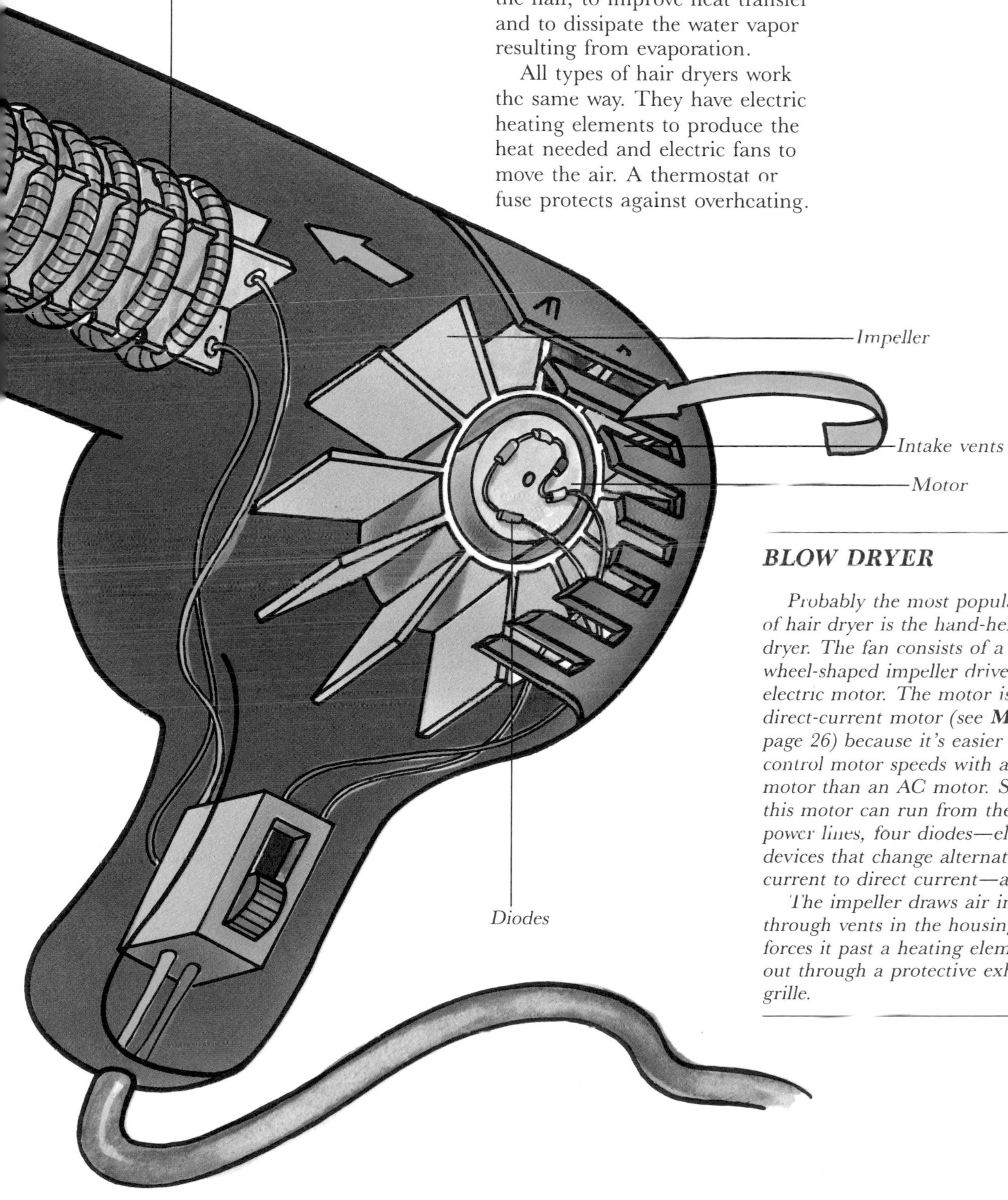

Heating element

A hair dryer has to do two things: heat the damp hair, so that the water will evaporate more quickly, and blow air over the hair, to improve heat transfer and to dissipate the water vapor resulting from evaporation.

All types of hair dryers work the same way. They have electric heating elements to produce the heat needed and electric fans to move the air. A thermostat or fuse protects against overheating.

Impeller

Intake vents

Motor

BLOW DRYER

Probably the most popular style of hair dryer is the hand-held blow dryer. The fan consists of a paddle-wheel-shaped impeller driven by an electric motor. The motor is a direct-current motor (see **Motors**, page 26) because it's easier to control motor speeds with a DC motor than an AC motor. So that this motor can run from the AC power lines, four diodes—electronic devices that change alternating current to direct current—are used.

The impeller draws air in through vents in the housing and forces it past a heating element and out through a protective exhaust grille.

Diodes

MICROSCOPES

Virtual image

How clearly the human eye can see an object depends on the size of its image on the eye's retina and whether the eye can bring it into sharp focus (see **Eyeglasses,** page 62). Light waves from a nearby object enter the eye at a wider angle than light waves from the same object at a distance, and they form a larger image on the retina. But extremely small objects are hard to see even when brought very close to the eye, because the visual angle is still too small. Even if the image were large

enough, the unaided eye can't form a sharply focused image of any object closer than about six inches.

The purpose of a microscope is to increase the visual angle. A small object can be brought very close to the eye, thereby increasing its visual angle and image size, and still be seen distinctly. And with the aid of electronics, we can see images of objects that are so small they are invisible to the naked eye.

MAGNIFYING GLASS

The simplest form of microscope is a magnifying glass, which is a **converging** *lens (see page 63). The focal length of a lens is the distance at which parallel light rays passing through the lens from an object converge into an image. By placing a magnifying glass so that the object you are studying is within the focal length of the lens, the glass will produce an enlarged* **virtual image** *(see* **Mirrors,** *page 60) at a distance your eye can readily focus on.*

The number of times the image is larger than the object is called the magnifying power of the lens. A lens that magnifies 20 times (20X) is called a 20-power lens. Hand-held magnifying glasses are usually limited to 20X because images larger than that are difficult to hold steady.

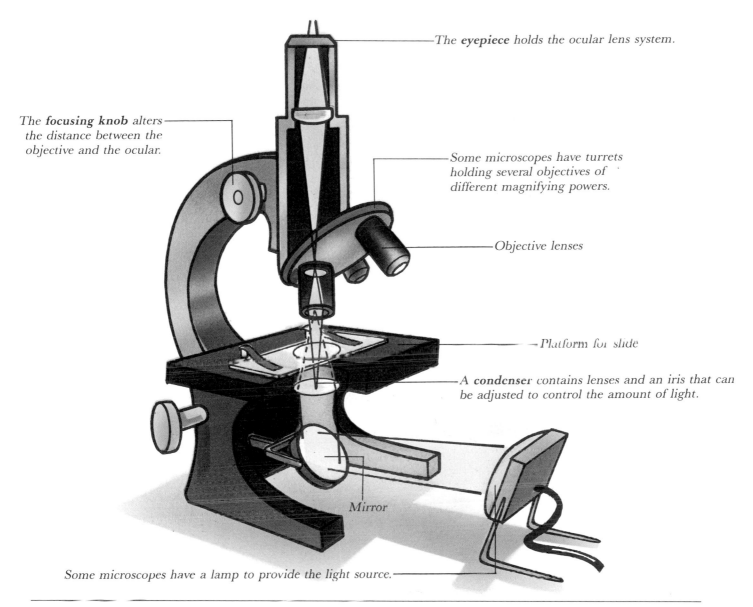

The **eyepiece** holds the ocular lens system.

The **focusing knob** alters the distance between the objective and the ocular.

Some microscopes have turrets holding several objectives of different magnifying powers.

Objective lenses

Platform for slide

A **condenser** contains lenses and an iris that can be adjusted to control the amount of light.

Mirror

Some microscopes have a lamp to provide the light source.

COMPOUND MICROSCOPE

For larger magnifications, it is possible to place one lens far enough from another so that the second lens magnifies the image produced by the first. That is essentially what happens in a compound microscope—the type of microscope used in biology classes. The first lens, the one closest to the object you're looking at, is the **objective;** the second lens, the **ocular,** magnifies the image produced by the objective. The objective system and the ocular system are actually groups of lenses.

The specimen to be studied, which is sliced thin enough so that light can pass through it, is mounted on a glass slide. The slide is placed on a platform under the objective. Light passes through the slide to the objective, which forms one magnified image. The ocular lens system, in the eyepiece, magnifies that image further, forming an enlarged virtual image. This image enters the eye at a wide visual angle. The subject's apparent size is increased, and you can see it clearly and in detail.

The right combination of objective and ocular lenses can provide magnifications of 2500X, which theoretically would enable you to see objects as small as .00001 inch. However, just as your finger is too thick to feel small bumps on a pane of glass, light waves can't distinguish details much smaller than their wavelengths—.000016 to .000028 inch. For practical purposes, optical microscopes can't magnify objects more than about 2000X or distinguish detail closer together than the wavelength of light. Atoms, molecules, and viruses are so small that they can't be seen with an optical microscope.

(continued)

MICROSCOPES

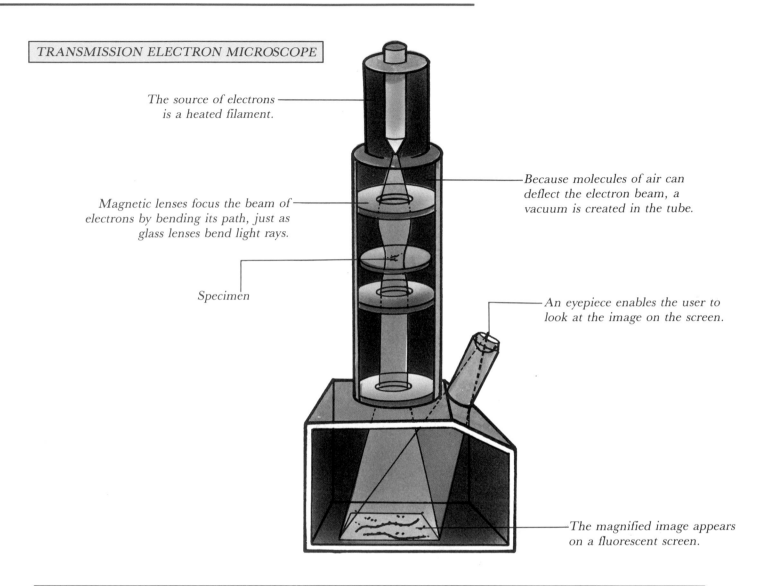

The source of electrons is a heated filament.

Magnetic lenses focus the beam of electrons by bending its path, just as glass lenses bend light rays.

Specimen

Because molecules of air can deflect the electron beam, a vacuum is created in the tube.

An eyepiece enables the user to look at the image on the screen.

The magnified image appears on a fluorescent screen.

ELECTRON MICROSCOPE

An electron microscope uses a beam of electrons instead of a beam of light. Electrons are negatively charged particles that are parts of all atoms. Since the electron beam has shorter wavelengths than those of light, the electron microscope can distinguish smaller items than the optical microscope can. It works on the same principle as an optical microscope except that the "lenses" are not shaped pieces of glass, but electrostatic fields or magnets that bend the path of the electrons. (See **Electromagnet,** page 24.)

There are two types of electron microscopes. The **transmission electron microscope** transmits the beam of electrons through the specimen, just as in an optical microscope the beam of light goes through the specimen. Objects as small as bacteria or molecules can be seen with a transmission electron microscope, and with further enlarging of the images by photography, magnifications of 100,000X to 500,000X are possible.

The **scanning electron microscope** can show the three-

dimensional surface of the object. The beam of electrons is highly focused and sweeps across the specimen, dislodging a shower of secondary electrons. The number of secondary electrons varies according to the structure of the specimen. These secondary electrons are collected and a display is created on a television tube. The narrow beam moves over the specimen, building an image on the television tube that shows the three-dimensional nature of the specimen.

DENTIST'S DRILL

Everybody's least favorite tool is surely the dentist's drill. But the modern dentist's drill is a lot closer to "painless dentistry" than those of years ago. Nearly every improvement in dental drills has addressed one or more of the factors that contribute to pain: heat, vibration, and lengthy sessions. All of these can be lessened by increasing the speed of the drill.

Before 1850, most dentistry was limited to extractions. Simple files, chisels, and scoops were used for what little dental reconstruction was done. The advent of anesthesia, along with an increased appreciation of all areas of medicine, encouraged dentists to seek ways of fixing teeth rather than pulling them.

In 1850, the first flexible-shaft dental drill was developed; it was cranked by hand. A foot-treadle model was introduced in 1870. Both of these were low-speed devices—hand-cranked tools operated at about 300 revolutions per minute (RPM), and the treadle models could turn a little more than double that speed. Not only were they painful for the patient, they were tiring for the dentist, because of both the effort required to power them and the length of time required to complete a session.

By 1920, electrically powered drills operated at 1,000 RPM with little physical effort from the dentist. Higher speeds were possible, but the grinding wheels and steel burrs wouldn't last at higher speeds. The development of the diamond drill—actually a steel disc or shaft with diamond dust bonded to it—and, later, the tungsten carbide burr effectively removed these speed limitations, and drill speeds reached 12,000 RPM. The introduction of miniature high-precision ball bearings in 1953 made drill speeds of 25,000 RPM possible, and a drill powered by a small water turbine, introduced in 1954, brought speeds up to 40,000 RPM.

At higher speeds, frictional heat is a serious problem; water jets were added that directed cooling sprays of water onto the drill tip.

AIR-POWERED TURBINE DRILL

Introduced in 1957, the air-powered turbine drill has brought drill speeds up to 800,000 RPM. A nozzle directs a high-speed stream of air at a small windmill-like turbine; the turbine rotates the drill, and the used air returns through a separate duct. Rather than ball bearings, the air itself is used to reduce friction. The turbine floats within a cushion of the air that rotates it.

This ultra-high-speed design allows the drill to cut very rapidly, which reduces working time and is less disturbing to the patient. Studies have found that vibrations from drills operating at more than 250,000 RPM cannot be sensed directly—the patient is conscious only of a high-pitched whistling.

The high-speed air-turbine drill actually works too fast for the fine control required for the last stages of most dental work. The dentist goes back to a more conventional, lower-speed drill.

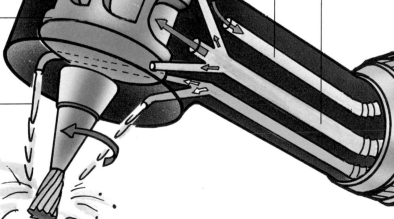

Air cushion

Turbine

Water jets

Air out

Air in

X RAYS

The ability to take a photograph of the inside of our bodies seems like magic, and in fact X rays were a mystery for their discoverer, W. K. Roentgen. He called them X rays because he did not understand their nature. Even though much has been learned about X rays since, the name has stuck.

X rays, like radio signals and visible light, are **electromagnetic waves** (see page 50). The wavelengths of X rays are so short that the rays can pass through some solid materials, such as paper or human skin and tissues. However, they are absorbed by denser materials such as bone and metal. X rays can also expose photographic film. If you put your hand between a source of X rays and a plate of film, the X rays go right through your skin to expose the plate, but your bones absorb the rays, preventing the film from being absorbed. The denser the material, the more X rays it absorbs, and the less the film will be exposed.

X rays can be harmful in large doses; to guard against overexposure, they are not used indiscriminately. Although the X rays used in medical diagnosis are low-energy X rays (those with relatively longer wavelengths), great care is taken to shield both the patient and the operator from scattered radiation. X-ray examinations that are used as screening tests are generally recommended only at intervals of one year or more.

Industry makes use of X rays. Internal cracks and evidence of stress in metal parts will show up on an X-ray photograph. Airplane welds can be examined with X rays for invisible faults.

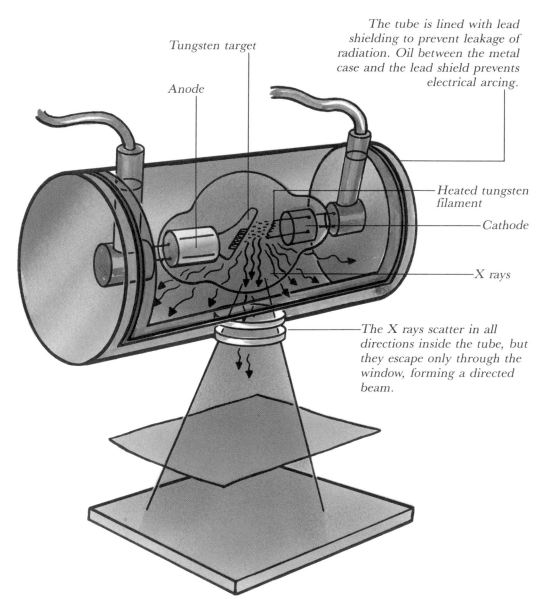

Tungsten target

Anode

The tube is lined with lead shielding to prevent leakage of radiation. Oil between the metal case and the lead shield prevents electrical arcing.

Heated tungsten filament

Cathode

X rays

The X rays scatter in all directions inside the tube, but they escape only through the window, forming a directed beam.

Some radioactive atoms and stars produce X rays, but for practical purposes they are produced in a vacuum tube. Inside the tube are two metal plates called electrodes—an anode (the positive electrode) and a cathode (the negative electrode). A heated tungsten filament in the cathode gives off electrons. They stream toward the anode, which is made of copper with a tungsten target set into it.

The electrons strike the tungsten target and bump the electrons of the tungsten atoms in such a way that energy is emitted. When this happens, X rays are produced. The amount of voltage applied to the vacuum tube determines the wavelength, and thus the energy level, of the X rays. Hospital X-ray machines use a relatively low voltage of 30,000 to 100,000 volts.

MEDICAL ULTRASONICS

One of the earliest and still most common uses of ultrasonics in medicine is to "see" unborn babies in the womb. Ultrasound images of the fetus help the physician determine its age, whether it is growing properly, and whether there is more than one baby.

Ultrasonic scanning in medicine works on the same principle as sonar (see page 110). Sound pressure waves that are ultrasonic, or too high in frequency for humans to hear (see **Sound,** page 48), are sent into the body. When they meet any changes, such as bones or organs, some of their energy is reflected back.

The medical profession finds many uses for ultrasonic waves— for example, to take "pictures" of body organs and bones, to measure the velocity of blood flow, and to detect tumors in various parts of the body. Ultrasonics is also being used in research and therapy. Ultrasonic microscopes can give a peek at the structure of living cells unobtainable with electron microscopes. Ultrasonic waves can provide heat to ease the pain of muscle strain. Focused ultrasonic beams can destroy some malignant tumors without open surgery.

A **piezoelectric transducer** (see **Ultrasonic humidifier,** page 145) generates the ultrasonic waves. A specially cut crystal or ceramic is electrically charged and therefore vibrates to produce the ultrasonic waves.

The transducer is held in contact with the patient's skin. To eliminate air, a gel or oil is applied. The ultrasonic waves penetrate the womb and are reflected back to the transducer, which acts as a detector. The returned signals cause the crystal to move, which generates electrical signals that can be analyzed. The signals can then be displayed as moving images on a cathode-ray tube like a television tube.

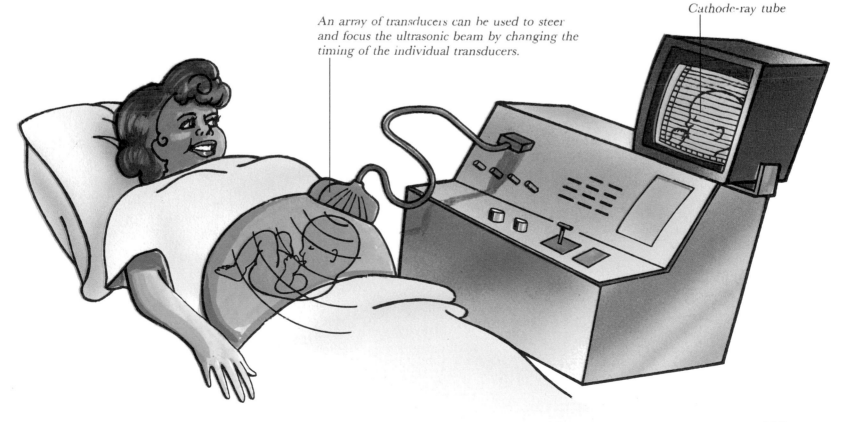

An array of transducers can be used to steer and focus the ultrasonic beam by changing the timing of the individual transducers.

Cathode-ray tube

CAT SCAN

A conventional X-ray photograph (see **X rays, page 286**) shows dense areas inside the body such as bones, but the view is from one angle only. Any structure behind or inside that dense area is still hidden. The development of Computerized Axial Tomography, or the CAT scanner, has made the interior of the body much more visible. A CAT scan uses X rays, but in a different way, to produce a sort of cross section of an area of the body. It can detect deeply imbedded tumors and abnormalities in body tissue and bone that ordinary X-ray examinations do not show. CAT scans are widely used for detecting cancerous areas or tumors in the brain and spine, but whole-body scanners are also used. They can easily identify hemorrhages, abscesses, and swellings that previously were almost impossible to detect.

A narrow X-ray beam, or a fan-shaped array of beams, sweeps around the patient's head or body, striking the target site from many angles. The transmitted X rays are not directed to photographic film, however; they go to detectors that measure the rays that were not absorbed. This data is fed to a computer. The computer can build up a picture of the area being studied using X-ray information from many angles. The result is a "slice," or cross section, of the scanned area that shows great detail and, with some systems, produces three-dimensional images.

A dramatic new development in medical technology is Nuclear Magnetic Resonance Testing (NMR). In this process, the subject is placed in a cylindrical tunnel and surrounded by a doughnutlike electromagnetic ring. The electromagnetism created causes the atoms in the body, especially hydrogen atoms, to align themselves. Radio-frequency signals are then introduced from a direction other than that of the magnetic field, and the hydrogen atoms realign themselves in that direction. When the radio signals are suddenly removed, the atoms flip back to their original position. This is a "relaxation" period; the time it takes for the tissue to relax and the amount of energy the atoms absorb can be analyzed by a computer. The computer can then produce a picture with shapes and colors that can reveal the chemical makeup of the area being examined.

Multiple X-ray exposures of the same area, taken from different angles, build up a detailed cross section of the area.

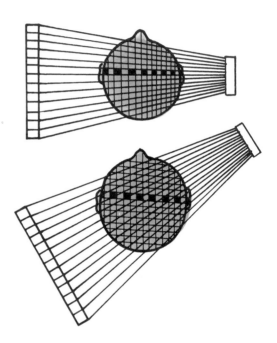

CAT SCANNER

The source of the X rays is contained in the **gantry,** a tunnel-shaped housing that also contains the detectors. The patient lies on a trolley, which is moved into the gantry.

A fan-shaped group of X-ray beams is directed at the area of the body to be scanned. The denser the material in the body, the more X rays will be absorbed. What X rays are not absorbed or scattered are picked up by detectors, which convert the beam to an electric current that reflects the amount of X rays transmitted. The gantry is rotated very slightly, so that another set of X-ray readings of the same area can be taken from a different angle.

The computer measures the amount of X-ray transmission at each angle and reconstructs a section, or "slice," of the area as a black-and-white image on a monitor. Like an ordinary X-ray image, light areas in the image indicate a dense substance. The image can be photographed for further study.

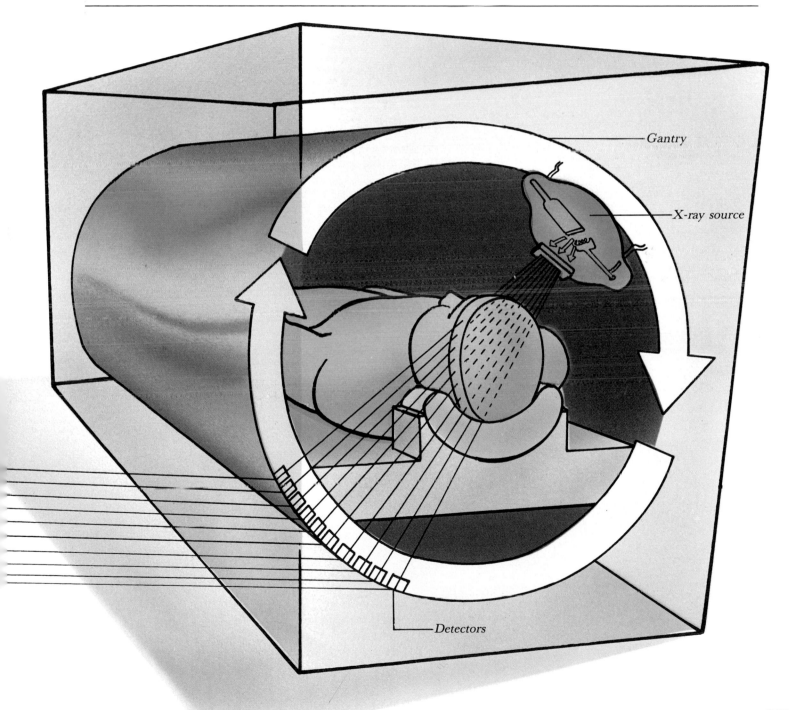

Gantry

X-ray source

Detectors

EKG MACHINE

As the heart pumps blood around the body, it is giving off a minute electrical current that can be detected on the skin. The electrocardiograph (ECG or, more commonly, EKG, from the German term *elektrokardiogramm*) senses these electrical voltages and displays them on a printed graph, a fluorescent screen, or magnetic tape.

Much can be learned about the condition of the heart from an EKG. The rhythm and rate of heartbeat shows on the graph. If one part of the heart is enlarged, it will produce currents that are stronger than normal. After an accident or illness, the condition of the patient's heart can be monitored continually.

Portable EKG machines can be worn by a patient for a full day, so that the performance of the heart during the patient's daily routine can be observed. The signals are recorded on magnetic tape and are converted to a paper readout.

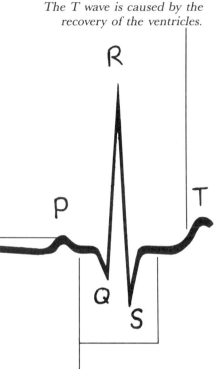

The T wave is caused by the recovery of the ventricles.

The P peak represents electrical activity in the upper chambers of the heart.

The QRS outline represents activity in the ventricles, from which the blood is forced into the arteries.

*Electrical conductors are smeared with a special paste to improve conductivity and then attached to carefully selected parts of the body. The small electrical currents that have traveled to the skin can then be amplified. The signal is printed out by a **galvanometer,** which is an electromagnet (see page 24) that moves a stylus across a graph paper. In electronic instruments, the signal is directed to a cathode-ray tube, which is like a television tube.*

The peaks and valleys of the electrocardiogram's familiar jagged line are commonly labeled from P to T, representing electrical impulses from each part of the heart's cycle.

EEG MACHINE

As we go about our daily business and even as we sleep, bundles of nerve cells in the outer layers of our brains are producing miniature electric currents. This electrical activity provides a clue to the operation of the brain and can be recorded with an electroencephalograph (EEG). Records of these "brain waves" are valuable in diagnosing strokes, epilepsy, muscular dystrophy, and other diseases of the brain and the central nervous system.

Certain parts of the brain are devoted to specialized activities within the body. For this reason, current-conducting electrodes are attached to about 20 areas of the skull, and one is attached to the ear to provide a reference. The electric current from the outer layer of the brain is picked up by the electrodes and fed to a master recording device with several channels. Each channel has an electromagnetic pen that moves in accordance with the electrical signal fed to it. The result is a set of wavy lines on paper that form a graph of the electrical activity of different parts of the brain.

One primary use of an EEG is to pinpoint the origin of epileptic seizures. Each line on an EEG shows a different type of brain activity. The wave patterns of healthy people are quite similar to each other. Normal waveforms show rhythmic fluctuations without extreme peaks and valleys. A sudden appearance of abnormal "spikes" or sharp waves is often caused by an epileptic seizure. The absence of any fluctuations confirms brain death.

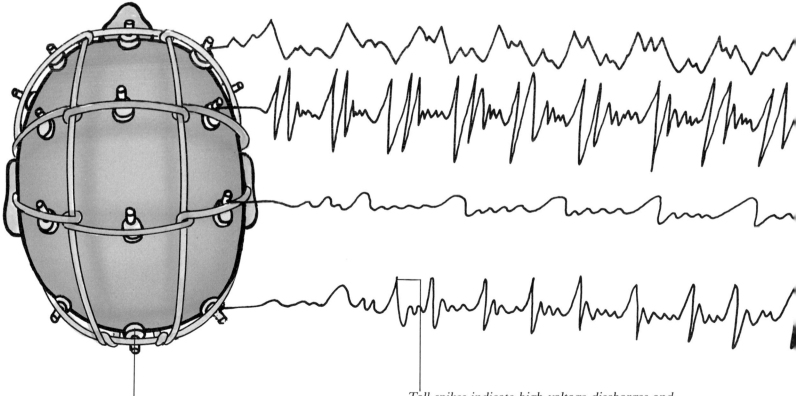

Electrodes

Tall spikes indicate high-voltage discharges and can result from an epileptic seizure.

ENDOSCOPE

Light waves are refracted by the layers of glass toward the center of the fiber.

With an endoscope, a physician can actually peer inside the human body. A narrow, flexible tube containing fiber-optic light guides is inserted through natural body openings such as the throat or into a small incision. Through the endoscope, the doctor can examine internal organs—the stomach, intestines, lungs, esophagus—and detect diseases that otherwise might not be seen without exploratory surgery.

Light sent through the fiber-optic glass fibers illuminates the interior organs and is reflected back through other light guides to magnifying lenses. The fibers of glass, each one smaller in diameter than a human hair, are so flexible that they bend around curves and corners, following the twists and turns of the body's interior passages. What the physician sees is a picture made up of tiny dots, each dot consisting of the light leaving one fiber.

The endoscope can include small instruments to gather samples of tissue for microscopic study or even to do minor surgery, such as removing growths in the colon.

FIBER OPTICS

A ray of light entering one end of an optical glass fiber doesn't shine out the side of the fiber; it continues all the way to the opposite end so that no light is lost. This is because of the special construction of the fiber.

Light waves passing through glass are **refracted,** or bent (see **Waves,** page 47). Different types of glass bend the light waves to different degrees. The fiber is made of concentric layers of glass that progressively refract the light waves more and more. The effect is that a light wave entering at one end is continually being bent toward the center of the fiber, and the light waves do not leak out the side of the fiber.

Another consideration is the purity of the glass used. Ordinary window glass absorbs light. If a window were three feet thick, very little light would get through it. Optical fibers are made of glass so pure and transparent that light can travel through them for miles.

PACEMAKER

The human heart has a natural pacemaker, the sinus node, which produces an electric signal that stimulates the heart to beat. If this electrical system is damaged, causing an irregular heartbeat, an artificial pacemaker can be installed as a substitute. A device that has maintained life for countless people, the pacemaker is an application of space-age technology and miniaturization to everyday living.

Despite its mysterious nature and demonstrable importance, the basic pacemaker is a relatively simple gadget. A pulse generator, powered by a battery, builds up a tiny electric charge. The electric charge goes to an electrode, which is in contact with the heart. The electric charge is transferred to the heart muscle, which immediately contracts and maintains the heart's pumping action.

The most common type of pacemaker is the "demand" type. It senses when the heart is beating normally and shuts off; if the electronic circuit determines that the heart is beating too slowly or not rhythmically, it immediately turns the pacemaker back on. The heart is thus encouraged to beat under its own force.

The pacemaker consists of three parts: the **pulse generator,** which houses the battery and associated circuits; the **pacing lead,** which carries the output of the pulse generator to the electrode; and the stimulating **electrode,** which is the tip that actually comes in contact with the heart. The battery builds up a tiny electrical charge in the circuitry, which delivers the charge to the electrode at precisely timed intervals.

The pulse generator, no bigger than a matchbox, is surgically placed in the chest cavity between the patient's breast and collarbone. The pacing lead runs through a blood vessel toward the heart, and then attaches to the heart. The lithium batteries used in the pulse generator can last ten years or more.

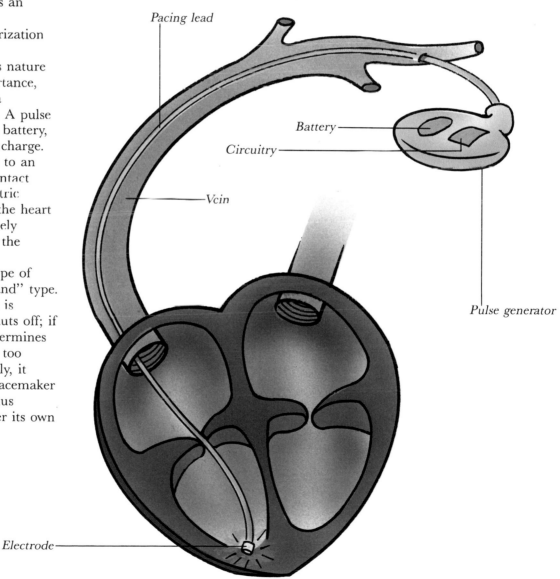

Pacing lead

Battery

Circuitry

Vein

Pulse generator

Electrode

LASER SURGERY

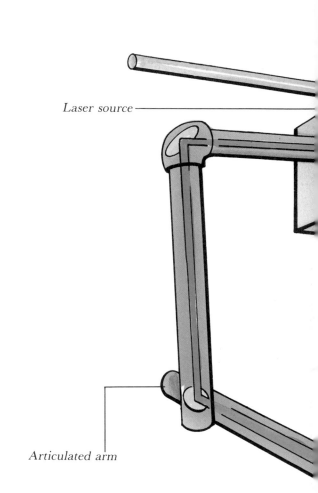

Laser source

Articulated arm

Surgery was the first application and is still the most common use of lasers in medicine. The coherent light from a laser can be used to cut body tissue like a scalpel, to cauterize—or sear—blood vessels and coagulate blood, to weld tissue, and to evaporate diseased tissue.

Laser light can be used for surgery because the light rays are parallel (see **Laser,** page 55), and they can be focused by lenses to a very fine point. Even though the total energy in the rays is less than that from an ordinary light bulb, it can be concentrated on a very small spot, increasing the intensity at that spot and permitting very precise cutting.

The principle underlying laser surgery is that the intensity of laser light delivered in a very short time to a very small area will cause body tissues to absorb heat and vaporize. As the beam cuts, its heat cauterizes and seals blood vessels, stemming the flow of blood. Thus the surgeon can see the area better than if conventional scalpels were used. Focused to a small point, the beam causes less damage to surrounding tissues than conventional cauterizing knives, which contain a heating coil that may burn more of neighboring tissue and cause excessive pain and swelling. In cancer surgery, the use of a scalpel exerts mechanical pressure on the tissue, which could push cancer cells into the bloodstream. The laser beam exerts no mechanical pressure, and the sealing of blood and lymph vessels lessens the spread of cancer cells.

Lasers are used to repair ulcers in the gastrointestinal tract and to "spot-weld" retinas that have become partly detached from the back surface of the eye. Laser surgery is used successfully to remove or destroy cancer cells in both soft tissue and bone; the area to be treated is exposed by conventional scalpel surgery.

To reach areas within the body, such as the throat, lungs, stomach, bladder, or colon, the laser beam is often directed through an **endoscope** (see page 292). Lasers are also used to open up blood vessels that are blocked by plaque. A hair-thin glass fiber is inserted into the artery or vein in a leg and is threaded up to the blocked section. The laser pulse sent through the fiber burns a path through the plaque. Similar techniques are used to unblock fallopian tubes and to remove tumors from the brain or spinal cord.

Sophisticated equipment directs and focuses the laser radiation and measures and controls the amount of energy that is emitted by the laser and absorbed by the tissue. An articulated arm with mirrors directs the laser beam to a laser scalpel held by the surgeon. For precision work, the beam can be delivered through a binocular microscope with a TV camera and micromanipulator controls. The surgeon can watch what he or she is doing either directly or on a TV monitor. For surgery inside the body, the laser scalpel is replaced by an eyepiece and endoscope instrument.

Frequently laser scalpels are linked with computers to help guide the beam more precisely. When paired with new imaging methods such as CAT scanners (see page 288) and Nuclear Magnetic Resonance Testing (NMR—see page 288), the surgeon has a display of the area to be operated on and can "see" precisely where to direct the laser beam.

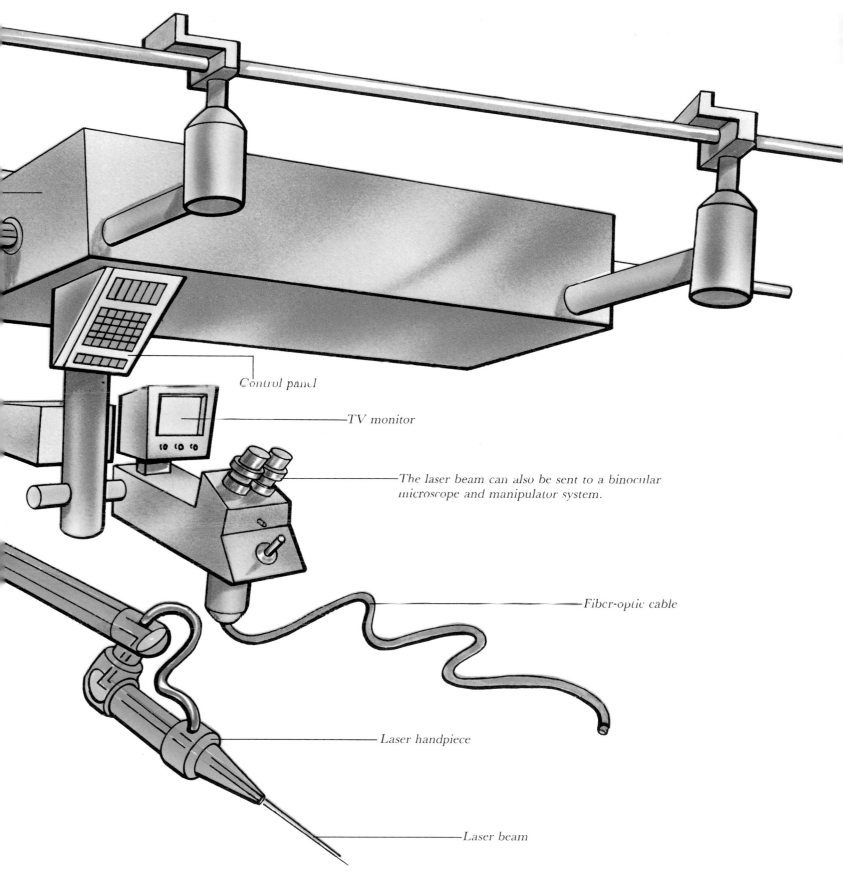

Control panel

TV monitor

The laser beam can also be sent to a binocular microscope and manipulator system.

Fiber-optic cable

Laser handpiece

Laser beam

ARTIFICIAL LIMBS

Knee block

Every year, tens of thousands of people lose a limb as victims of accidents, war injuries, and diseases such as diabetes that affect blood circulation. Although no artificial limb yet devised is a worthy substitute for the original, a good modern **prosthesis** (artificial limb) can enable its wearer to perform most everyday activities.

There are roughly ten times more amputations of legs than of arms, largely because the legs are more likely to be affected by poor blood circulation than the arms. The difficulty of constructing a fully functional artificial leg derives from the considerable complexity of the act of walking. Walking upright has been described as a controlled fall. Trying to imitate the balance and coordination of walking with a mere machine is a challenge.

The conventional prosthetic leg was carved from willow wood and had a foot and ankle attached at the bottom and a socket at the top. It attached to the wearer with straps and laces. More recently, metal alloys and plastics that are stronger and lighter than wood are being used. Newer designs sometimes incorporate a

Piston

full-contact suction socket, which fits snugly over the stump and does away with a cumbersome harness.

Artificial arms can be powered so that the wearer can grasp and lift things. The most advanced of these powered prosthetic arms operate by **myoelectricity,** the tiny electric current that is produced by muscle contraction. Electrodes in the prosthesis pick up these electric impulses from the stump; they are amplified and used to operate a small motor in the arm. Myoelectric arms have been in use since the 1960s, but myoelectric legs are not often used. The complexity—and cost—of myoelectric limbs is very high, and it is possible that a myoelectric leg would not be a significant improvement over a conventional artificial leg.

Shin piece

ARTIFICIAL KNEE JOINT

If a leg amputation is above the knee, some form of artificial knee joint is necessary. Since a leg's two major functions are weight-bearing and force-acceleration, an artificial knee must lock when the wearer is standing but flex for walking or stair-climbing. An advanced design uses a hydraulic piston (see **Hydraulics & pneumatics,** *page 45), which works like an automatic door closer. The smooth movement of the piston gives a soft, natural motion to the gait.*

MYOELECTRIC ARM

Contractions in the muscle of the shoulder or the remaining part of the arm generate a small electric current. This current is picked up by electrodes in the artificial arm and amplified. The current triggers a battery-powered motor, which drives a system of gears to make the thumb and first two fingers move. With practice, the wearer can move the arm and hand by sheer will—impulses in the brain automatically contract the correct muscles. Even the strength of the grip can be controlled by the strength of the muscle contraction.

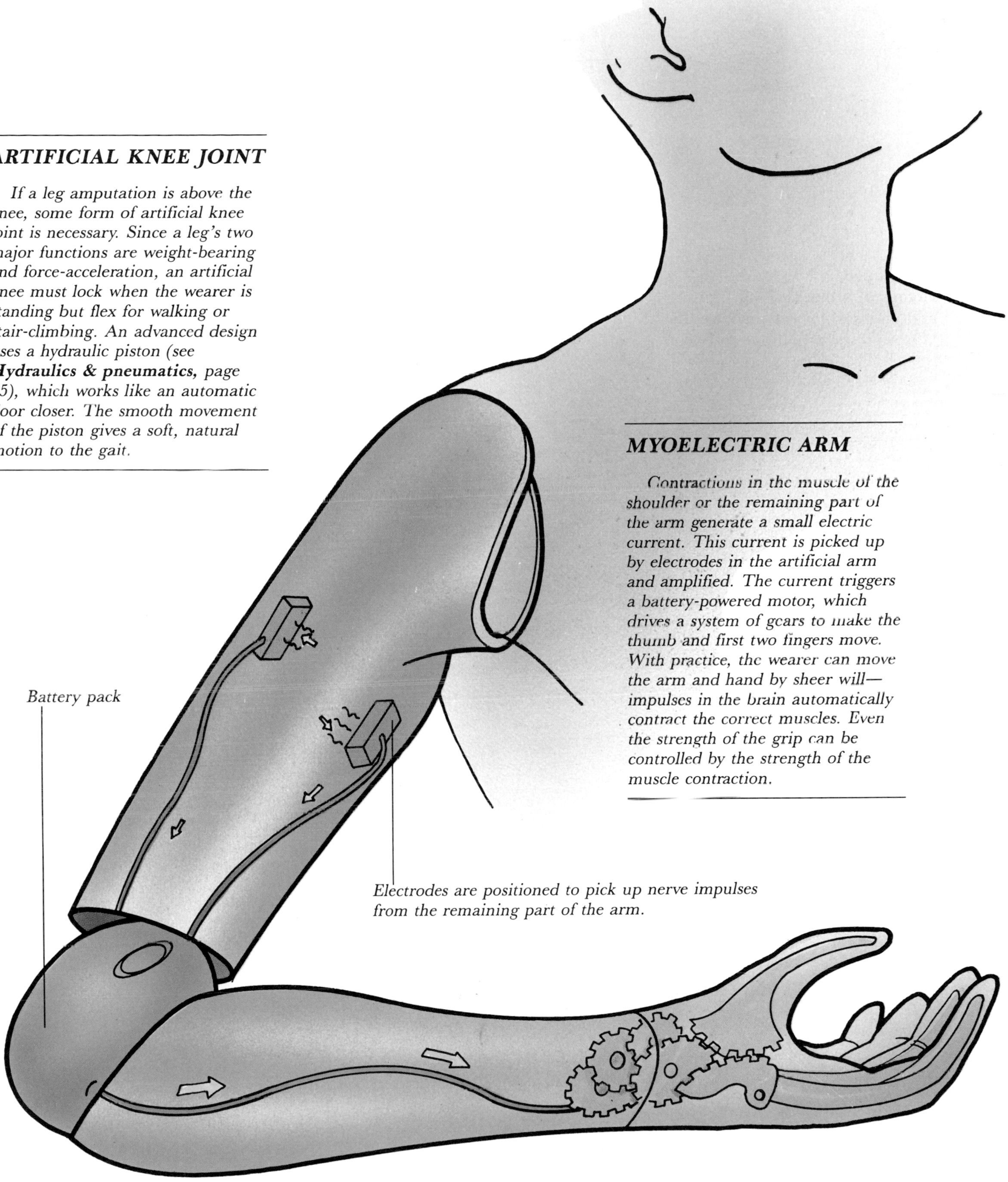

Battery pack

Electrodes are positioned to pick up nerve impulses from the remaining part of the arm.

297

ARTIFICIAL HEART VALVES

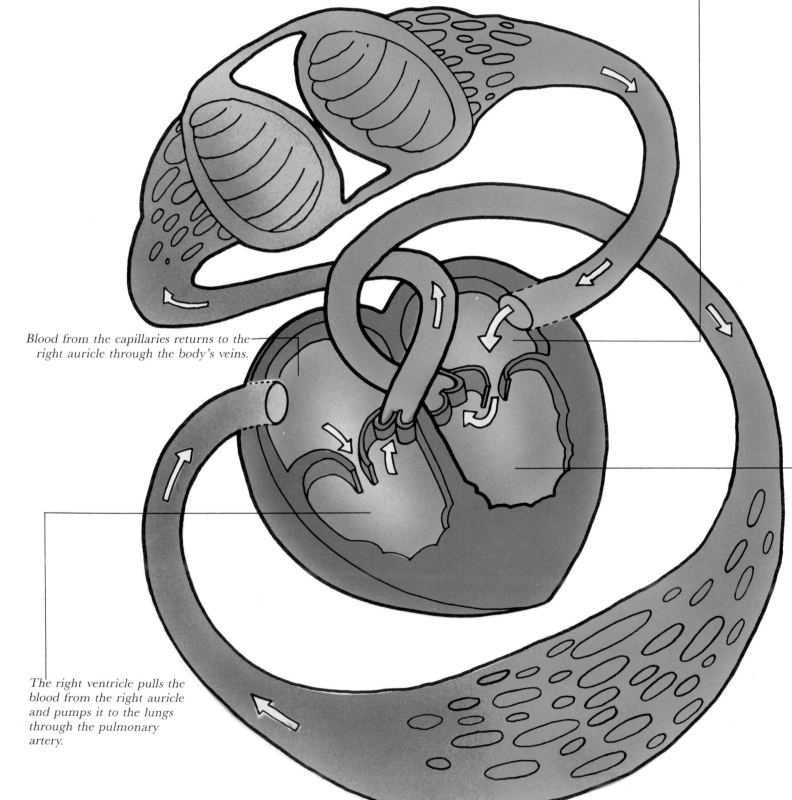

Oxygenated blood from the lungs returns to the left auricle through the pulmonary vein.

Blood from the capillaries returns to the right auricle through the body's veins.

The right ventricle pulls the blood from the right auricle and pumps it to the lungs through the pulmonary artery.

The human heart is an elaborate pump that circulates blood through the body. It is divided into four sections, or chambers, that keep this circulation going. The two chambers that do the actual pumping—the **ventricles**—are powerful muscles that contract and relax rhythmically; the other two—the **auricles,** or **atria**—act primarily as receiving chambers. Since each chamber has an entrance and an exit, valves are required to keep the blood flow going in one constant direction. Otherwise, the contraction of the ventricle would pump some of the blood backward. There are valves between each chamber of the heart and in the blood vessels leading out of the heart. These valves are the type that normally permit the blood to flow one way only.

A natural heart valve consists of thin, light, fibrous tissue that grows in a funnel shape pointing in the direction of blood flow. Various diseases and infections can damage them. If heart valves malfunction, the heart can't circulate blood efficiently. The patient becomes weakened from lack of oxygen reaching the body.

In the United States, more than 50,000 people each year have surgery to replace one or more defective heart valves with artificial ones. They mimic the action, if not the construction, of natural valves. They have a moving section that opens when fluid presses against it from one direction but shuts tightly when the pressure comes from the other side.

Artificial heart valves are made of inert materials that aren't rejected by the body. Nonetheless, they do introduce problems. Blood platelets—small, flat components of the blood that aid in blood clotting—tend to coat the inside of a mechanical valve. Since an excessive buildup of platelets would eventually clog the replacement valve, patients take anticoagulant medication to prevent this. People with artificial heart valves can lead an otherwise normal life and enjoy a normal life span.

—*The left ventricle pulls blood from the left auricle and pumps it through the body's arteries to blood vessels, or capillaries, throughout the body.*

THE HEART

The heart circulates blood through two loops: in and out of the lungs, and then through the rest of the body. It pumps oxygen-rich blood to the blood vessels throughout the body and takes it up again after it has traveled through those vessels and given up its oxygen. The heart then pumps the blood to the lungs, where it absorbs new oxygen, and takes it back in order to start the cycle again.

ARTIFICIAL HEART VALVES

*Artificial heart valves come in several types. A **caged-ball valve** consists of a ball that lifts away from the valve whenever pressure is applied from the valve side and presses tightly against the valve to seal it when the direction of pressure is reversed. **Disk valves** work similarly, except that the moving part is a disk that can swing open or shut. The frames of artificial heart valves are mounted in flexible rings of inert material that can be surgically sewn into the large blood vessels of the heart.*

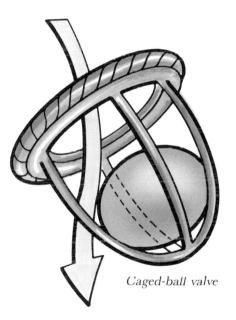

Caged-ball valve

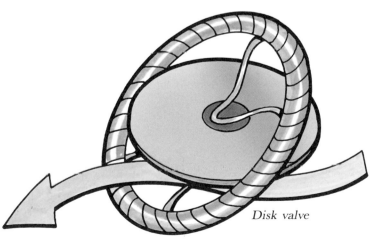

Disk valve

A Time Line of Discoveries

3500-3000 B.C.	The earliest known numerals are in use in Egypt Wheeled vehicles are in use in Mesopotamia
3000-2500 B.C.	Iron objects are probably being manufactured for the first time
2500-2000 B.C.	Skiing is pictured on an ancient rock carving in Norway
2000-1500 B.C.	The decimal system is in use in Crete
800-700 B.C.	Spoked wheels are used in Europe
600-500 B.C.	The invention of the lock and key is attributed to Theodorus of Samos Thales of Miletus discovers that magnets attract iron
500-450 B.C.	Dams are constructed in India
250-200 B.C.	Eratosthenes theorizes that the Earth moves around the sun; he also estimates closely the Earth's circumference
200-150 B.C.	Gears are in use, making possible the ox-driven water wheel
50-1 B.C.	An early form of the oboe is being played in Rome
A.D. 300	The triangular sail is used on boats on the Arabian Sea
527	Paddle-wheel boats are powered by animals
600	Books are being printed in China
650	The windmill is used in Persia
678	A missile weapon, called "Greek Fire" and made of sulfur, petroleum, rock salt, and resin, is invented by a Byzantine
748	In Peking, the first printed newspaper appears

800	The first mention of the crank handle
1045	The Chinese are using moveable type
1090	A water-driven mechanical clock is constructed in China
1115	The first recorded use of the magnetic compass, by Chinese navigators
1290	The invention of spectacles
1453	The Gutenberg Bible is printed with moveable type
1480	Leonardo da Vinci designs the parachute
1500	Pencils made of black graphite are used in England
1510	Da Vinci develops the horizontal water wheel, which is the principle of the water turbine
1535	The first diving bells are in use
1546	Gerardus Mercator claims that the Earth has a magnetic pole
1553	The violin appears in its present form
1581	Galileo discovers the principle of the pendulum
1596	Galileo invents the thermometer
1600	The telescope is developed by Dutch opticians
1658	Robert Hooke invents a balance spring for watches
1666	The color spectrum is discovered by Isaac Newton
1694	In Germany, Gottfried Liebniz creates a mechanical calculator that can multiply and divide

1709	The piano is invented
1710	German engraver Jakob Christoph Le Blon invents three-color printing
1712	Thomas Newcomen builds the first steam engine in England
1714	Gabriel D. Fahrenheit of Germany develops the mercury thermometer
1729	English scientist Stephen Gray discovers that certain materials conduct electricity and other materials are nonconductors
1743	A private elevator is built for Louis XV at the Palace of Versailles
1752	American printer Benjamin Franklin invents the lightning rod
1775	James Watt perfects the steam engine
1777	The torpedo is invented by American engineer David Bushnell
1782	The Montgolfier brothers, in France, construct a hot-air balloon
1785	The seismograph, for measuring earthquakes, is invented
1786	German chemist M.H. Klaproth discovers uranium
1797	French balloonist André Garnerin makes the first parachute descent
1798	Lithography is developed
1800	The existence of infrared solar rays is discovered by William Herschel Alessandro Volta produces electricity with the first battery made of zinc and copper plates
1801	Robert Fulton develops the submarine *Nautilus*
1802	John Dalton formulates the atomic theory
1803	The steamboat is invented by Robert Fulton
1814	The first practical steam locomotive is built by George Stephenson in England
1819	Danish physicist Hans C. Oersted discovers that a magnetic compass needle is influenced by electric currents
1823	English mathematician Charles Babbage attempts to construct a calculating machine Waterproof fabric is invented by Charles Macintosh
1829	L.J.M. Daguerre constructs a camera An electromagnetic motor is built by American physicist Joseph Henry William B. Burt takes out the first U.S. patent for a typewriter
1832	The Frenchman Hippolyte Pixii builds an electric generator
1834	Charles Babbage invents the principle of his "analytical engine," the forerunner of the computer The refrigerator compression system is developed by American Jacob Perkins
1835	The earliest photographic negative is taken by William Henry Fox Talbot, in England
1836	The screw propeller is developed by Swedish-American inventor John Ericsson
1837	The telegraph is patented by Sir Charles Wheatstone and W.F. Cooke
1839	American inventor Charles Goodyear makes the commercial use of rubber possible by developing the vulcanization of rubber Ozone is discovered by German-Swiss chemist Christian F. Schönbein

1839	The first electric clock is made by Swiss physicist Carl August Steinheil
	Scottish inventor Kirkpatrick Macmillan invents the pedal-operated bicycle
1846	A patent for a sewing machine is taken out by American Elias Howe
1848	American locksmith Linus Yale develops the cylinder pin-tumbler lock
1849	The safety pin is invented by American Walter Hunt
1854	Elisha G. Otis invents the safety elevator
1855	George Audemars takes out a patent for rayon
1859	The first practical storage battery is developed by R.L.G. Planté
1860	The first successful internal combustion engine is produced by French engineer Étienne Lenoir
	American inventor Christopher L. Sholes produces a practical typewriter
1862	French physicist Leon Foucault successfully measures the speed of light
1866	Alfred Nobel invents dynamite
1873	Color photographs are produced for the first time
1876	Alexander Graham Bell invents the telephone
1877	Thomas A. Edison invents the phonograph
1878	The microphone is invented by David Hughes
1880	Edison and J.W. Swan independently develop the first practical electric lights
	The existence of semiconductors is known

1884	Sir Charles Parsons develops the first practical steam turbine engine
	L.E. Waterman of New York makes the first easily refillable fountain pen
1885	Karl Benz builds a single-cylinder engine for an automobile
1888	An alternating-current motor is constructed by U.S. inventor Nikola Tesla
	Radio waves are discovered to be in the same family of waves as light waves
1891	The zipper is invented by American W.L. Judson, although it is not in common use until 1919
1892	German Rudolf Diesel patents the diesel engine
1893	Henry Ford builds his first motor car
1895	Wilhelm Roentgen discovers X rays
	Guglielmo Marconi develops radio telegraphy
	Auguste and Louis Lumière invent the motion picture camera
	The safety razor is invented by King C. Gillette
	Konstantin Isiolkovski develops the principle of rocket propulsion
1896	French physicist A.H. Becquerel discovers radioactivity
1897	The electron is discovered by J.J. Thomson
1898	An airship is constructed by German Count Ferdinand von Zeppelin
1900	American scientist R.A. Fessenden transmits speech by radio waves
1901	In Britain, Hubert Booth invents the vacuum cleaner
1903	Orville and Wilbur Wright are the first to fly a powered plane successfully
	The electrocardiograph is invented by Wilhelm Einthoven

1905	Albert Einstein develops his theory of relativity
1909	Bakelite, the earliest synthetic plastic, is commercially manufactured
1911	Charles F. Kettering develops the electric self-starter for automobiles
1919	Experiments are being made with shortwave radio
1920	Water skiing is introduced at Lake Annecy in France
1923	Col. Jacob Schick patents the electric razor
1924	John Harwood patents the self-winding wristwatch
1925	Scottish inventor J.L. Baird transmits recognizable images by television
1926	The first liquid-fuel rocket is fired, having been developed by Robert H. Goddard of the United States
1928	J.L. Baird demonstrates color television

George Eastman exhibits the first color film |
1929	Quartz crystal clocks are introduced by W.A. Morrison
1930	The analog computer is developed by American Vannevar Bush
1935	Englishman Robert Watson-Watt builds a radar detection device
1936	Dr. Alexis Carrel develops the artificial heart
1937	In England, Frank Whittle builds the first jet engine
1938	Chester Carlson, a worker in the patent department of an electronics firm, invents the xerographic copying process

The ballpoint pen is invented by Hungarian Lajos Biró |

1939	Igor Sikorsky constructs the first reliable helicopter

Nylon stockings are introduced

Frequency modulation (FM) is developed by American Edwin H. Armstrong |
| *1940* | Scientists at RCA demonstrate the electron microscope |
| *1942* | Enrico Fermi, in the United States, splits the atom

Magnetic tape recording is developed

The first jet plane is built in the United States |
| *1945* | The first atomic bomb is detonated near Alamogordo, New Mexico |
| *1947* | The transistor is invented by scientists at Bell Laboratories

The hologram is invented by Dennis Gabor, who cannot put his idea into use because the laser has not been invented yet |
| *1948* | The long-playing record is invented |
| *1949* | The Raytheon corporation develops the first commercial microwave oven |
| *1955* | In Schenectady, New York, atomically generated power is used for the first time

British engineer Christopher Cockerell invents the hovercraft |
| *1957* | *Sputnik I*, the first artificial satellite, is launched by the Soviet Union |
| *1958* | American engineer Jack Kilby invents the integrated circuit

Stereophonic recordings are introduced |
1960	American Theodore H. Maiman builds the first laser
1972	Japanese manufacturer Sony Corporation markets the Betamax videocassette recorder
1980	The compact disc is introduced

ACCELEROMETER
An instrument that measures how rapidly speed changes.

AILERON
A flap on the trailing edge of an airplane wing that can be moved to control the plane's position in the air.

AIRFOIL
A contoured construction (such as an airplane wing) that is shaped to produce lift when moved through the air.

ALGORITHM
A set of specific rules and procedures for solving a mathematical problem or achieving a goal.

ALTERNATING CURRENT (AC)
Electric current that periodically reverses its direction of flow.

AMPERE
The unit for measuring electrical current. One ampere (or **amp**) is equal to the amount of current flowing through a circuit produced by one volt across a resistance of one ohm.

AMPLITUDE
One half the full extent of vibration of a wave or an oscillation. The amplitude of a wave is the distance from its halfway position to its most extreme position.

ANALOG
Relating to devices (such as telephones or some tape recorders) that represent information in a way that varies continuously relative to variations in the information. Compare **Digital.**

ANGSTROM
A very small unit of measurement (abbreviation Å), equal to one ten-millionth of a millimeter, that is used to measure wavelengths.

ANODE
The positive electrode through which an electric current goes into an electric device. See **Cathode; Electrode.**

ARMATURE
The part of an electric motor, usually a metal core wound with coils of wire, that moves or rotates between the opposite poles of a magnet.

ATOM
The smallest particle that a chemical element can be divided into and retain its identity. Atoms consist of electrons orbiting around a nucleus of protons and neutrons. See **Electron; Element; Neutron; Proton.**

AUDIO-FREQUENCY SIGNAL
An alternating electrical current that represents sound vibrations because it changes direction at frequencies corresponding to the frequencies of the sound waves.

AXLE
The shaft on which a wheel or pair of wheels rotates.

BALLAST
Heavy substances arranged so as to give stability to a ship or other floating vehicle.

BALL BEARING
A bearing in which the shaft rotates against a set of freely revolving metal balls in order to reduce friction.

BEAM
A stream of particles or waves, often composed of several rays.

BEARING
The unit that supports a revolving part or wheel.

BIMETALLIC STRIP
A strip or coil composed of two different metals layered together. When heated, the two metals expand at different rates, which makes the strip bend.

BINARY CODE
A system in which data is represented by a series of binary numbers for use in computers and other digital devices. See **Digital.**

BINARY NUMBER
A number expressed in terms of only two digits, 0 and 1, rather than the ten digits of decimal notation.

BIT
One of the digits—0 or 1—in a binary number. A bit forms a unit of computer information.

BRUSH
A piece made of a conducting material such as carbon that, when rubbing against a commutator in a motor, provides a passage for electric current.

BYTE
A group of bits, usually eight, used as a measure of computer memory.

CAM
An irregularly shaped wheel that rotates against a rod called a cam follower. The arrangement is used to change a rotary motion into a back-and-forth motion.

CANTILEVER
A horizontal beam with one unsupported end.

CAPACITOR
A device used to store an electrical charge.

CARRIER WAVE
A radio wave whose frequency or amplitude is modulated so as to transmit a signal for radio or television broadcasting.

CATENARY
The curve assumed by a flexible cord or cable when it is suspended and hangs freely from two fixed points.

CATHODE
The negative electrode through which an electric current passes out of an electric device. See **Anode; Electrode.**

CATHODE-RAY TUBE
A vacuum tube in which a cathode produces a stream of high-speed electrons—a cathode ray—toward a fluorescent screen; the cathode ray

forms a bright spot on the screen. Cathode-ray tubes are used in TV sets and computer monitors.

COAXIAL CABLE
A cable consisting of an insulated conducting core surrounded by a conducting tube that is also insulated. The conductors thus have the same axis—they are coaxial. The signals carried in a coaxial cable resist disturbance from electromagnetic fields outside the cable.

COIL
A number of turns of wire, often in spiral form, used in electromagnets.

COMMUTATOR
An arrangement of metal bars that is the part of an electric motor that reverses the direction of current.

COMPOUND
A substance made up of two or more elements chemically combined in definite proportions. The compound has properties different from those of the separate elements. See **Element.**

CONCAVE
Having a surface that curves inward.

CONDUCTOR
A substance that transmits some form of energy such as electricity.

CONVEX
Having a surface that curves outward.

COUNTERWEIGHT
A weight in a machine that is attached to a moving part so as to balance the weight of the moving part.

CRYSTAL
A solid that by nature has flat surfaces and in which the atoms or molecules are arranged in a regular pattern.

CYLINDER
A chamber in an engine or pump in which a piston moves back and forth.

DICHROIC
Showing transmitted or reflected light in two different colors.

DIFFERENTIAL
The unit in an automobile that will drive both rear axles at the same time but will allow them to rotate at different speeds on turns.

DIFFRACTION
A change in the direction of waves after they pass an obstacle or go through an opening.

DIGITAL
Relating to devices (such as computers or certain sound recorders) that represent information in the form of a sequence of numbers. Compare **Analog.**

DIODE
An electronic device having two electrodes, especially one that permits the flow of electricity in one direction only.

DIPOLE
A molecule or other object with two equal, opposite electric charges or magnetic poles separated by a short distance.

DIRECT CURRENT (DC)
Electric current that flows in one direction only.

DISTRIBUTOR
A unit in an automobile designed to distribute the high-voltage electric current from the ignition circuit of the engine to the proper cylinder at the correct time.

ELECTRET
A material that retains a permanent electric polarization, so that one end of it is positively charged and the other end is negatively charged. Shortened form of *electricity* and *magnet.*

ELECTRIC CHARGE
A condition of being able to discharge electricity. A gain of electrons results in a negative charge; a loss of electrons results in a positive charge.

ELECTRIC CURRENT
The continuous flow of electrons through a conductor such as wire.

ELECTRIC FIELD
An area in which an electric charge produces a force.

ELECTRODE
A conductor through which electric current either goes into or goes out of an electrical device. See **Anode; Cathode.**

ELECTROLUMINESCENCE
Emission of light that is induced in certain materials by passing an electric current through them.

ELECTROLYTE
A nonmetal conductor of electricity.

ELECTROMAGNET
A core of iron wound with wire, which becomes magnetized as long as electricity is passed through the wire.

ELECTROMAGNETIC SPECTRUM
The family of electromagnetic waves, ranging from radio waves of low frequency and long wavelength to X rays and cosmic rays of high frequency and short wavelength. Visible light waves are a part of the electromagnetic spectrum.

ELECTROMAGNETIC WAVE
An energy wave generated by electric and magnetic fields that vary simultaneously.

ELECTROMOTIVE FORCE
The force or pressure necessary to cause the flow of an electric current. See **Voltage.**

ELECTRON
An elementary particle that has a negative electric charge. Atoms have one or more electrons surrounding their nuclei. The movement of electrons through a conductor such as a wire is electricity.

ELECTRON BEAM
A stream of electrons that are

moving at the same speed in the same direction.

ELECTRON GUN
The part of a cathode-ray tube that directs and focuses a stream of electrons to the screen.

ELEMENT
A substance that consists entirely of atoms that have the same number of electrons and protons. Hydrogen and oxygen are elements; their chemical combination, water, is a compound. More than 100 elements are known to exist; some of them, such as plutonium, are synthetic and do not occur in nature. Elements are the fundamental substances from which all matter is made up.

EXHAUST MANIFOLD
The part of an automobile's exhaust system that collects gases from the engine and transmits them to the exhaust pipe. A manifold is a pipe with many fittings that connects multiple pipes to one inlet or outlet.

EXTERNAL COMBUSTION ENGINE
An engine (such as a steam engine) that gets its power from fuel consumed outside the engine's cylinders.

EYEPIECE
In an optical instrument, the lens or set of lenses that is closest to the user's eye.

FISSION
The splitting of the nucleus of an atom into smaller nuclei by bombarding the atom with a neutron, which results in a release of energy.

FLUORESCENCE
The ability of some substances to give off light when absorbing radiation from another source such as X rays.

FLYWHEEL
In an automobile, a large, heavy wheel attached to the crankshaft of the engine. Its inertia keeps the crankshaft rotating smoothly despite fluctuations in the movement of the engine.

FREQUENCY
The number of complete cycles per second of a wave or of alternating current.

FUEL CELL
A device that produces electricity directly from a chemical reaction between oxygen and a fuel.

FULCRUM
The support or pivot that a lever rotates on.

FUSION
The combining of the nuclei of two atoms to create a larger atom, which results in a release of energy.

GAMMA RAY
An electromagnetic wave of very high frequency that is given off spontaneously by radioactive substances.

GIMBAL
A support that permits the object it is supporting to incline in any direction, such as in a gyroscope.

HALOGEN
Any of the group of nonmetallic elements that includes fluorine, chlorine, bromine, iodine, and astatine. They display similar properties and react with other elements to form numerous compounds, principally metallic salts.

HEAT EXCHANGER
An arrangement of pipes that allows a hot liquid or gas to warm a cool one, or a cold liquid or gas to cool off a hot one.

HELICAL
Shaped in the form of a helix, which is, roughly, the shape of a spiral staircase.

HERTZ
A measurement of frequency (abbreviation Hz) that is applied to electromagnetic waves. One hertz equals one cycle per second. One kilohertz (kHz) is 1,000 hertz; one megahertz (MHz) is 1,000,000 hertz.

HYDRAULIC
Operated by means of water pressure or other liquid pressure.

HYDRODYNAMIC LIFT
The lift provided by motion through water or along its surface.

HYDROMETER
An instrument for determining the specific gravity of a liquid and, therefore, the strength of a solution such as battery acid.

IMPEDANCE
In an electric circuit, the apparent opposition to the flow of alternating current, including not only resistance but also the amount of electromotive force induced in the circuit and the amount of electric charge that can be stored.

INCANDESCENCE
The emission of light when heated.

INERTIA
The tendency of all matter to remain motionless if already motionless or, if already in motion, to resist changes in the speed or direction of the motion unless an outside force is applied.

INFRARED
Referring to the invisible electro-magnetic waves whose wavelengths are longer than those of visible light and shorter than those of microwaves. Infrared rays include heat rays.

INFRASONIC
Referring to sound waves with a frequency below the range of human hearing.

INSULATOR
A material that is a poor conductor of electricity or a material that prevents the passage of heat or sound.

INTEGRATED CIRCUIT
See **Microchip.**

INTERNAL COMBUSTION ENGINE

An engine (such as a gasoline or jet engine) that gets its power from fuel consumed inside the engine itself.

INTERNAL GEAR

A gear whose teeth are not on the outside of a disc, but are arranged on the inside of a circle facing inward toward the center.

ION

An atom that has a positive electric charge because of having lost electrons or a negative electric charge because of having gained electrons.

IONIZATION

The formation of ions.

ISOTOPE

Any of different forms of an element that have different numbers of neutrons in the nucleus and therefore have slightly different physical properties, though they have nearly identical chemical properties.

KELVIN

A unit of temperature measurement in the Kelvin scale, which is widely used in science. Water freezes at 273.15 kelvins and boils at 373.15 kelvins.

KILOHERTZ

See **Hertz.**

LATENT IMAGE

In photography, an image that is chemically present in film but invisible until the film is developed.

LIFT

The force acting on an airfoil that constitutes the upward force counteracting the pull of gravity.

LINKAGE

A system of links and levers connected together to transmit motion or force.

LUMINESCENCE

An emission of light that is produced not by heat but by chemical or electrical activity.

MAGNETIC FIELD

The space around a magnet or a current-carrying wire in which a magnetic force can be detected.

MAGNETISM

The property of certain objects that causes them to attract or repel other objects.

MECHANICAL ADVANTAGE

The advantage in effort gained by using a mechanism. The mechanism multiplies the force applied to it and produces a larger force that does the work. The number of times the force is multiplied is the mechanical advantage.

MEGAHERTZ

See **Hertz.**

MICROCHIP

A very tiny assemblage of electronic components on a slice of silicon.

MICROPROCESSOR

The central processing unit of a computer; it controls data flow and performs arithmetic and logic calculations.

MICROWAVE

An electromagnetic wave with a wavelength longer than those of infrared heat waves and shorter than those of radio waves.

MODEM

A device used to convert signals from one type of equipment to signals that can be used by another type of equipment.

MOLECULE

The smallest particle into which an element or compound can be divided and still retain its chemical and physical properties. A molecule of water (H_2O) consists of two hydrogen atoms and one oxygen atom bound together.

MYOELECTRICITY

The electric current that is produced by muscle contraction.

NEBULIZER

A device that converts a liquid to a fine spray.

NEUTRON

A particle with no electric charge that is one of two types of particles present in the nucleus of an atom. The other type of particle is the proton.

NUCLEUS

The center of an atom, consisting of one or more positively charged protons and a nearly equal number of neutrons.

OBJECTIVE

In an optical instrument, the lens or set of lenses that is closest to the object being studied.

OHM

The unit for measuring electrical resistance. One ohm is equal to the resistance of a conductor in which one volt produces a current of one amp.

OSCILLATOR

A device that changes an electric current to one that oscillates, or alternates, at a specific frequency.

PARABOLIC

Having the approximate shape of a bowl.

PAWL

A machine part that pivots to engage with a notch on a ratchet wheel to permit movement in only one direction.

PHONEME

A distinctive, basic sound; the smallest unit of speech.

PHOSPHOR

A substance that emits light when subjected to certain kinds of radiation.

PHOSPHORESCENCE

An emission of light that not only occurs while the substance is subjected to a source of radiation, but continues even after the source has been removed.

PHOTOCONDUCTOR
A substance that is able to conduct electricity upon being exposed to light.

PHOTODETECTOR
A device that detects and measures radiant energy.

PHOTODIODE
A diode made of semiconducting material that varies the amount of electric current flowing through the photodiode depending on the amount of radiant energy present.

PHOTOELECTRIC CELL
An electronic device that produces a varying amount of electricity depending on the amount of radiant energy present.

PHOTON
A very small unit of radiant energy.

PIEZOELECTRICITY
The generation of an electric current by certain crystals when subjected to stress, and also the reverse: the vibration of a crystal when it is exposed to an electric current.

PINION
A small gear whose teeth engage with a larger gear or a toothed rack.

PISTON
A sliding plug that moves up and down inside a cylinder. In an engine, the piston is moved by fluid pressure against it; in a pump, the piston compresses a fluid.

PIXEL
One of the separate, individual elements that make up the image on a television screen.

PNEUMATIC
Operated by means of air pressure or the pressure of other gases.

POLARITY
The condition of having two opposite poles like those of a magnet; or the condition of having a positive or negative electric charge.

POLYMERIZATION
A reaction that combines many small, simple molecules into a larger, more complex molecule.

PROSTHESIS
An artificial limb or other part of the body.

PROTON
A positively charged particle that is one of two types of particles present in the nucleus of an atom. The other type of particle is the neutron.

PULLEY
A wheel with a grooved rim to hold a rope or chain, used to change the direction or the location of a pulling force.

QUARTZ
A hard, colorless, transparent, crystalline mineral composed of silicon dioxide.

RACK AND PINION
A machine part to convert rotary to linear motion or vice versa, consisting of a small gear intermeshed with a toothed bar.

RADIATION
Energy transmitted in the form of rays or waves; a stream of particles or waves emitted by certain elements as their nuclei decay.

RADIOACTIVITY
Radiation emitted by certain elements as the result of nuclear decay.

RADIO WAVE
An electromagnetic wave with a wavelength longer than that of a microwave and used to transmit signals.

RAM
In computers, random-access memory, the memory that uses or temporarily stores data or programs.

RAREFACTION
A less dense condition; an area where a passing sound pressure wave has caused a decrease in air pressure.

RAY
A thin line of light or other radiation.

RECEIVER
An electronic device that receives incoming electromagnetic signals and converts them to a form that can be seen or heard.

RECIPROCATING MOTION
A back-and-forth motion.

REFRACTION
The bending of a wave as it passes between two mediums with different densities.

RELAY
An electric device that uses a small amount of current to control a much larger flow of current that operates other electric devices.

RESISTANCE
The opposition existing in some substances to the passage of energy, such as electricity, through them. Electrical energy that meets resistance changes into heat or other forms of energy.

RESISTOR
A device used in an electric circuit to provide resistance and control the flow of current.

RETINA
A membrane in the eyeball that receives the image formed by the lens and is connected by the optic nerve to the brain.

ROM
In computers, read-only memory, the memory that stores data or programs permanently.

ROTOR
A part of an electric or mechanical device that rotates within a stationary part.

SEMICONDUCTOR
A substance such as silicon that conducts electricity more efficiently than an insulator but not so efficiently as a true conductor. Semiconductors can be used to control the flow of electricity.

SERVO
An automatic device that uses a sensing element and a motor to control a mechanism.

SIGNAL
An electric voltage or current that is varied in order to form a code for transmitting information.

SILICON
A nonmetallic element whose compounds are found extensively in the Earth's crust and which is used in semiconductor devices.

SOLENOID
An electromagnetic device that is used to convert electrical energy into mechanical energy.

SPECIFIC GRAVITY
The ratio of the weight or mass of a substance to that of an equal volume of water.

SPECTRUM
A continuous range of values, such as the range of colors formed by a beam of white light passing through a prism, or the entire range of electromagnetic waves in order of wavelength.

SPROCKET
One of the toothlike projections from the circumference of a wheel that mesh with a chain.

STATOR
The stationary part of a motor or machine about which the rotor rotates.

SUPERCONDUCTOR
A metal or alloy that, under temperatures at or near absolute zero, will conduct electricity with no resistance.

TELEMETRY
The use of devices that measure speed, heat, radiation, or other attributes and transmit the data over a distance.

TEMPLATE
A pattern or guide for making something.

THERMOPLASTIC
Able to be softened when heated and hardened when cooled.

THROTTLE
In an internal combustion engine, the valve that controls the amount of vaporized fuel delivered to the cylinders.

TORQUE
A turning or twisting force.

TRANSCEIVER
A piece of equipment that both transmits and receives radio waves.

TRANSDUCER
A device that converts energy from one form, such as the energy from a sound wave, to another form, such as electrical energy.

TRANSFORMER
A device that transforms an electric current from one voltage to another voltage.

TRANSMISSION
In an automobile, a device that uses gearing or torque conversion to change the ratio between the rotational speed of the engine and that of the drive wheels.

TRANSMITTER
An electronic device that generates and modulates a carrier radio wave and transmits the signal to a receiver.

TRANSPONDER
An electronic detection device that receives a predetermined signal and then sends a radio signal of its own.

TURBINE
A rotary engine consisting of a rotating shaft with blades on it that is activated by the pressure of a gas or a liquid turning the blades.

ULTRASONIC
Referring to sound waves with a frequency above the range of human hearing.

ULTRAVIOLET LIGHT
The invisible electromagnetic waves with wavelengths shorter than those of visible violet light and longer than those of X rays.

UNIVERSAL JOINT
A joint that allows freedom of motion in any direction. Also, a joint that allows power from a rotating shaft to be transmitted to another shaft that is at an angle to it.

VALVE
A mechanical device that regulates the flow of a liquid or gas through a pipe or other structure.

VELOCITY
The rate of motion in a given direction within a certain amount of time. (Speed is the rate of motion irrespective of direction.)

VIRTUAL IMAGE
An image, such as that in a flat mirror, formed where light rays appear to come together but where it is physically impossible for them to actually come together. A virtual image is an image where no real image could possibly be.

VOLTAGE
Electric potential, expressed in volts. One volt is the amount of electromotive force that will cause a current of one ampere to flow through a wire with a resistance of one ohm.

WATT
A unit of electrical power equal to the current in amperes times the potential in volts.

WAVE FRONT
The corresponding parts of waves that vary in amplitude and frequency by the same amount at the same time and are traveling in the same direction.

WAVELENGTH
The distance between corresponding parts of a wave.

X RAY
An electromagnetic wave with a wavelength that is extremely short.

INDEX